Storey's Guide to Raising Miniature Livestock

Storey's Guide to

RAISING MINIATURE LIVESTOCK

Health • Handling • Breeding

Sue Weaver

Storey Publishing

The mission of Storey Publishing is to serve our customers by
publishing practical information that encourages
personal independence in harmony with the environment.

Edited by Sarah Guare and Deborah Burns
Art direction and book design by Cynthia N. McFarland
Cover design by Kent Lew
Text production by Erin Dawson

Cover photograph by © Jason Houston
Interior photography credits appear on page 452
Illustrations by © Elayne Sears

Indexed by Christine R. Lindemer, Boston Road Communications

Storey Publishing
210 MASS MoCA Way
North Adams, MA 01247
www.storey.com

Printed in the United States by Sheridan Books, Inc.
10 9 8 7 6 5 4

Library of Congress Cataloging-in-Publication Data

Weaver, Sue.
Storey's guide to raising miniature livestock / by Sue Weaver.
p. cm.
Includes index.
ISBN 978-1-60342-481-3 (pbk. : alk. paper)
ISBN 978-1-60342-482-0 (hardcover : alk. paper)
1. Miniature livestock. I. Title. II. Title: Raising miniature livestock.
SF61.W24 2010
636—dc22

2009041819

To Gib and Melba Mullins

of Ridgerunner Miniature Horses in Caulfield, Missouri,

who raise the best little horses in the world.

Thanks for your friendship, guys!

Contents

Why Raise Miniature Livestock?

ASK A HOST OF ESTABLISHED HOBBY FARMERS and most will agree that there is little (if any) money to be made in commercial, full-size livestock. Feeder cattle, market hogs, and standard lamb-and-wool operations are faltering. With the cost of keeping farm animals hurtling skyward, steadily increasing numbers of owners and producers are turning to raising smaller breeds of livestock, and wisely so. Miniature livestock require less housing space, pasture, fencing, and feed than do their full-size counterparts. According to American Miniature Horse Association figures, you can feed, house, and maintain five miniatures for the cost of keeping a single standard-size horse. Beef cattlemen can stock two or three miniature Herefords or Lowline Angus to one garden-variety cow (Hereford, Angus, Holstein). And a pair of miniature sheep or goats or even a pig can lodge happily in a doghouse in a fenced backyard.

Minis are easier to handle and less intimidating than commonplace livestock, especially for beginners, children, old folks, and the physically challenged, not only due to their smaller stature but also because many miniatures are specifically bred for calm disposition and tractability. Chores such as hoof trimming, shearing or clipping, giving shots, and administering dewormers are infinitely easier, as is training peewee livestock for show or pleasure.

In most states, keeping miniature livestock (especially cattle and heritage pigs, often goats and sheep, but rarely equines) qualifies landowners for valuable agricultural-use land tax exemptions, even on relatively small parcels of land. And minis are sometimes acceptable where zoning laws prohibit full-size barnyard pets.

Enthusiasts exhibit pint-size livestock and poultry at species-specific shows, state and county fairs, and as 4-H projects. Most species are locally transportable in a van or SUV. Scaled-down equines, cattle, goats, and llamas handily pull carts and wagons, even with adults at the reins. Country kids dress their mini friends in costume for the kiddies' day parade.

MINIATURE LIVESTOCK: WHAT'S THAT?

There are three kinds of miniature livestock. For this book's purposes we'll refer to all three groups as miniature livestock, though owners of traditional breeds in the first two groups are quick to point out that their favorite breeds weren't miniaturized by man.

- Naturally diminutive breeds that evolved as small animals to better survive the conditions nature handed them. Think Soay sheep, San Clemente Island goats, and British Shetland ponies.
- Small breeds that retained their original breed character when their parent breeds were selected for greater size. These include Miniature Jersey cattle and Babydoll Southdown sheep.
- Breeds that were deliberately miniaturized by breeders who selected for smaller stature, often through outcrossing to an established smaller breed. Miniature Highland and Miniature Longhorn cattle spring to mind.

Minis' irresistible "cute appeal" and their easygoing natures make them unrivaled tagalongs for nursing home, school, and hospital visitations. What better traffic-stopper than a 30-inch pint-size Zebu bull; an 18-pound miniature horse foal; or a teensy Pygmy goat kid. Minis are fun to own!

And the bottom line: Quality miniatures tend to pay their way. Miniatures cost less to raise, and because they are in short supply, a ready market awaits the conscientious breeder. They can fetch good money at annual sales. And customers tend to pay a premium for value-added products in lieu of the usual kind (goat cheese or yogurt, handspun yarn, and heritage beef or pork). If you want to raise livestock and show a profit (or at least break even), think small! This is the best time in history to invest in miniature cattle, horses, donkeys, mules, sheep, goats, and llamas: breeders abound, and they are eager to welcome newcomers into the fold. Breeders on one side of the United States can easily ship miniature lambs, kids, and piglets by air, two to a standard large-size dog crate.

Whether you're considering a pair of miniature fiber llamas, a herd of meaty Lowline cattle, or a flock of tiny sheep, this book is designed to help you find, select, buy, and take good care of your tiny charges, and if you wish, market them to others who would like to do the same.

SECTION 1

Raising Miniature Livestock

1

Before You Begin

BEFORE LAUNCHING any animal-related enterprise, be certain you understand its demands. Raising animals will impact your lifestyle, your physical and emotional health (typically for the better), and your bank account. Not everyone is cut out to keep livestock, however, miniature or otherwise. Are you?

Lifestyle Choices

Are you willing to be on call 24 hours a day, 7 days a week, 365 days a year, whenever your livestock need you? Will you camp in the barn when foals are due? Roll out of bed at two in the morning to feed a bottle lamb? Retrieve errant cattle and patch their flattened fences under a noonday sun, missing that long-anticipated televised ball game in the process? Animals rarely get hungry, sick, loose, or injured at convenient times.

Is reliable relief help available when you need it? If not, would forgoing dinner invitations, overnight trips, and well-deserved vacations upset you (or your family)? Keeping livestock invariably ties you down.

Discuss your venture with your veterinarian. (Or if you don't yet have a veterinarian, talk with a few local vets to see who might be available.) Is she qualified to doctor Pygmy goats, miniature llamas, or whatever else you've chosen? Does she want to? If not, are you willing (and able) to transport sick or injured animals to a specialty practice and learn to perform routine maintenance procedures yourself?

Physical and Emotional Considerations

Can you weather the inevitable livestock keeper's lows? When your favorite cow dies of bloat or a coyote slays that special lamb, what then? Animals die. They

REQUIRED ITEMS FOR ANIMAL RAISERS

Before you bring any livestock home, you will need to have the following. Are you prepared?

- Safe shelters (see chapter 6 for information on facilities and fencing), proper fencing, bedding, feed, feeders, water containers, and a consistent source of clean drinking water.
- If you plan to breed your livestock, you also need safe birthing quarters and a well-stocked birthing kit and, if you maintain breeding males, strong, secure quarters in which to house them.
- Halters, leads, hoof or toenail trimming equipment, coat care necessities, and any type of specialized tack (carts, harnesses, show halters, and so on) your breed and species requires.
- Well-stocked first-aid kits, one for the barn and one for your towing vehicle (if you have one).

And most important:

- Phone numbers of at least three reliable veterinarians who treat your species. (Introduce yourself in advance so that they are familiar with who you are; what species you have; and, if they make farm visits, where you live.)

injure themselves and each other. Things go terribly wrong. Some stockkeepers neatly handle these stressors — could you?

Do you have the patience or the means to wait for the best animals? Some miniature breeds are in exceedingly short supply. With location and price factored into the equation, newcomers must often bide their time until the right individuals come along.

Are you capable of selling your livestock? Can you send the steer to slaughter; could you sell the sow you love? Are you willing to pull out the stops to market your wares? You'll need to continually monitor market trends to stay on the cutting edge, advertise, and maintain a Web site. Are you ready to haul your livestock to expos, demonstrations, shows, and sales? If not, think "pets," not "produce." Don't become a breeder. That way lies heartache if you find you can't (or won't) sell what you produce.

One animal or one hundred, you should genuinely enjoy working with the species you select. And you must also like the people associated with it. Buying, selling, co-op marketing, or showing, you'll be dealing with them on an ongoing basis.

Are you prepared to handle dwarfs? Some (but not all) miniature cattle, horse, and donkey breeds produce occasional dwarfs. Because of serious limb and internal abnormalities, dwarf calves and foals that survive birth seldom live happy, pain-free lives, so it behooves breeders not to knowingly produce them. The gene responsible for dwarfism in cattle has been identified, but miniature equine dwarf genetics are poorly understood. Miniature equine dwarfism is more prevalent than most new breeders imagine, so it's wise to research this issue before you commit.

Is the species you choose suited to your climate, your temperament, and your physical capabilities? You could raise Miniature Highland cattle in South Texas and Miniature Zebus in northern Minnesota — but why would you want to? For the animals' sake, choose something adapted for the weather where you live. You will need to do your homework to determine what your options are.

Interacting with some animal species requires more brawn than with others, and with some species you'd better be swift on your feet. Don't take on a species or breed you physically can't handle. It will be frustrating and dangerous if you do.

Loud, abrupt individuals rarely resonate with flighty, reactive livestock. "Do it my way or else" humans and headstrong, aggressive animals are bound to clash. Assess your mind-set carefully, and choose a compatible species. It will save a world of upset for all concerned.

Economic Considerations

Do you have the wherewithal to purchase good-quality animals of the breed you want? Since miniature livestock tend to command top prices, getting started with certain breeds of certain species can be a pricey venture. Even so, start with the best animals you can find and afford. One calf or foal from outstanding parents will fetch a higher price than half a dozen mediocre animals, and it will cost far less to produce the single good one.

Are there enough dollars in your pocket to support your animals when things go awry? Markets falter. Disease rips through your herd. Expect the unexpected. Keeping livestock can be a pricey proposition. Do you have the financial resources to see yours through those bumpy times?

Will your animals be pets? Freezer fodder? A means of producing offspring to raise or to sell? If you keep livestock to claim a lower-cost agriculture land tax assessment, your venture must eventually turn a profit. How much profit is enough? And would you be content if you lost money or if your animals simply paid their way?

Make certain your facilities are adequate before you bring any livestock home. If they aren't safe and appropriate for the species of your choice, do you have enough land, financial resources, and know-how to make the necessary improvements? Can you obtain the building permits to do it? If you need chutes and squeezes or tall fences to repel predators, can you build them or hire someone else to build them? If not, you might need to choose a different species. Factor in injuries, losses, and breakouts; it's the safe and economical thing to do.

Before purchasing livestock of any kind, acquire any licenses and owner/breeder permits required by state and local authorities, and make certain your property is zoned for the type of livestock you plan to keep.

BEFORE YOU CHOOSE AN ENDANGERED BREED

The feeling of helping a rare livestock breed survive is priceless. Some breeds, however, are so rare that every individual in their population counts, so be sure you know the best way to care for your breed.

Be prepared to be a stellar steward. Find a mentor before you begin and thoroughly understand the breed's needs and peculiarities before you buy. Hone your skills on a similar but more common breed before tackling endangered livestock. For example, before you try Kunekunes, get your feet wet keeping pet Potbellied pigs, or try raising Pygmies before you purchase San Clemente Island Goats.

Don't dabble in rare breed livestock; these animals need committed breeders. Don't crossbreed females of rare breeds. Carefully and permanently identify each individual. Plan for your breed's future; make certain your heirs know how precious they are.

2

Which Species?

THIS CHAPTER WILL COMPARE SPECIES and present basic information so that you can choose which species best suits your needs and resources. See section 2 for more in-depth information about the species you select. Whether you keep farm animals for pleasure, food, or profit, there are miniature livestock options designed for every pocketbook, lifestyle, and taste. Which of them beckon to you?

When choosing the species and breeds you might like to raise, factor in personal preference; time constraints; ease of handling; space, fencing, and facilities needed to maintain the species; start-up costs; and profit potential. The chart on pages 8 to 11 may help.

Miniature Cattle

Hobby farmers seeking a diminutive, moderately priced, American Livestock Breeds Conservancy heritage breed (see box on pages 16 and 17) choose Dexters. Dexter cows stand 36 to 42 inches (91 to 107 cm) tall and tip the scales at roughly 750 pounds (340 kg); bulls can be 2 inches (5 cm) taller than cows. The Dexter is a good dual-purpose breed: steers dress out at a respectable 55 to 60 percent live weight, and cows give one to three gallons of 4 percent butterfat milk per day. And Dexters are fairly readily available in North America.

Where's the beef? On miniature steers! Pint-size Herefords registered with the American Hereford Association come in miniature (43 inches [109 cm] and under) and classic (44 to 48 inches [112 to 122 cm]) sizes. Stockier and shorter-legged than everyday Herefords, they mature 20 percent faster and dress out at 65 percent live weight. At 40 to 70 percent less than the size of standard Angus,

Lowlines require one-third the amount of feed to produce 70 percent as much rib-eye, and Lowlines dress out at up to 75 percent live weight.

If it's milk you want, miniature Jerseys are a good choice. Only 38 to 42 inches (97 to 107 cm) tall and weighing 600 to 650 pounds (272 to 295 kg), miniature Jersey cows give two to four gallons of butterfat-rich milk every day — enough for most families and then some.

Or choose Miniature Zebus, Miniature Highlands, Miniature Longhorns, or any of the array of cattle breeds created by Richard Gradwohl of the International Miniature Cattle Breeds Registry. Prices vary widely based on breed, availability (expect to be put on a waiting list), quality, age, and sex.

Miniature Longhorn

THINK BEFORE YOU BUY

Good things come in small packages, but please don't take miniatures for granted. Although small by commercial standards, these animals aren't toys. Miniature cattle sometimes weigh more than half a ton, and miniature equines tip the scale at 200 pounds (90 kg) and more. Even 100-pound (45 kg) miniature sheep can bowl over unwary adults with ease. We'll talk about handling livestock later in this book; in the meantime, take your strength and agility into consideration when choosing which species you'd like to raise.

By the same token, small livestock don't automatically make great pets. Folks are often surprised and dismayed when the miniature pig purchased from an unethical breeder grows to twice the size it's supposed to be, and people who buy sheep for suburban backyard pets are disappointed when their charges behave like sheep instead of dogs. It's perfectly possible to raise a house pig, sheep, or goat, but just because the critter's small doesn't mean raising it will be easy.

FACTORS TO CONSIDER WHEN CHOOSING A SPECIES

Species	Availability	Ease of care	Fencing*
MINIATURE CATTLE	3–5 (depending on breed)	Beef cattle and nonlactating dairy cattle: 1 Lactating dairy cattle: 3–4	Any standard cattle fencing: multiple strands of barbed, high-tensile, or electric fencing; welded-wire mesh cattle panels; wood, plastic, or pipe post-and-rail construction
MINIATURE HORSES	1–3 (depending on quality)	1	Wood, plastic, or pipe post-and-rail construction; multiple strands of high-tensile or electric wire (but not barbed) fencing; welded-wire mesh cattle panels; woven-wire horse or sheep fencing
MINIATURE DONKEYS AND MULES	2–3 (depending on quality)	1	Wood, plastic, or pipe post-and-rail construction; multiple strands of high-tensile or electric wire (but not barbed) fencing; welded-wire cattle panels; woven-wire horse or sheep fencing
MINIATURE LLAMAS	4–5	1	Wood, plastic, or pipe post-and-rail construction; multiple strands of high-tensile or electric wire (but not barbed) fencing; welded-wire mesh cattle panels; woven-wire horse or sheep fencing
KEY:	1 = readily available 5 = rare or demand currently exceeds supply	1 = very easy 5 = time consuming or difficult	* It's always best to house and pasture miniature livestock and poultry of all species inside tall, stout, predator-proof fencing. Also consider keeping a livestock guardian dog or a full-size guard llama or donkey with miniature goats, miniature sheep, and with small groups of miniature equines.

Pasture and shelter	Compatibility with other species	Profit potential
Allow approximately ⅓ the pasture-stocking rate recommended for a conventional cow, based on local conditions (consult your county extension agent for particulars) Communal loose housing: allow 25–35 square feet of space per animal Box stall: allow 64–80 or more square feet of space per animal (ideal sizes: 8' × 8' to 8' × 10')	All other livestock species of all sizes (barring horses that habitually chase cattle)	3–5 (depending on breed)
Allow approximately ¼ to ⅕ the pasture-stocking rate recommended for a full-size horse, based on local conditions (consult your county extension agent for particulars) Communal loose housing: allow 30–40 square feet of space per animal Box stall: allow 64–80 or more square feet of space per animal (ideal sizes: 8' × 8' to 8' × 10')	All other livestock species except aggressive full-size horses	1–4 (depending on quality)
Allow approximately ¼ to ⅕ the pasture-stocking rate recommended for a full-size horse, based on local conditions (consult your county extension agent for particulars) Communal loose housing: allow 25–35 square feet or more of space per adult donkey or mule Box stall: allow 64–80 or more square feet of space per donkey (ideal sizes: 8' × 8' to 8' × 10')	All other livestock species except aggressive full-size horses	1–4 (depending on quality)
Communal loose housing: allow 25–30 square feet of space per llama Box stall: allow 48–80 or more square feet of space per llama (ideal sizes: 6' × 8' to 8' × 10')	All other livestock species except aggressive full-size horses	3–5 (depending on quality)
		1 = very unlikely 5 = high

continued on next page

FACTORS TO CONSIDER WHEN CHOOSING A SPECIES *CONTINUED*

Species	Availability	Ease of care	Fencing*
MINIATURE DAIRY GOATS	1–5 (depending on breed)	Non-lactating: 1 Lactating does: 3–4	Goats require secure fencing to keep them in and predators out: welded-wire cattle or tall sheep panels; woven-wire horse or sheep fencing; multiple strands of high-tensile or electric fencing
MINIATURE FIBER GOATS	3	2–4 (1 on an everyday basis, but Pygoras and Nigoras require annual or biannual stripping or shearing, depending on fiber type)	Same as for miniature dairy goats
PYGMY GOATS	1–3 (depending on quality)	1 (3–4 if milking lactating does)	Same as for miniature dairy goats
MINIATURE SHEEP	3–5 (depending on breed)	2 (1 on an everyday basis, but most miniature sheep also require annual shearing)	Sheep require secure fencing to keep them in and predators out: welded-wire cattle or tall sheep panels; woven-wire horse or sheep fencing; multiple strands of high-tensile or electric fencing
MINIATURE PIGS	Potbellied pigs: 1 Kunekune and heritage pigs: 5		Welded-wire hog panels work best

KEY:	1 = readily available 5 = rare or demand currently exceeds supply	1 = very easy 5 = time consuming or difficult

Note: Figures and comments are subjective and based on the author's experience and that of veteran livestock owners surveyed through species-specific e-mail forums.

Pasture and shelter	Compatibility with other species	Profit potential
Communal loose housing: allow 12–20 square feet or more per goat Box stall: allow 16–24 square feet or more of space per goat (ideal sizes: 4' × 4' to 4' × 6', depending on breed)	All miniature livestock species; full-size goats, sheep, llamas	2–4 (depending on breed)
Communal loose housing: allow 16–20 square feet or more per goat Box stall: allow 16–30 square feet or more of space per goat (ideal sizes: 4' × 4' to 4' × 6')	All miniature livestock species; full-size goats, sheep, llamas	3
Communal loose housing: allow 12–16 square feet or more per goat Box stall: allow 16–24 square feet or more of space per goat (ideal sizes: 4' × 4' to 4' × 6')	All miniature livestock species; full-size goats, sheep, llamas	1–3 (depending on quality)
Communal loose housing: allow 12–16 square feet or more per sheep Box stall: allow 16–24 square feet or more of space per sheep (ideal sizes: 4' × 4' to 4' × 6')	All other livestock species	3–4 (depending on breed)
Communal loose housing: allow 20–25 square feet of space per pig Box stall: allow 24–48 or more square feet of space per pig (ideal sizes: 4' × 6' to 6' × 8', depending on adult size and breed) Smaller miniature pigs can also be housetrained and kept indoors as pets	All other livestock species with one caveat: pigs have been known to kill and eat small, weak lambs and kids, so they shouldn't be kept with sheep or goats with newborn offspring at their sides	Potbellied pigs: 1 Kunekune: 4–5 Heritage breeds: variable (based on whether they're raised for breeding stock or pork)
		1 = very unlikely 5 = high

Miniature Horses

Miniature horse

Once the playthings of royalty and the ultra-rich, today's miniature horses are priced within the means of people with moderate incomes.

Miniature horses are registered by the American Miniature Horse Association, which requires mature AMHA horses to measure 34 inches (86 cm) and under, and the American Miniature Horse Registry, which registers in two divisions: 34 inches (86 cm) or less and 34 to 38 inches (86–97 cm) in height. It's best to buy breeding stock recorded with both groups; other factors being of equal quality, the tiniest minis, the flashiest individuals (pintos, Appaloosas, Pintoloosas), and the double-registered horses sell dearer than the rest.

Size, type, training (minis make outstanding driving horses, singly or in tandem), show record, color, and bloodlines play a big part in determining mini horse prices, so if you plan to breed to show a profit, study the market before you invest. Depending on your needs, you could spend $200 or $20,000 for a mini horse and in both cases get precisely the horse of your dreams.

Or consider breeds less frequently encountered in North America, such as Falabella miniature horses from Argentina (most are also registered with one or both of the American miniature horse registries listed above), British Shetland ponies from the United Kingdom, or Caspian horses from Iran.

ONE SPECIES OR SEVERAL?

Folks who keep miniature livestock tend to start with one breed of a single species and later expand to raising several breeds and species. Read the breeder profiles scattered throughout this book to get a feel for raising multiple species. For your sake and the animals' too, however, start small, learn everything you can about your first breed and species, and expand later on if you like.

Miniature Donkeys

In 1958, Bea Langfeld of Danby Farm in Nebraska incorporated the Miniature Donkey Registry (MDR) to record the pedigrees of Sicilian and Sardinian imports and their offspring. Ownership eventually passed to the American Donkey and Mule Society (ADMS), which still maintains the original MDR studbooks. More than 40,000 miniature donkeys measuring 36 inches (91 cm) and under at the withers have been duly recorded. Another group, the International Miniature Donkey Registry (IMDR), records donkeys in two divisions: Class A (36 inches [91 cm] and under) and Class B (36.1 to 38 inches [91.7–97 cm] tall). The American Council of Spotted Asses registers spotted miniature donkeys of all sizes.

Mini donkeys are arguably the world's most charming pets. Most are hefty enough to pull an adult or several children in a cart or to pack up to 50 pounds (22 kg) of camping gear. They're a hit at parades, and donkey and mule shows offer a wide array of classes for bantam brayers.

Miniature donkey prices vary widely, from $200 or so for an unregistered, plain-Jane gelding to $7,500 or more for a well-bred young jack or jennet with a show record, popular bloodlines, and color du jour. Tinier donkeys tend to garner the most cash.

Miniature Mules and Hinnies

Miniature mules are created by breeding miniature donkey jacks to miniature horse mares; miniature hinnies (they are much rarer beasts) are the offspring of miniature stallions and miniature donkey jennets. The American Miniature Mule Society and the American Donkey and Mule Society both register miniature mules and hinnies; the American Council of Spotted Asses registers spotted mules and hinnies of all sizes.

Miniature mules and hinnies can do anything miniature donkeys do, and they're crowd pleasers wherever they go. Not a lot of folks are breeding these winsome creatures — yet. If you love equines and want something different, miniature mules might be just the thing.

Miniature Llamas

Some folks think miniature llamas and alpacas are one and the same, but they aren't. Llamas (*Lama glama*) and alpacas (*Vicugna pacos*) both belong to the *Lama* genus of the Camelid family, but they are two separate species.

The American Miniature Llama Association is allied with the International Llama Registry, the governing body that registers llamas of all sizes.

For a llama to be registered with the American Miniature Llama Association, it must also be registered with the International Llama Registry, three years of age or older, and no more than 38 inches (97 cm) tall (foundation stock llamas can be 38.1 to 40 inches [96.7–102 cm]). Immature llamas are conditionally registered until they turn three, but they will be officially registered only if they stand 36 inches or smaller, their mothers are registered miniature or foundation stock llamas, and their sires are registered miniature llamas.

Miniature llama

Miniature llamas make fine pets; many yield wonderful fiber; and major

THIS FARMER'S TAKE

The best piece of advice I can give you is this: Know what you need before you buy. By determining precisely what you require regarding breed, size, color, coat style, popular bloodlines, and other incidentals, you can avoid the loss of time and momentum, cost, and aggravation of upgrading later on.

llama shows host classes for "vest-pocket" llamas. The future of tiny camelids seems rosy. Fewer than 1,000 llamas in the United States meet American Miniature Llama Association standards and not all of those are registered, which means finding breeding stock can be difficult indeed.

Miniature Goats

Whether you choose tiny mini goats like plump, perky Pygmies or svelte, colorful Nigerian Dwarfs — or slightly larger miniatures like Kinders, Pygoras, and Miniature Dairy Goat Association breeds, you're sure to fall in love with these fey and charming creatures.

Consider the familiar, friendly Pygmy, a short-legged, meat-type goat descended from dwarf stock imported from West Africa in the 1950s. Just 16 to 23 inches (41 to 58 cm) tall, good-natured, gregarious Pygmies make huggable, fun-loving pets. A bonus: Lactating Pygmy does give about a quart of rich, 6 to 10 percent butterfat milk per day — enough for a small family's table.

Nigerian Dwarf does give up to twice as much milk as the Pygmies. Elegant, refined, and with good dairy conformation, Nigies can be up to 23 inches (58 cm) tall, and they come in a staggering array of colors and patterns; some even have pretty blue eyes. The American Dairy Goat Association, the American Goat Society, and the Nigerian Dwarf Goat Association register these minute caprine cuties.

Pygora and Kinder goats were developed using Pygmy goat crosses — the former with Angora goats to create miniatures with marketable, soft fleeces and the latter with Nubians to create a scaled-down, dual-purpose meat and dairy goat breed. Nigora fiber goats are Nigerian Dwarf and Angora goat crosses.

Miniature goats come in an array of shapes, sizes, and coat types, such as smooth-coated Mini-Nubians, hairy Pygmies, and Pygora fiber goats.

Breeders developed the Miniature Dairy Goat Association breeds (Mini-Alpines, Mini-LaManchas, Mini-Nubians, Mini-Oberhaslis, Mini-Saanens, and Mini-Toggenbergs) by crossing standard-size purebreds with Nigerian Dwarf goats. All produce about two-thirds as much milk as a standard dairy goat while consuming one-third as much feed. Preferred heights are 21 to 25 inches (53–64 cm) for does and 23 to 27 inches (58–69 cm) for bucks.

Unregistered pet-quality miniature goats, especially wethers (castrated males) and bucklings (immature intact males), can be purchased for $100 or less in some locales, but expect to spend from $300 to $1,000 and up for quality registered breeding stock from certified, disease-free herds.

Miniature Sheep

Kept for pets, fleece, or meat, petite Shetlands and Soay, teddy bear–faced Babydoll Southdowns, and elegant Miniature Cheviots are among the most versatile and productive breeds of small farm sheep.

RAISING HERITAGE BREEDS

The American Livestock Breeds Conservancy (ALBC) is a nonprofit membership organization devoted to the promotion and protection of more than 150 breeds of livestock and poultry. Now in its third decade of service, it's the primary organization in the United States working to conserve rare breeds and genetic diversity in heritage livestock.

The ALBC breeds that are included in its Conservation Priority List fall into the following categories:

- **Critical.** Breeds that have fewer than 200 annual registrations in the United States and whose global population is estimated at less than 2,000.
- **Threatened.** Breeds that have fewer than 1,000 annual registrations in the United States and whose global population is estimated at less than 5,000.
- **Watch.** Breeds with fewer than 2,500 annual registrations in the United States and a global population estimated at less than 10,000. Breeds that present genetic or numeric concerns or that have limited geographic distribution are also included.

The Soay is an ancient breed shaped by nature on the Isle of Soay in the St. Hirta archipelago, 41 miles off the windswept coast of Scotland, where it has dwelled since Neolithic times. The lithe, lean, elfin sheep stand about 22 inches (56 cm) tall. Mature rams weigh 85 to 90 pounds (38–41 kg), and ewes run about 30 pounds (14 kg) lighter.

Shetlands descend from sheep carried to the Shetland Islands by Viking settlers during the first millennium AD. They come in a startling array of colors and patterns. Rams tip the scale at 90 to 125 pounds (41–57 kg); ewes weigh 75 to 100 pounds (34–45 kg).

Babydoll Southdowns are the traditional British Southdowns of yore (after World War II, the British Southdowns were crossed with larger animals to satisfy the hunger for larger cuts of meat). The woolly, sweet faces of these short, chunky sheep endear them to most everyone they meet. Colors include off-white, black (often fading to gray), and spotted. Babydoll Southdowns stand 18 to 24 inches (46–61 cm) tall.

- **Recovering.** Breeds that were once listed in another category and have exceeded Watch category numbers but are still in need of monitoring.
- **Study.** Breeds that are of genetic interest but either lack definition or lack genetic or historical documentation.

Several of the miniature livestock breeds showcased in this book are old-fashioned heritage breeds listed as Critical or Threatened breeds by the ALBC, including cattle registered by the Florida Cracker Cattle Association and the Pineywoods Cattle Registry and Breeders Association; Caspian horses; Myotonic (fainting) and San Clemente Island goats; and Ossabaw Island and Guinea hogs. Others, like Highland and Dexter cattle, Nigerian Dwarf goats, Shetland sheep, and Miniature Mediterranean donkeys, are listed as Recovering breeds. These breeds all need the help of dedicated breeders if they're to survive.

Other organizations at home and abroad are dedicated to promoting and protecting rare breeds, among them the Rare Breeds Survival Trust (UK), Rare Breeds Trust of Australia, New Zealand Rare Breeds, Rare Breeds Canada, and the Equus Survival Trust. All field inquiries from new, potential breeders. Check the resources section for contact details.

Miniature Cheviot sheep

As Babydolls are to modern Southdowns, Miniature and Classic Cheviots are to British Border Cheviots of the nineteenth century and before. They may be black (usually fading to gray or tan), white, and even spotted. Easy-care, wool-free faces and legs; elegant arched profiles; and erect, horselike ears are their hallmarks. Naturally polled Miniature and Classic Cheviots are strongly built, broad, and short-legged, ideally standing 24 inches (61 cm) tall or less.

Miniature Pigs

Miniature pigs fall into two categories: pets (Kunekunes and Potbellied pigs) and heritage swine (Guinea Hogs and Ossabaw Island Hogs) — though the heritage swine make fine pets, too.

Potbellies descend from the Í breed of Vietnam. Most purebred Potbellies run 80 to 200 pounds (36–91 kg) and stand 18 to 36 inches (46–91 cm) tall at three to four years of age. Piglets can be housetrained, but indoor pig parenting is not for the faint of heart. Many former house pigs languish in rescue facilities and pig sanctuaries awaiting adoption. Rescued and surrendered pigs make fine pets, and many such pigs need homes, so please consider adoption before purchasing a Potbellied piglet.

The new pet pig in North America is New Zealand's Kunekune, a cute, colorful, hairy pig with an upturned snout. Many have wattles (also called tassels or pire pire) dangling from their lower jaws. Big ones stand 30 inches (76 cm) tall and weigh up to 250 pounds (113 kg); most are considerably smaller. They are gaining in popularity but are still rare.

Two of the critically endangered breeds listed by the ALBC are miniature pigs: Guinea Hogs and Ossabaw Island Hogs. Guinea Hogs (also called Guinea Forest Hogs) originated on the Guinea coast of Africa. They were carried to the

American South with slave traders, where they were once common farmstead pigs. Guinea Hogs are black and hairy and stand 15 to 20 inches (38–51 cm) tall. Ossabaw Island Hogs descend from Spanish pigs brought to Ossabaw Island (off the Georgia coast) during the 1600s. They are hairy, long-snouted pigs. Wild ones weigh about 100 pounds (45 kg); they grow somewhat larger in captivity. Both breeds need help if they're to survive.

Kunekune pig

Guinea Hog

THINK BEFORE YOU BREED

Unless you're breeding for a terminal market, such as producing grass-fed beef from Miniature Herefords or home-cured hams from Guinea Hogs, think about this before you breed.

Animal rescues and sanctuaries are bursting at their seams trying to house and feed an unending supply of unwanted, excess pets including thousands of hapless farm animals: Pygmy goats, miniature horses and donkeys, and, especially, Potbellied pigs. Granted, these are not the crème de la crème of their species, but they need homes, too. You might find the animal you want at one of these shelters. So unless you are reasonably certain the animals you produce won't end up in the same situation, please refrain from breeding them at all.

Don't contribute mediocre stock to a saturated market. Breed only the best to the best, be responsible about finding good homes for young stock, and consider lending a hand to organizations that work hard to find homes for unwanted examples of the species you care for. As one leading llama information site succinctly says, "If you don't rescue, don't breed!" By contributing time, talent, and financial support to rescue groups and sanctuaries, you can help offset the damage perpetrated by breeders less conscientious than yourself.

BREEDER'S STORY

Gib and Melba Mullins
Ridgerunner Miniature Horses, Missouri

GIB AND MELBA MULLINS breed "tinies," the smallest of correctly proportioned American Miniature horses, on their Ozark Mountain ridgetop ranch near Bakersfield, Missouri. The Mullins' homebred senior stallion, Ridgerunner's Moonlight Bay, only 26 inches (66 cm) tall and better known as Bubba, shares his domain with a herd of teensy mares of many colors, including Appaloosas, pintos, and an array of dilutes, including rare cremellos.

We asked Melba about their horses and breeding program.

"We bought our first three minis in 1998, just to play with," she replied. "I'd always wanted horses, and when Gib retired and we moved back to the farm, we were on a trip to Rogersville to buy guinea fowl and accidentally stumbled upon a breeder who was selling all of his horses. We didn't realize at first that we bought very nice stock, but as our little herd grew and we met others in the business, we began to realize that others wanted what we were raising, so we thought, 'We must be doing something right!' I've *never* regretted getting into the mini business!

"When I think 'miniature horse,' I think of the tiny ones. They're my first love. Even though a registered mini can be up to 34 inches tall in one registry and 38 inches tall in the other, to my way of thinking, the tinier, the better! But you can't just breed for size — you have to consider conformation above all else. And I love Appaloosas and dilutes, but as I've heard many times, 'There's no bad color on a good horse.'

"Bloodlines are important to some people, and I have to admit I'm a sucker for a couple of them myself, but actually they mean very little. It's just nice to have that World Champion stallion's name on your foal's papers! Had we known what we had when Bubba was born, we'd have shown him, and I have no doubt he'd have that Champion title in the 28 inches and under class by now.

"The market for any horse right now is not the best, but as a rule, there's a good market for the tinies. But I'm pretty careful about who I sell my tiny mares to. They have to have experience breeding and foaling tiny ones. I haven't seen that the mini mares have any more trouble foaling than their bigger sisters, but if they do have a problem, there's not much room for you to work. You *have* to have a good vet.

"Double-registered horses are worth the price, in my opinion. Why cut your market in half by only using one registry? Some folks show at American Miniature Horse Association shows and others prefer American Miniature Horse Registry shows, so having your horses registered with both makes them attractive to everybody.

"There is certainly room in the miniature horse business for new breeders. We love to see new folks discover the wonders of owning minis! This economy has forced quite a few breeders out of the business and lowered prices across the board, so this is a super-good time for new folks to get started! There are some really good bargains on some really nice horses right now. In fact, I know people that are getting into the business and buying horses that they couldn't have afforded a few years ago. That won't last forever, though, so if you've ever entertained the notion of buying a mini — or three — now's the time!

"Miniature livestock in general is the hot thing to have. As we Baby Boomers age, we still love our animals but may not be as spry as we once were. Miniature horses and cattle are the right fit: they're much easier to handle and eat a LOT less than their full-size counterparts. Lots of people that used to love riding now love driving minis. A 'horse person' will always long to own a horse of their own, and with these little fellers, they can continue to do so in their retirement years!

"I would advise anybody who's thinking about buying a mini, be it a horse, donkey, or cow, to do their research, take a knowledgeable friend along to visit some farms, and buy the very best they can afford. Less is more! It is better to have one really good mare than two or three mediocre ones.

"I tell people that miniature horses are the true quarter horse: a quarter the size, a quarter the feed, a quarter the space, but four times the love. Especially the foals! They're like puppies with hooves. Once you've had a mini foal sitting in your lap, you're hooked for life!"

3

Getting Started: Education, Vets, and Where to Buy

ONCE YOU HAVE DECIDED which species and breeds to raise, the next step is to learn all you can before buying. It is also important to locate a good veterinarian at this time, before you bring home any animals. Then, when you're prepared, begin seeking your stock in the right places.

Educate Yourself

When starting a new venture, there is usually a strong learning curve to overcome, so it's best to gain as much knowledge as possible before you begin. Make certain you know your target species to the nth degree. Don't charge headlong into any livestock enterprise on the basis of hearsay; educate yourself before you buy.

Seek out breed associations. A useful first step is to join appropriate breed associations and scope out their local affiliates. Attend meetings. Volunteer to work on committees. You'll meet people "in the know" and learn while having a great time.

Attend livestock shows and expos. Examine livestock, watch the judging, and ask questions. Approach exhibitors when they aren't engaged in important tasks like last-minute grooming or watching their animals being shown.

Read books. If you don't know what's available, visit your library and ask to speak with the research librarian. Rare books and books that aren't in your

THE INSIDE SCOOP FOR FREE!

Joining e-mail groups (also called e-mail lists and Listservs) is a fun, free, and invaluable way to learn from folks already walking the path you're embarking upon. The best of these services is YahooGroups (see Resources).

Once you've subscribed to a group, you can choose whether to receive individual e-mails or daily digest e-mails or to read the posts at the Web site. Posting is easy: simply send an e-mail to the group's e-mail address. All of the subscribers who signed up for e-mail delivery will receive a copy and it will also be posted online.

These groups maintain photo albums and file directories; the latter are often packed with useful items, such as how to build equipment for your species or treat its various ailments, so be sure to check them out when you join.

library's collection can be ordered via interlibrary loan. Read everything. Don't overlook juvenile nonfiction; Storey's guides for young adults (*Your Goats*, *Your Calf*, *Your Sheep*) are jam-packed with easily accessible information.

Utilize the Internet. If you've never learned to use a search engine, this is the time. I recommend Google. If you need help, use the great tutorials at Google Guide. There's a world of easily accessible, free information on the Internet (see Resources for a list of Web sites to get you started).

Use ATTRA's services. Appropriate Technology Transfer for Rural Areas is a National Sustainable Agriculture Information Service (see Resources) jointly funded by the National Center for Appropriate Technology and the USDA Rural Business Cooperative Service. In 2002, its agricultural agents, working out of offices in Fayetteville, Arkansas; Davis, California; and Butte, Montana, fielded more than 30,000 requests from ranchers, farmers, researchers, and educators — people just like you — and all of ATTRA's services and publications are provided free of charge.

First-time callers receive ATTRA's bulletin "Sustainable Agriculture: An Introduction," which is a complete list of more than 200 detailed publications, and a two-year subscription to ATTRAnews. A program specialist takes every call. If existing publications address your concern, he'll send them; if not, you'll be mailed a custom report in two to four weeks.

Find a teacher. "When the student is ready, a teacher will come," so states an ancient adage. And any time you break into a new venture, be it raising miniature sheep for pets or Guinea Hogs for meat, an experienced mentor will pave your way to success.

Don't, however, choose the first mentor who happens down the pike. It's important to learn from someone truly knowledgeable; "experience" doesn't always equate with wisdom. Find someone who produces the type of livestock you admire, not just the "right" breed or species but animals you'd like to have in your barn. This person would ideally be engaged in the same sort of venture you're embarking on, be it raising grass-fed beef or milking dairy goats. She doesn't have to live close by so she can mentor you in person (though that's an asset); telephone and e-mail contact works well, too.

Once you have found someone, politely approach this person and ask if you may ask questions as they occur. Some mentors gladly talk newcomers through scary midnight birthings by means of cell phone, but others don't. Establish contact protocols early on, and please don't overstep your bounds.

Finding the Right Vet

Always choose a veterinarian for your livestock *before* bringing any animals home. The process isn't as easy as you might think. There is a serious (and growing) shortage of large-animal vets nationwide (though minis are relatively small animals, they still need to be seen by large animal vets, as they specialize in the needs of livestock species).

Developing the Initial List

To find a *good* vet, ask other folks who breed your species for recommendations. Narrow down your selection to several vets most people like, then call their offices and talk to the person in charge. Here are some questions to ask.

- How many veterinarians are associated with your practice? Do any of them specialize in large animal practice?
- Are your veterinarians familiar with the species I'm raising? Do any of them raise this species themselves?
- What range of medical services does your practice provide?
- Can I stipulate which of your vets I want to treat my livestock?
- How are phone calls handled? When I call, may I speak to a vet?
- Do your vets make routine farm calls? Will someone come to my farm in an emergency? If not, do you have facilities to board my species at your practice? What about after-hours, weekend, or holiday emergencies?

LARGE-ANIMAL VET SHORTAGE

Vets who operated large-animal and mixed practices for years are switching to small-animal practices in droves, while fewer graduating vets each year handle large stock. According to a survey by the American Veterinary Medical Association, in 1994 about 36 percent of new veterinary school graduates said they would treat large animals at least part time. In 2004, only 25 percent were willing, and some of those said they would devote no more than 20 percent of their time to large-animal practice.

The major reason for the shift away from large-animal practices: money. The average vet leaves college owing between $80,000 and $100,000. A large-animal practitioner who makes farm calls sees a fraction of the paying clients that a clinic-bound small-animal vet treats every day. With massive debt hanging over new vets' heads, there's little wonder small-animal practice makes sense.

Another reason vets cite: pet and horse owners are more willing (and perhaps more able) to pay what it takes to treat an ailing animal than many farmers are, who are reluctant to sink money into injured animals they can convert into meat. It's discouraging when clients give up on an animal that could be saved, say vets who treat them. Other reasons vets switch from large-animal to small-animal practice include the inability to recruit partner vets for their practices, the lack of qualified emergency practices to refer clients to during the vet's time off, and the simple fact that it's easier (and safer) to work with pets instead of livestock species.

- What are your hours? To whom do you refer clients when you're unavailable?
- How is payment handled? Do you offer payment plans for major procedures or collaborate with a lender that does?

Paying the Clinic a Visit

Then, arrange for a time to visit the clinic (don't just show up unannounced). Note the mileage from your door to the vet's. When you arrive, meet the staff; they are your link to the practice's vets, so it's important that you all get along. Are they friendly and knowledgeable? Will someone give you a walk-through tour of the treatment facilities? Do they allow you to speak with available vets who will be treating your animals?

Ask each vet some questions. How long has she been treating the species you plan to raise? Will she mind if you research a problem in books or online and bring her your results? (Some vets appreciate input; others take offense.)

How does she feel about clients performing routine health care procedures such as treating minor injuries and giving their own shots? Will she dispense prescription drugs such as painkillers and epinephrine if you need them? If you would like to investigate alternative or complementary holistic therapies, will she object?

What sorts of diagnostics are available in-house (X-rays, ultrasounds, blood work, endoscopies, and the like)? If they aren't done in-house, who does them?

Ask about policies and fees. Some vets require extensive, expensive diagnostics before they'll treat a beast. Can you afford them?

If the practice offers overnight or long-term, in-house care, ask to see the boarding facilities. Are they safe, clean, and arranged so patients can't physically interact with one another? What sort of feed are the animals eating? Can you provide your own feed if you want to? Is water readily available and is it clean?

Whatever questions you ask, you should never be made to feel inadequate or stupid. They're your animals and you'll be footing the bill, so if you don't like the way you're treated or how your questions are handled, by all means take your business elsewhere.

Test Out Your New Vet and Treat Her Right

Once you've selected a practice, don't wait for an emergency to try out the vet. Schedule a routine farm visit and see how that goes. Does she arrive promptly, or does someone from her office phone to inform you of delays? How does she interact with your livestock? Are you comfortable with her attitude and her work?

James Herriot, world-famous Yorkshire veterinarian, doesn't practice medicine any more; today's vets are a breed apart. It's unlikely you'll find a practice that satisfies all your needs, so prioritize depending on what matters most to you. Is it worth driving farther to work with a better vet? And which is better, the brusque, abrasive vet with years of experience with your species or the friendly, young vet who is eager to work with you and learn?

Once you decide, treat your new vet like a prince; a good vet is worth his or her weight in gold. If your vet makes farm calls, pay attention to the following:

- Have your animals caught, cleaned up, and ready for treatment when the vet arrives. Chasing your animals across a forty-acre field is not her job. Learn to handle emergencies until your vet arrives.

- Be there. Your animals know you, and they'll behave better if you're on hand to help. If you don't understand a treatment, ask questions. If follow-up care entails detailed instructions, write them down.
- Provide a comfortable, weatherproof, well-lit place for your vet to work. Provide a reliable means of restraint even if it's a makeshift chute or extra hands to help control a fractious animal.
- A cold beverage on a sweltering summer afternoon or a steaming cup of coffee in the winter makes a good vet feel appreciated.

If your vet doesn't make farm calls, she may be able to talk you through a procedure on the phone. Find out if your cell or wireless house phone works in the barn. Stock well-equipped first-aid and birthing kits and know how to use them. If you have to bring your animal to the vet's practice, phone ahead, even in an emergency. Know the fastest, safest route so you can arrive quickly and treatment can begin right away.

Don't disturb your vet during nonworking hours unless she asks you to do so. By the same token, don't wait until a minor problem escalates into an after-hours or weekend emergency. Know what you can and cannot do on your own and involve a vet as soon as one is needed.

If you have a problem with the vet or her staff, talk to them about it; don't bad-mouth the practice behind their backs.

And with no exceptions, settle your bill when payment is due.

Seeking a Holistic Vet?

Nowadays many people prefer a holistic approach to livestock management and care. When their animals become injured or ill, it's important to find a holistic vet or at least one who is open to alternative and complementary therapies. In rural America, such vets are often few and far between. So, it's doubly important to locate a vet before you need one. Here are some things to consider:

- The American Holistic Veterinary Medical Association (see Resources) offers a comprehensive Find a Holistic Vet feature on its Web site. Search by state, modality, or species.
- Mainline veterinarians are often willing to collaborate with holistic vets upon request, usually by phone. Ask.
- Many modalities, such as massage, treatment with homeopathic or flower remedies, herbs, and Reiki, can be used to complement your mainline veterinarian's allopathic treatments. Take some classes, read some books; many of these therapies aren't difficult to learn and they are very, very effective.

BE A DO-IT-YOURSELF VET (SOMETIMES)

In many parts of rural America, you had better learn to be your own vet at least some of the time. Fewer veterinarians are willing to make farm calls to handle everyday procedures such as birthing young stock, giving vaccinations, and treating minor injuries.

Everyone who keeps livestock should be able to handle routine medical procedures and minor emergencies. Many of us learn to handle bigger problems as well. Decide how much you feel comfortable taking on, then take a class or seminar, read some books, or ask your vet or mentor to help you learn the basics.

That said, never hesitate to call the vet if something seems not quite right or you're looking at a situation you know you can't handle. That's what he's there for.

Where to Buy

Once you have covered the preliminaries, it's time to start looking for livestock. In the next chapter we'll talk about what to look for. For now, let's discuss where to look.

Touching Base with Sellers

No matter what type of livestock you're buying, it's important to deal with reputable sellers of healthy stock.

- Start by contacting organizations involved with the species and breeds you choose. Most registries supply lists of breeders on request and most host directories on their Web sites.
- Check out online breeder directories. To find them, type the name of the animal you're searching for (for example, Lowline cattle, miniature horse, Pygmy goat) and the word *directory* in the search box at your favorite search engine (for example, Google).
- Visit breeders' Web sites. Find them the same way by substituting the word *sale* for *directory* in your search engine's search box. You could also qualify it by state (*Babydoll Southdown sale Montana*). If breeders' Web sites don't offer the precise animal you're searching for, e-mail sellers and ask if they have additional livestock for sale. If they don't have what you want, they may know someone who does.

- Join species- and breed-related e-mail groups in which subscribers post what they have to sell.
- Check out the periodicals listed in the resources section at the back of this book; they all run display ads, directories, and classifieds. Subscribe to your favorites or pick them up at newsstands and farm stores such as Tractor Supply Company.
- Take in a show, seminar, or expo. Visit information booths and chat with exhibitors between classes. State and regional breed associations sanction shows and host seminars and expos; e-mail or call these organizations for specifics.

If you want to breed miniature livestock, start with breeding stock from successful producers. Study show results, read pertinent material in magazines and online, and talk with other breeders who consistently produce the kind of animals you'd like to own.

TELEPHONE PROTOCOL

Chances are you'll make a lot of calls before you buy your first handsome beast. Sometimes you'll reach call waiting or an answering machine. When you do, make it easy for sellers to get back to you. In the last two months I've been unable to return three calls because I couldn't understand what the caller was saying!

- If possible, use a landline for making inquiry calls. In some rural areas, my own included, cell phone reception is erratic. Cell phones are notorious for fading and cutting out.
- Please, speak slowly and clearly, especially when leaving your name, a phone number, or a mailing address. People that hail from other parts of the country may not readily understand regional speech patterns. I live in Arkansas, and some Southern dialects still leave me scratching my head.
- When leaving a phone number, read the number through, pause, then read it through again. If the person you're calling doesn't understand you the first time, give him a second chance to get it right.
- If you don't receive a return call, don't assume the recipient isn't interested — call again. It can't hurt and may help you find just what you want.

If you can buy the stock you need close to home, this is good. Healthy animals acclimated to your region that are spared the stress of long-distance hauling tend to stay healthier than those trucked in from distant sources. If you plan to breed high-end show and breeding stock, however, you may not have a lot of local options.

Buying Online

There are many reasons to shop the Internet when you're looking for livestock: you can shop anywhere in the world, at any time of day, seven days a week, without leaving home; you can select from a vast pool of animals and breeders; and, you can research interesting animals and sellers before you deal, thus saving time and money on farm visits. However, be especially careful to deal with reputable sellers. Request buyers' references, and always follow through.

When you find something that piques your interest, check it out. Calls are better than e-mails (unfortunately, many times e-mails go unanswered). You can request video footage of the animal that interests you. If video isn't available, ask to see additional photos taken from many different angles. Examine these materials very carefully and address any issues before you buy.

When you've decided to purchase an animal, insist on a written guarantee and negotiate its terms before you commit. Be clear about how and when you will take delivery of the animal before you make a deposit. Work out all of the details: Who pays for interstate health papers and the tests they entail? How long will the seller hold the animals once payment is made and you're lining up transportation? Who foots vet bills incurred during that wait? What happens if an animal dies or is injured? Get *everything* in writing; never leave anything to chance.

Visiting Farms

Once you've narrowed the field to a group of people selling what you want, contact them, and if possible, arrange to visit their farms.

Arrive at the designated time. Look around. Farms need not be showplaces, but they should not be trash dumps, either. Are animals housed in safe, reasonably clean facilities? Are they in good flesh — neither skinny nor blubbery fat? You might spot individuals that are skinnier or fatter than the norm, but the majority should be in just-right condition.

If you ask the right questions and get the right answers, and if you like the animal you came to see, ask to see its registration papers and its health, vaccination, deworming, and production records. And ask about guarantees. Some producers give them, some don't. If there is one, insist on getting it in writing.

A FEW MORE WORDS ABOUT PAPERS

Let me repeat: Carefully check the registration papers of animals you buy to make certain you're getting what you're paying for.

Carefully compare the animal against the description on its registration papers. If it's tattooed or wearing an ear tag, this should be mentioned in the papers. Unscrupulous sellers have been known to provide unregistered stock with papers from an animal that died or to "upgrade" an animal with papers much better than its own.

The animal must be recorded in the seller's name. The last recorded owner is the only person who can transfer the animal to you. In rare cases, the seller may have a transfer form made out in his name but he's never gotten around to having the papers updated. Some registries will allow this person to issue a second transfer form transferring his impending ownership to you.

If you're buying a bred female, you'll usually need a breeder's certificate signed by both the male's and female's owners in order to register her offspring. Depending on the registry you do business with, the service memo may be a separate document or be part of the transfer slip. Know which type you need.

Unless you want a given animal very badly, don't accept irregularities in its paperwork. Even then, phone the appropriate registry to see what it takes to set things right. For example, in some cases late registrations can cost hundreds of dollars. Don't get bitten in the end.

Finally, trust your intuition. If a seller makes you feel uneasy for any reason, thank him for his time and leave. There are so many honest sellers out there; there's no need to deal with someone you don't quite trust.

Production and Dispersal Sales

Production and dispersal sales are first-class venues for buying quality animals at fair market prices. The best are publicized months in advance. These sales offer printed catalogs that highlight sale animals' pedigrees and production records. Production sales can be the perfect places to meet people and purchase quality livestock.

Major sales make provisions for absentee bidding (usually by phone, fax, or e-mail), but it's best to be there in person. Arrive before the sale starts and do hands-on inspections of the animals you think you might bid on. Another sensible ploy: mark your catalog, designating which animals you plan to bid on, and make a notation of your absolute top bid so you don't get carried away in the heat of a bidding war.

Unless special arrangements are made in advance, payment in full is expected on sale day; don't plan to pay by check unless you've cleared this in advance, too. Animals sell with registration papers, health certificates, and any other documentation needed for interstate shipment. Guarantees, if any, are stated in the sale catalog.

If you bring livestock home from *any* type of sale, plan to quarantine them away from your existing herd. House them in an easy-to-sanitize area at least 50 feet (15 m) from any other livestock but preferably where they can see other animals of the same species at a distance. Deworm them, vaccinate them, trim their hooves or toenails if needed, but keep the new animals isolated for at least three weeks. Don't forget to sanitize the conveyance you hauled them home in.

During the quarantine period, feed and care for your other livestock first, so you can scrub up after handling the new additions. Never go directly from the quarantine quarters to the rest of your animals. If you can prevent it, don't allow dogs, cats, poultry, or any other livestock to travel between one group and the other. When the new additions' time in quarantine is up, sanitize the isolation area and any equipment you've used while caring for them.

Sale Barns

The first rule of livestock buying is to buy from individuals or at well-run dispersal and production sales and *never* from neighborhood sale barns — they're dumping grounds for sick animals and culls.

If you buy at sale barns, you won't know for sure if the animal you buy has been vaccinated, how it's been handled, if it's pregnant and by what sort of male, or if difficult or impossible-to-eradicate diseases such as hoof rot, caseous lymphadenitis, tuberculosis, and Johne's disease are present in its flock or herd of origin. A male may be infertile or so dangerous that its owner is willing to see the last of it at any price. And animals that weren't sick or exposed to disease before they're sold through a sale barn will be by the time the sale ends.

If you attend sale barn livestock auctions, even just to look, you're sure to be tempted to buy. If you succumb to temptation, remember: Never, *ever* turn newly purchased sale livestock out with the rest of your herd. Always, without exception, quarantine them for at least three weeks.

And whether you buy or not, scrub your hands using plenty of soap and sanitize the clothes you wore to any sale before going near your other livestock. Use one part household bleach to five parts plain water in a fine-mist spray bottle to thoroughly spritz boots and shoes, and launder all other clothing in hot water and detergent. It sounds like overkill, but it isn't. Foot rot, soremouth, ringworm, respiratory diseases, and a host of other problems can hitchhike home on your hands and your clothes, so it's best to never take chances.

Exotic Sales

Miniature livestock frequently sell through exotic animal auctions — auctions that sell everything from lion cubs to exotic birds to gnus to more mundane species such as unusual breeds of cattle, sheep, and goats — often at bargain basement prices. But should you patronize them? It depends.

Some exotic sales are long-established, well-run affairs, but most aren't, and even the best are usually held at livestock sale barns where the animals will be exposed to the hazards named in the section above. The bottom line? Attend a sale before you go to buy. If you're comfortable with what you see, it might be worth it. However, it's always better to buy directly from individuals and from production or dispersal sales, because you're more likely to find healthy livestock. The dollars you save at an exotic sale may not go far toward paying a hefty vet bill.

BREEDER'S STORY

Bev Jacobs and Bill Lanier Jr.
Dragonflye Farms Miniature Livestock, Arizona

BEV JACOBS AND BILL LANIER JR. raise miniature Zebu cattle; miniature horses; and Pygmy, Miniature LaMancha, and Nigerian Dwarf goats at Dragonflye Farms, located 25 miles west of Phoenix between Buckeye and Goodyear, Arizona. The couple raise their little livestock to be multi-purpose animals, filling the niche for pets and companions, while also producing quality breeding stock and filling the family's dairy needs.

Their animals shine in the show ring as well, having won multiple Grand Champion ribbons. Dragonflye Farms is the breeder of the AGS Nationals 2008 National Grand Champion Buck, DF Farms HD National Hero! Their Miniature Zebu cattle are all foundation pure animals (an elite group within the Miniature Zebu registry), and their small herd of miniature horses includes Buckeye WCF Eternal Flame — an American Miniature Horse Registry Hall of Fame mare.

When we asked Bev about the Dragonflye Farms breeding program, this is what she said.

"We decided to breed miniature goats after getting our first few goats. We fell in love with them, began showing them, and owned mostly registered stock, so breeding soon followed. We started with a pair of Pygmy doelings and found the goats to be amazing, addicting, fun-loving little companions. Our two goats quickly turned into many more: first a Mini-Alpine and a full-size Nubian, then we added Nigerian Dwarf and Mini-LaMancha goats. Miniature Zebu cattle and miniature horses soon followed. We are absolutely pleased with our decision to raise miniatures!

"The market for miniature livestock can vary; some of it is local or regional. Some is based on 'the flavor of the moment.' It also depends on the economy and if the market is saturated or not.

"Offering more than one miniature species is definitely an asset. This enables you to target more than one audience and to hit many 'niches' in the miniature livestock market. For example, you are able to market to people wanting pets, along with those wanting breeding stock and those who want to show their animals. There are also people who are interested in a particular breed due to its milking ability, color, or how rare it is (so they can be 'the only one in the neighborhood with one' or because they want to help conserve an endangered breed).

"There is always room for new miniature livestock breeders, especially since people's lives change — some breeders leave, making room for up-and-coming breeders.

"My advice to new breeders is to seek out knowledgeable breeders and pick their brains. Go to shows so you can learn what good conformation looks like. Become knowledgeable about your breed or breeds — be able to identify their good points and any problems they're known for. The more you know about your breeds of choice, the better you'll be able to take care of your animals, the better your herd will be, and the more marketable your animals will be in the future.

"If targeting more than the pet market, you need to be up on your particular breed's standards. When you raise or sell registered animals, there will always be some that aren't good representations of the breed or that will never meet the breed standards. Unlike with dogs and cats, you can't always get your mini livestock spayed or neutered if they shouldn't be reproducing. You have to know what to do with animals that shouldn't be added to the gene pool, be it placing them in pet or nonbreeding homes or selling them for meat. Raising livestock is a wonderfully fulfilling endeavor, but along with the fun part comes the responsibility of dealing with life-and-death issues.

"You always want the best you can afford, and then plan to improve from there. If you start with quality you'll have fewer changes to make, making it easier for you and your pocketbook — and much easier to sell to the consumer."

4

Selecting Miniature Livestock

WHETHER SHOPPING FOR DEXTER CATTLE, miniature donkeys, or Pygmy goats, certain points remain the same. To choose healthy, sound, good-natured livestock, keep these thoughts in mind.

Don't buy other people's problems. Know which maladies affect the species you're buying so you can watch for them when evaluating livestock (we'll talk more about species-specific problems in section 2). However, some signs of sickness are the same across the board (see table on next page).

If in doubt about the soundness of the animals you're interested in buying, schedule a second visit to reexamine them. Better yet, arrange for a prepurchase veterinary exam at your own cost. This is always a good plan when buying expen-

NORMAL BIOLOGICAL VALUES*

Species	Temperature (Fahrenheit)	Pulse (beats per minute)	Respiration (breaths per minute)
Llamas	99.5–101.5	60–90	15–30
Equines (horses, donkeys, mules)	99.5–101.5	30–60	8–16
Cattle	100.5–103	50–80	10–30
Goats	101.5–104	60–90	12–20
Sheep	101.5–104	60–90	12–20
Pigs	100.5–104	60–100	25–35

*Average for an adult at rest; for more specific values, consult the species-specific chapters to come.

WHAT DOES HEALTHY OR SICK LOOK LIKE?

Healthy animals	Sick animals
Are alert and curious	Are dull and disinterested in their surroundings Tend to isolate themselves from their peers
Have bright, clear eyes	Have dull, depressed-looking eyes May have fresh or crusty opaque discharges in the corners of their eyes
Have dry, cool noses (cattle have cool, uniformly moist noses) May have a trace of clear nasal discharge (this isn't cause for concern)	May have thick, opaque, creamy white, yellow, or greenish nasal discharge
Have regular and unlabored breathing	May wheeze, cough, or breathe heavily and/or erratically
Have clean, glossy hair coats and pliable, vermin- and eruption-free skin	Have dull, dry hair coats; skin may show evidence of external parasites or skin disease
Move freely and easily	Move slowly, unevenly, or with a limp
Weigh the average amount for their breed and age	Are thin or emaciated; these individuals should always be suspect Are chronically obese; these animals are more prone to fertility and birthing problems, so you may want to avoid these, too
Have healthy appetites	Refuse food; usually a sign of illness
If a ruminant (llama, cattle, sheep, or goat), ruminates (chews cud) after eating	If a ruminant, doesn't chew cud; this animal is very ill
Have firm droppings (relative to their species); tail and surrounding areas are clean	Matted tail, tail area, and hair on hind legs, with fresh or dried diarrhea; may have scours (a.k.a. diarrhea)
Have normal temperatures (see page 36)	Run high or low temperatures; subnormal temperatures are generally more worrisome than fevers

sive animals, but doing so can be cost-effective any time you're introducing new stock to your farm. It's much better to have a vet bill than to bring home difficult-to-eradicate infectious diseases such as caseous lymphadenitis or hoof rot.

Avoiding Dwarfism

"Miniature" does not equate with "dwarf." Miniatures are proportionate, scaled-down versions of full-size members of their species. Dwarves carry one or more crippling genes that cause their bodies to grow in nontypical manners. Dwarfism is a serious problem in miniature equines and in some breeds of cattle. Anyone considering raising these species *must* be fully informed before selecting livestock for their farms.

GENETICS GLOSSARY

Allele. Genes come in pairs (one from the father, one from the mother). An allele is either of the two paired genes affecting an inherited trait.

Autosomal recessive. A condition that appears only in individuals who have received two copies of an autosomal gene — one copy from each parent. The gene is on an autosome, a nonsex chromosome. The parents are carriers who have only one copy of the gene and don't exhibit the trait because the gene is recessive to its normal counterpart.

Carrier. An animal that has a recessive gene in its genetic makeup.

Chromosomes. Cell material that transports genes during cell division.

Codominant. An allele that causes a heterozygous animal to look like an intermediate between homozygous recessive (dwarf) and homozygous dominant (normal) forms. This animal is a carrier. In other words, all three forms — dwarf, carrier, and normal — look different from one another.

Dominant. An allele that causes a homozygous animal and a heterozygous animal to look the same. For example, you cannot visually tell a homozygous curly-coat horse from a heterozygous curly-coat horse. They are both curly-coated, because that allele trumps the allele for straight-haired coats in the heterozygous animal.

Dwarfism in Miniature Horses

To date, no definitive studies have been conducted into the heritability of miniature horse dwarfism, so what follows is what science presently believes.

Miniature horse dwarfism appears to be inherited as a non–sex-linked autosomal recessive. What that means, broken down into everyday terms, is that dwarf miniature horses inherit one copy of the mutated allele that causes dwarfism from each of their "carrier" parents. Furthermore, a dwarf can inherit more than one type of dwarfism, each caused by a separate defective gene.

A dwarf may also be the offspring of a dwarf miniature horse and a normal-looking carrier parent. The dwarf parent may possibly be a "minimally expressed dwarf," meaning it appears almost normal but it manifests characteristics that indicate mild dwarfism, although this is more typical of codominant genetic expression. Or the dwarf parent could be a visibly affected dwarf.

No responsible breeder ever uses a full-fledged dwarf for breeding purposes. Unfortunately, that wasn't true during the breed's early history when dwarfism

Gene. The unit of heredity responsible for the expression of a given trait. Genes normally occur in pairs.

Genotype. The genetic makeup of an animal. The animal may hold genes to control traits that are not physically manifest. Compare with phenotype.

Heterozygote. An organism that is heterozygous for a particular trait.

Heterozygous. Having two nonmatching paired alleles for a genetic trait.

Homozygote. An organism that is homozygous for a particular trait.

Homozygous. Having two identical paired alleles for a genetic trait.

Lethal gene. A gene that causes the death of a developing fetus, usually before or soon after birth.

Mutation. An unexpected, spontaneous change in DNA.

Phenotype. The way an animal looks. Compare with genotype.

Probability. The likelihood of any particular event occurring.

Recessive. An allele that affects an animal's appearance only if it's paired with another identical recessive allele (the homozygous state). For example, a miniature horse carrying a single dwarfing allele will appear normal; it must have two matching dwarfing alleles to be a dwarf. The "normal" gene is dominant; the dwarfing gene is recessive.

Sex-linked. A term used to describe a gene or disease whose inheritance is related to an animal's sex chromosomes.

was poorly understood and small stature was the yardstick by which most minis were measured. Vintage images show many a dwarf stallion or mare whose name appears, sometimes many times over, in today's best pedigrees. Bond Tiny Tim, for example, was severely dwarfed yet was bred extensively, as were many of his dwarf and carrier offspring.

I came to miniatures from a background in Morgan, Arabian, and Thoroughbred horses — breeds that incurred heavy inbreeding early on to establish desirable characteristics — and I bought my first Miniature Cheviot sheep from an inbred flock (see page 56 for a discussion of the common misconceptions about inbreeding) that I continue breeding today. So when we bought our first registered miniature horses, I bought with the idea of setting the type that I preferred: strongly built, cob-type miniatures with wavy coats.

My best mare was sired by a stallion of the type I wanted to produce and whose line was known for wavy coats. After a time I bred her to her paternal

Normal foals' heads are in proportion to their bodies; brachycephalic dwarf foals (the most common type) have highly domed foreheads, deeply dished faces, and undershot jaws.

half-brother, a wonderful little stallion that exemplifies everything I dreamed of creating in my little herd.

Their foal, a charming little soul named Thorfinn, was a dwarf. And he manifested two types of dwarfism, indicating both parents carried both of those genes. Thorfinn was an unusually happy youngster and was greatly loved. He got around well in his own way, but when he was two, he was badly injured in a freak accident, and sadly, we had to euthanize him. His birth and short, sweet life inspired me to study dwarfism in miniature horses. This is what I've learned:

- Many, many breeders quietly euthanize dwarf foals because they don't want the stigma of dwarfism associated with their breeding programs.
- Dwarfism is present in most, if not all, miniature horse bloodlines.
- There are no genetic tests available and to date, neither major miniature horse registry has sponsored substantial research that might lead to the development of such tests.
- If a miniature horse breeder tells you his stock is free of dwarfism, take the statement with a grain of salt. While it might be true, a surprising number of owners and breeders don't recognize minimally expressed individuals (those with borderline dwarfing characteristics like mildly undershot jaws and short necks), and they unknowingly continue using them in their breeding programs; others prefer to simply look the other way. Until reliable DNA testing becomes a reality, dwarfs *will* happen, and they can happen to anyone who breeds miniature horses.

Both brachycephalic and achondroplasic dwarfs have uncommonly short legs; achondroplasic dwarfs' heads are more in proportion to their bodies.

Also, although I remain a proponent of inbreeding and linebreeding (we'll talk about these breeding tools later on), I would never recommend doing so in a population of animals carrying a serious recessive condition such as dwarfism.

Common Characteristics of Dwarfism in Miniature Horses

This is an adaptation of a chart issued by the American Miniature Horse Association to help breeders and buyers evaluate miniature horses for dwarfism. Horses exhibiting any two of these characteristics cannot be registered with AMHA, and the presence of only one characteristic might prohibit registration, depending on its severity.

1. A foal may appear "cute" at birth, but as his head and body mature, his legs do not grow in length. The adult dwarf appears to have an oversized head and body for his overall height.
2. Dwarf foals are often born with contracted tendons, club feet, and buck knees that cannot be straightened out. Joint enlargements and joint deviations (epiphyseal growth irregularities) are common. Extreme cow hocks, extremely short gaskins, and severe sickle hocks, all with varying degrees of visible joint looseness and/or joint weaknesses, are also common. Premature arthritic processes take place in most dwarfs, resulting in progressive ambulatory disabilities.
3. Undershot jaws are rampant among dwarfs. Their molars, therefore, are also misaligned, making it necessary to float their teeth much more often than one would for a normal horse.

4. The brachycephalic dwarf has a large, bulging forehead with an extremely dished (concave) face and up-turned nose. Overly large and bulging eyes sometimes placed at uneven angles and nostrils placed too high up on the face are typical as well. A second, similar type of dwarf has a more normally shaped head, but the head is still much too large for the body; this type does not usually have an undershot jaw.
5. The achondroplasic dwarf has extremely short legs. The bone develops unevenly at the joints, causing gross leg deformities.
6. A dwarf's head is usually much longer than its neck. In some dwarfs, the neck is so short that the head seems to come directly out of the animal's shoulders.
7. Males often have enlarged genitals. Internal organs may also be enlarged and pronounced; potbellies are common.
8. Scoliosis, kyphosis, and/or lordosis (vertebral deviations) are common.
9. Dwarfs are often unable to rear or stand on their hind legs. Many exhibit an odd "tilting backward" gait, with shoulders markedly higher than their rumps.
10. Less obvious characteristics include heart and internal organ defects, shortened life span, and depression due to pain.

How Dwarfism in Miniature Horses Works

An autosomal recessive is a mode of inheritance of a specific genetic trait. Every mammal, including humans, has two copies of every gene on its autosomal (non-sex-determining) chromosomes — one from its mother and one from its father. The dominant allele (one of these two copies) of a gene will always be expressed, but the recessive allele will only be expressed if the animal inherits two recessive forms — one from each parent. When that occurs, the trait becomes physically apparent, and the animal is considered homozygous for that trait.

Carriers appear normal; however, they possess one mutated (defective) allele and one normal copy. When a carrier is mated to a normal individual, none of the offspring will be dwarfs, but 50 percent of the offspring will be carriers and 50 percent will be normal.

Trouble begins when two normal-looking carriers are mated. There is a 25 percent chance that both will transmit the mutated gene, resulting in a dwarf foal; a 50 percent chance that one parent will transmit the mutated gene and the other will transmit the normal gene, resulting in a carrier foal; and a 25 percent chance that both carrier parents will transmit the normal gene, resulting in a normal foal that will not be a carrier. In this last case, dwarfism stops here; if this foal is mated to another normal type, it will not produce any dwarfs.

AUTOSOMAL RECESSIVE INHERITANCE AT A GLANCE

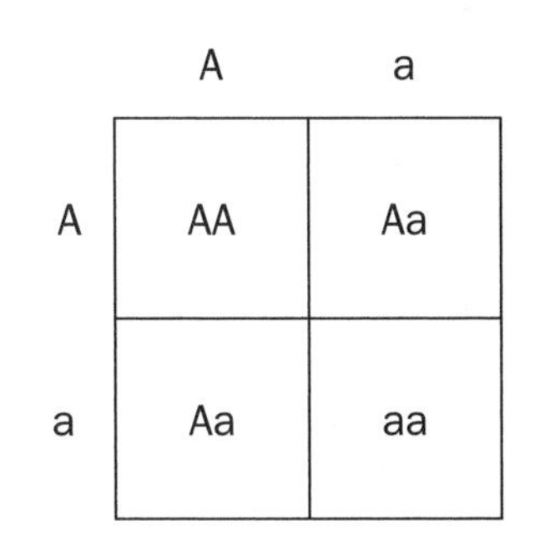

	A	a
A	AA	Aa
a	Aa	aa

This Punnett square illustrates the possible genetic combinations of a foal produced by two carrier (Aa) parents. There is a 25 percent chance it will be a dwarf (AA); a 50 percent chance it will be a carrier (Aa); and a 25 percent chance it will be a normal, noncarrier foal (aa).

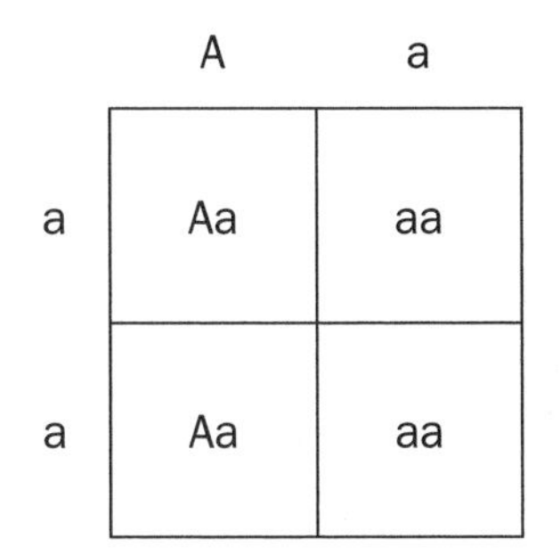

	A	a
a	Aa	aa
a	Aa	aa

This combination illustrates the genetic possibilities of a foal produced by one carrier (Aa) parent and one normal (aa) parent. There is a 50 percent chance a foal wili be a carrier (Aa) and a 50 percent chance it will be normal (aa).

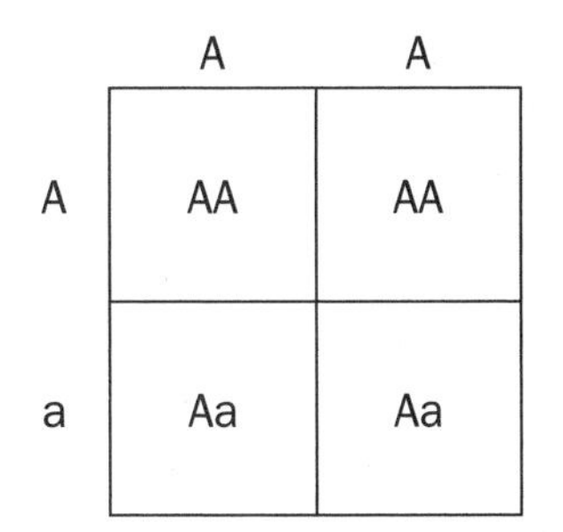

	A	A
A	AA	AA
a	Aa	Aa

When a dwarf (AA) is bred to a carrier (Aa), the foal has a 50 percent chance of being a dwarf and a 50 percent chance of being a carrier.

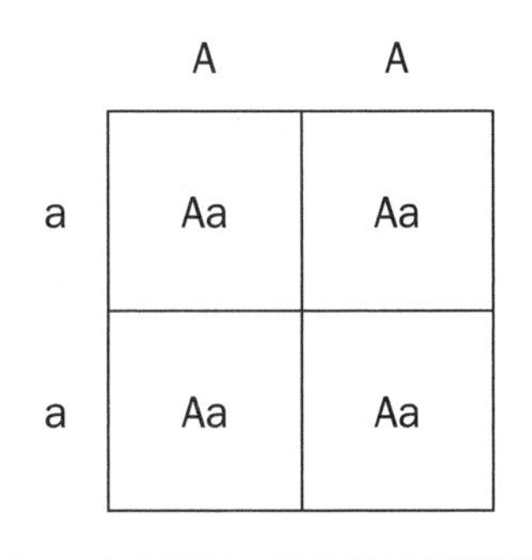

	A	A
a	Aa	Aa
a	Aa	Aa

All of the foals produced by a dwarf (AA) and a normal (aa) horse will be carriers.

Preventing Miniature Horse Dwarfism

Some dwarfs are so mildly affected by the dwarfing gene that they pass for normal in herds belonging to people not aware of subtle signs of dwarfism; others are born so crippled that they're unable to stand or nurse. The breeder must decide what to do with these unfortunates. Many can be saved, and with care, they can lead quality lives. The birth of a dwarf in your herd is nothing to hide. Until reliable genetic testing is developed, it's going to happen, but there are a few things to do in the meantime to lessen the odds.

Develop an eye for subtle signs of dwarfism. Some people believe that horses that display subtle signs are carriers; others think they're minimally expressed dwarfs. Whichever is true, these individuals are almost certainly carrying one or more dwarfing genes, and the wise breeder doesn't use them for breeding purposes. Here are some things to watch for:

- Incorrect occlusion. Overbite (parrot mouth) is rarely seen in miniature horses but underbite (sow mouth, monkey mouth) is common. Never buy a miniature horse without checking his bite. If buying sight unseen, ask for close-up photos of the horse's mouth from the front and side. The upper and lower teeth should match perfectly, but up to one-quarter of a tooth's difference is acceptable. If the difference is more than that, the horse should not be used for breeding purposes.
- Very short necks, with or without unusually large heads.
- Strongly domed foreheads or deeply dished faces, especially in adults.
- Extremely short legs under bodies of substantial size. Cob-type miniature horses are short-legged and stockily built, as are British Shetland ponies, but their legs are long enough to allow them to move like normal small equines. Suspect the individual with legs so short it can't trot, much less canter, or whose small stature derives from short legs alone.

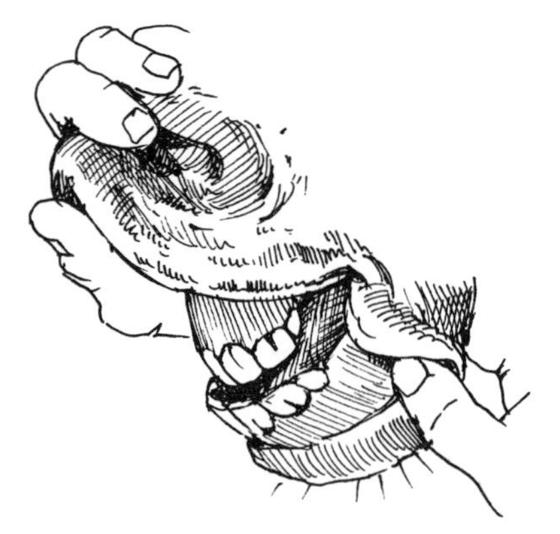

A very undershot jaw showing improper occlusion (sow mouth)

All of these traits are serious conformation faults and shouldn't be propagated in any case, but still, some breeders select for size alone. They use borderline dwarfs due to their small size, so expect to see some on shopping forays.

Before buying, ask "Has this stallion sired dwarfs? Has this mare produced one? How about their close relatives?" Some breeders who don't automatically offer that information will answer truthfully if directly asked.

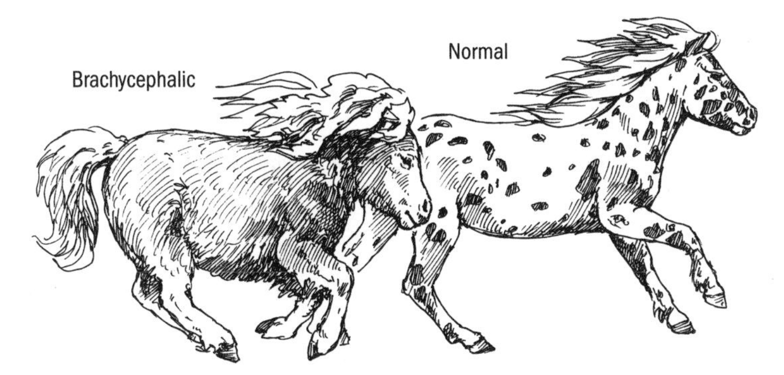

Compared to normal miniature horses, brachycephalic dwarfs have uncommonly short necks.

Cob-type miniature horses and achondroplasic dwarfs both have deep bodies, but cobs' legs are in proportion to their bodies; short-legged dwarfs' legs aren't.

SMALLEST HORSE

According to the *Guinness Book of World Records*, the smallest horse on earth is Thumbelina, a 17.5-inch, 57-pound dwarf born at Goessling's Goose Creek Farm in St. Louis, Missouri. Thumbelina has her own Web site (see Resources), has been featured on *Good Morning America*, and hosts her own Thumbelina Friends Club. She travels the United States each year visiting children's hospitals, special camps and schools, group children's homes, and animal shelters from coast to coast.

Most dwarf miniature donkeys have large heads set on short necks, with full-size bodies and short legs.

What about Carriers?

Should normal-looking miniature horses that have sired or produced dwarfs be used for breeding? If they are otherwise fine examples of the breed and they aren't rebred to the mate with which they produced the dwarf or to another one carrying much the same bloodlines, possibly so. Until DNA testing becomes the norm, breeding these great little horses will remain a crap shoot. It would be a shame to remove some of the breed's best animals from the gene pool and breed lesser animals in their stead.

Dwarfism in Miniature Donkeys

Even less is known about dwarfism in miniature donkeys, but it does occur. Dwarf miniature donkeys have large heads set on relatively short necks and full-size bodies with unusually short legs that are often a little longer in back than in front. It's believed that donkey dwarfism is inherited in the same manner as miniature horse dwarfism; that is, as a simple autosomal recessive.

When buying miniature donkeys, look for properly proportioned individuals, and be especially vigilant about head and leg length.

Dwarfism in Miniature Cattle

The main type of dwarfism of interest to miniature cattle fanciers is the one found in Dexters and various other miniature breeds developed using Dexter genetics. Dwarfism was once a major problem in full-size cattle breeds, too. See page 49 for a full explanation.

Cattle Dwarfism in the Midtwentieth Century

"Cattle breeders are in a fluster about dwarf calves, which are being born in ever-increasing numbers in the U.S. and Canada," begins an article titled "The

Sinister Gene" that appeared in the Monday, January 4, 1956, issue of *Time* magazine. "Cattle have always produced a few dwarfs, just as humans do," it continues, "but dwarf cattle were never common until about seven years ago, when they began to increase alarmingly in all three leading beef breeds — Shorthorn, Hereford, and Aberdeen Angus. The increase has continued ever since."

During the midtwentieth century, show-winning beef cattle were short-legged, thick-bodied, super-wide animals very unlike today's Angus and Herefords. Producers culled to perfect this type, and in doing so, they unknowingly selected carriers of the dwarfing gene as breeding stock; actual dwarfs were quietly dispatched as calves.

The *Time* article goes on to say, "Estimates of the proportion of dwarfs in beef breeds have gone as high as seven percent. Some unfortunate herds have

Short-legged, misshapen snorter dwarfs were common in the midtwentieth century.

INSULAR DWARFISM

Insular dwarfism (not true dwarfism) is also associated with animals of small stature, whereby the reduction in size of larger animals occurs when their gene pool is small due to the limitation of a very small environment, such as on an island, in a cave, or in an isolated valley. Several factors play into this process, including a gene-encoded response to environmental stress; selection of only the smallest and fittest as food declines to a borderline level; and inbreeding without the culling of unsuitable individuals.

Species and breeds that developed in this manner include the anoa and tamaraw (scaled-down wild water buffalo) of Southeast Asia; tiny deer in the Philippines and Crete; and diminutive Shetland sheep and ponies on Great Britain's Shetland Islands.

produced 12 percent and the figures might be higher if cattlemen did not conceal their monsters."

L. P. McCann, director of research for the American Hereford Association, was hired to track down the origin of dwarfism in Herefords. Based on extensive pedigree research, he traced the genetic material to an imported bull named St. Louis Lad, who was born before 1900. McCann recorded his experiences in a book called *The Battle of the Bull Runts* (see Resources) in which he explains that Hereford dwarfism spread so widely and so quickly because cattle produced by three wildly popular seedstock ranches, particularly the famous Wyoming Hereford Ranch, carried the gene. The Wyoming Hereford Ranch bull, WHR Royal Domino 51st, who carried the deadly gene, was also *the* bull of his day, and his bloodlines were widely dispersed across North America.

Once the culprits were found, cattlemen worked hard to eliminate dwarfism from their favorite breeds. Today, all duly American Hereford Association registered Miniature Herefords and registered Lowline cattle are *virtually free of dwarfism*, but it does occasionally occur in full-size Angus and Herefords. Therefore, it behooves the miniature cattle breeder to avoid buying unregistered "miniature" Herefords and Angus or to learn to tell true miniatures from full-size dwarfs.

Two types of dwarfism prevailed during the midtwentieth century: snorter dwarfs and long-headed dwarfs. Both types have low survival rates and are susceptible to bloat. Their carcasses are of poor quality, and as the *Time* article reports, "Most that go to market are ground into sausage."

Snorters are brachycephalic dwarfs with short faces and prominent lower jaws, bulging foreheads, large abdomens, and short legs. Because of their facial

DWARFISM IN COMPANION ANIMALS

A surprising number of breeds of dogs were developed by selecting for genetic dwarfism. Among the most obvious are English bulldogs, basset hounds, dachshunds, and corgis. Unintentional genetic dwarfism also occurs in a score of other breeds, including Chesapeake Bay retrievers, Labrador retrievers, Scottish terriers, Great Pyrenees, Irish setters, and Alaskan malamutes.

Cat breeders, too, have developed dwarf breeds, most notably the short-legged, achondroplasic Munchkin and its derivatives (Minskins, Skookums, Napoleons, and Kinkalows).

TYPES OF DWARFISM IN CATTLE

Type of dwarfism	Description	Breeds affected
Compressed	Heterozygotes (carriers with one gene for compressed type and one gene for normal type) demonstrate compact, thick type; homozygotes (both genes are the compressed type) are crooked-legged dwarfs that typically die at or soon after birth. The gene is believed to be inherited as an incomplete dominant.	Hereford
Dexter (bulldog)	Heterozygotes are short-legged and short-headed but otherwise healthy; homozygotes are "bulldog" lethals.	Dexter; guinea-type Florida Cracker and Pineywoods cattle
Midget	Heterozygotes are generally indistinguishable from normal individuals; homozygotes are 50 to 60 percent of normal size.	Brahman
Snorter	Abnormally shortened head causing difficulty breathing; shortened legs. Inherited as an autosomal recessive.	Angus, Hereford
Long headed	Abnormally small, but bone growth of the head is not affected. Inherited as an autosomal recessive.	Angus, Hereford

Source: Adapted from a chart in the research paper "Inheritance of Coat Coloration and Spotting Patterns of Cattle: A Review" by T. A. Olson (see Resources).

structure they have trouble breathing, hence their name. They are approximately half the size of full-size Herefords or Angus.

Long-faced dwarfs were more common among the Angus breed than the Hereford. They resembled the snorter dwarfs in every respect except they had full-size heads.

The Dexter Bulldog Gene (chondrodysplasia)

Unlike many miniature horse and donkey breeders, Dexter breeders acknowledge the taint of dwarfism in their breed and are working hard to prevent the birth of bulldog calves.

To address the problem of chondrodysplasia, Dexter Cattle Australia initiated and supported research to develop a DNA-based diagnostic test to identify carrier animals. According to a paper presented to the Second World Congress on Dexter Cattle in 2002, researchers at ReproGen, the Centre for Advanced Technologies in Animal Genetics and Reproduction at the University of Sydney in Australia, found the gene and developed a reliable DNA test to root it out. That test is inexpensive and now available in North America. Dexter breeders can use it to identify carriers and cull them if they like, though not all breeders are likely to do so. Why? Because the bulldog trait, chondrodysplasia, is a codominant trait (meaning carriers look distinct from normal individuals and dwarfs) and affected individuals are precisely what some breeders prefer.

Dexter breeders recognize two types within their breed: long-leg (Kerry type) and short-leg (classic) Dexters. Short-leg Dexters are cute, compact, short-legged, short-faced cattle — and they are affected carriers of chondrodysplasia (see box). They are the Dexters that historically won in Britain's show rings and the type most associated with Dexter cattle. Many breeders would like to keep it that way.

Long-leg Dexters are still quite small, but they are more proportionate than short-leg Dexters. They're called Kerry types after the Kerry breed they resemble that originated in Ireland at about the same time as Dexters. Long-leg types don't carry the gene for chondrodysplasia and when bred together can never produce a bulldog calf.

Though codominant genetic transmission occurs a bit differently from autosomal recessive transmission, the Punnett square examples (see page 43) apply to breeding Dexter cattle, too. Breeding an affected (carrier) bull to an affected cow has a 25 percent probability of producing a severely affected (bulldog) calf, a 50 percent probability of producing an affected calf, and a 25 percent probability of producing a normal (noncarrier calf). Mating an affected individual of either sex to a normal individual of the opposite sex has a 50 percent probability of producing an affected calf and 50 percent probability of producing a normal calf but no chance of producing a severely affected calf. Mating two normal individuals produces only normal calves.

Dexter cattle come in two types: the longer-legged Kerry-style Dexters and the short-legged classic Dexters.

CHONDRODYSPLASIA TERMINOLOGY

Normal. Homozygous normal (HN); a long-leg, Kerry-type, dairy-type, proportionate Dexter. This Dexter has two normal alleles for the gene that otherwise causes chondrodysplasia, and it appears normal.

Affected. Heterozygous for chondrodysplasia (HC); a short-leg, beef-type, classic Dexter that is a carrier and chondrodysplastic dwarf. This Dexter has one mutant allele and one normal allele. It is affected by the mutation and will show some signs of chondrodysplasia, such as short legs and a choppy gait. Some suffer from congenital defects such as tracheal malformations, poor feet, and arthritis.

Severely affected. Homozygous for chondrodysplasia (SA); a bulldog. This calf is typically spontaneously aborted during the third trimester. It has a shortened muzzle, domed forehead, and extremely short legs. Its intestines probably protrude through a pronounced abdominal hernia, and the calf will have a cleft palate. Its spine will be shortened and its rib cage small, which may compress the lungs, causing edema in calves carried close to full term. Although calving difficulties do occur, most bulldog fetuses are delivered without assistance.

We'll talk more about the differences between long-leg and short-leg Dexters in chapter 14. For now, keep in mind that when breeding an affected Dexter to a normal individual of another breed, the resulting calf has a 50 percent chance of carrying the gene for chondrodysplasia. If this crossbred carrier calf grows up to be bred to another crossbred (half Dexter, half another breed) that also carries the gene for chondrodysplasia, the odds for perpetuating the gene rise dramatically, and from this mating a bulldog calf could be born. This is why understanding Dexter dwarfism is important for anyone raising cattle with Dexter genetics in their immediate background.

Dwarfism in Other Livestock Species

Because miniature breeders usually select for size, they should be aware that dwarfism can appear as a mutation in any mammalian race, and in most instances it should not be allowed to perpetuate.

A mutation is defined as a spontaneous and random change in the genetic makeup of an organism that causes it to have a trait not possessed by its ancestors;

this "new and (possibly) improved" individual is known as a *sport*. In many cases the sport's new trait will be handed down to its own descendants. Mutations are essential to biological evolution because they produce new species and new varieties within species. However, not all mutations are in the best interest of the sport or its progeny; geneticists estimate that 99 out of 100 mutations decrease, rather than increase, an individual's ability to adapt to its environment, making survival more difficult. Less than 2 percent of all mutations cause visible variations such as changes in color or shape; more common are invisible variations that affect fertility or the ability to live to maturity.

Ancon Sheep — A Cautionary Tale

In 1791, the story goes, a Massachusetts farmer named Seth Wright owned a flock of 15 ewes and a ram of the "ordinary" kind. These sheep, as sheep are wont to do, were fond of leaping over the stone walls that fenced them in. Once out on their own, they raided Wright's neighbors' fields. So imagine Farmer Wright's surprise (and delight) when one of his ewes produced a most uncommon ram lamb, long of body and incredibly short of leg. Sheep like this, Wright correctly deduced, could not jump over stone fences.

So Farmer Wright culled his normal ram, the short-legged ram's sire, and bred the short-legged ram to his flock of ewes. The first year only two short-legged lambs were born. However, when he began breeding short-legged sheep to short-legged sheep, the animals began to reliably reproduce their kind. And thus the Otter (later to be known as Ancon) breed of sheep was born.

Charles Darwin considered Ancon sheep a perfect example of macroevolution. In *Origin of Species*, first published in 1859, he referred to the Ancon thus: "It is a rare thing for a striking variety to spring as suddenly as this into existence, and it is singular that the peculiarity should be preserved unmixed in the cross or half-breed; but in other respects it is common-place enough and only represents what men do every day with their cattle, poultry, horses and dogs, and what is done by every nursery gardener in rearing plants. Whenever a breeder sees any peculiarity appear amongst his animals which he considers valuable, he carefully preserves the

Short-legged Ancon sheep had so many genetic abnormalities that the breed became extinct.

MUTATIONS HAPPEN

Ancon sheep are an example of a problematic mutation. The name, derived from the Greek word for elbow, was applied to Ancon sheep because their deformed front legs turned in so badly that they resembled elbows.

The bones of the first Ancon-type achondroplastic sheep were uncovered in the excavation of a sixteenth-century archaeological site in Leicester, England. Ancon-type mutations recur from time to time. For example, inherited chondrodysplasia of Texel sheep was recently discovered in New Zealand.

individual that shows it, and by pairing it with other individuals that manifest a tendency towards it, and selecting such of the offspring as have most perfectly inherited it, he succeeds in perpetuating and greatly improving it."

But was it an improvement? Ancons were achondroplastic dwarfs, a type of genetic dwarfism characterized by slow limb growth relative to the rest of the skeleton. Their condition was caused by a mutation that results in the failure of the cartilage between their joints to develop. Numerous other abnormalities existed, including abnormal spines and skulls, flabby subscapular (deep shoulder) muscles, loose leg joint articulations, and badly deviated inward forelegs. Adult Ancons were clumsy cripples that could neither run nor jump like other sheep; that they suffered from crippling arthritis is a given. The breed had so many major health problems that it became extinct decades ago.

So, as the adage goes, "It's not nice to fool Mother Nature." Should cute, mutated dwarfs happen along, it's wise to choose not to propagate them. There are scores of breeds of small and miniature livestock available for producers and hobbyists to raise. Let's do our best to raise animals that enjoy a good-quality life.

Choosing Good Pedigrees

When selling registered miniatures, pedigree counts. If you want top dollar for the livestock you produce, spring for popular bloodlines up front. Then, keep abreast of the trends and adjust accordingly. Most breeders also breed to achieve a personal goal: a certain type, colors, or another trait or set of traits that better the breed and bring personal satisfaction. Therefore, it's important to do the research early on and discover which bloodlines deliver the sort of results you're aiming for.

BREEDER'S STORY

Kathie Miller
Southern Oregon Soay Sheep Farms

KATHIE MILLER AND VAL DAMBACHER met after each purchased Soay sheep. They became business partners and friends. Val, with expertise in sheep husbandry and marketing, and Kathie, with a background in historical research, began learning all they could about what has become known in the United States as British Soay sheep.

In 2000, Kathie and Val imported from Canada the single remaining purebred flock of Soay left in North America. In 2008, after eight years of trying, Kathie was able to import semen from a historic flock in the UK with the hope of introducing important characteristics seen on St. Kilda but missing in North America (Val had since retired). Kathie's British Soay breeding schemes are based on pedigrees. She uses a professional database and received guidance from longtime Soay keeper Christine Williams in Great Britain (who worked tirelessly to get her the semen) and retired geneticist Steven Weaver. All of her sheep are recognized by the Rare Breeds Survival Trust as true Soay sheep and are registered in its Combined Flock Book.

We asked Kathie, whose boundless enthusiasm for these sheep is highly infectious, to share her thoughts about Soay. This is what she told us.

"I became involved with Soay in 1996 and with my former partner Val Dambacher in 1994. We had both been sold what we were told were pure Soay sheep from St. Kilda, a place about which neither of us had ever heard. At the time the only information we could even find out about the place was a scuba diving site on the Internet. In exchanging stories about the sheep we had each purchased, we realized these were not true Soay sheep, and they had not come directly from St. Kilda. There was more to the 'story' than we had been told.

"Through contacts in the UK we learned that there was a single remaining flock of true Soay sheep in North America. The flock was in Canada and had been imported from England in 1990. I traveled to Montreal in the summer of 1998 to see the sheep. The caretaker was so taken with Val's and my friendship and our commitment to the sheep that he eventually sold his entire flock to us in 2000.

"Since that trip in 1998 I have had extraordinary experiences researching the sheep in the UK. These included four trips to St. Kilda, a primitive sheep breed conference in Orkney, research at the archives of the National Trust for Scotland in Edinburgh, and the opportunity to visit a great number of flocks on the mainland and outer islands of the UK. Being able to see the sheep in the wild has given me a unique understanding of Soay sheep.

"The Soay of Scotland, known as British Soay in the United States, is the oldest surviving British livestock breed and possibly the oldest surviving domestic breed anywhere. It is certainly the most primitive breed of sheep in existence today. Because of its isolation it has survived unchanged into the twenty-first century. In this, the breed is unique, and its history alone makes it worthy of preservation. Because Soay sheep retain the genetic diversity of their wild ancestors, they are extremely adaptable and resistant to many of the diseases that decimate other breeds of sheep.

"Among the Soay's strongest points are its small size and ease of handling. Its smaller carcass is a good size for families. Soay are thrifty, surviving on land that is too marginal for more improved breeds. They lamb easily and are excellent mothers. (In 12 years I have had one assist, and that was for a twin whose leg was caught.) Because they are quite hardy they are ideal for people who are new at raising sheep — they survive most beginners' mistakes. They are intelligent, shy, but very curious and have lovely gentle dispositions — even most of the rams. They are irresistible and simply fun to live with.

"Their downside is that at this point British Soay are hard to come by; however, that is changing as flocks get larger. Presently there are fewer than 300 registered ewes in the United States, so the number of lambs available annually is rather small, and enthusiasts need to be patient.

"For those interested in raising British Soay as a conservation project, I recommend they read Lawrence Alderson's book *The Chance to Survive, Rare Breeds in a Changing World*. It is a wonderful explanation of the importance of preserving these ancient breeds. Those who are not interested in conservation I refer to American Soay breeders. American Soay are more readily available and a little bit cheaper."

PEDIGREE PERFECT

Livestock breeders utilize written and printed pedigrees when planning matings, while conducting advanced pedigree research, and for advertising purposes. Although preprinted forms are certainly available, it's easy to plot and print custom pedigrees online. The best free pedigree generators suitable for all species are at Pedigree Online Thoroughbred Database, SitStay, and The Boer & Meat Goat Information Center.

Breeding Strategies

Before investing in any type of livestock, learn all you can about breeding theories so you can choose the type of breeding program you want to pursue. Terminology varies widely between species, breeds, and breeders, but for our purposes there are four basic breeding strategies: inbreeding, linebreeding, outcrossing, and outbreeding or hybridization.

Inbreeding

Inbreeding is the mating of closely related animals; for example, mother to son, father to daughter, sibling to sibling, and half-sibling to half-sibling. It's probably the most misunderstood aspect of animal breeding.

A tongue-in-cheek saying often bandied about when breeders meet is, "If it works, it's linebreeding; if it doesn't, it's inbreeding." This is because many breeders associate inbreeding of animals with incest in humans. It's not the same thing! One common misconception is that it causes genetic diseases. It doesn't (see box on page 58), but it will certainly bring hidden recessive traits to the fore, as I learned when Thorfinn was born. If you don't know which hidden recessives reside in your breeding stock's gene pool, and you aren't willing to experiment to find out, inbreeding probably isn't for you.

Conversely, because it promotes homozygosity (an increase in the frequency of paired alleles in similar genes, good or bad, recessive or dominant), inbreeding also brings stellar traits to the fore. The inbred animal's reproductive genes are more uniform in their makeup; thus its offspring tend to uniformly display the characteristics of the parents. It's the fastest, most reliable way to "set" or "fix" type. This is why it was used so extensively in the formation of virtually all recognized breeds of livestock and companion animals.

The closest type of inbreeding is father to daughter and mother to son. Pedigree 1 (next page) shows the result of one such mating. This was, in fact, a fortuitous but accidental mating accomplished through an adjoining fence (yes, it happens!) that resulted in the best Miniature Cheviot ram (Baamadeus) we've ever seen. Baasha, his dam and paternal grand-dam, is our foundation ewe and a stellar example of the breed. She came out of retirement at age 12 to produce Baamadeus. We're very glad she did.

Wren is the nicest ewe we've produced to date (see Pedigree 2). She's the product of a half-sibling to half-sibling mating with an added dash of her common ancestor (Bear Farm 95) through her sire's side of the pedigree.

We chose to inbreed the Bear Farm 95 bloodline because we bought our first few sheep from a flock in which Bear Farm 95 was already being successfully bred to his daughters and granddaughters, so we knew it was likely that no undesirable recessive traits were present. We saw Bear Farm 95 and were taken with his type. We want our Miniature Cheviots to look like what they are: old-fashioned, pre-"improvement" British Border Cheviots. We want long bodies; short, sturdy legs; beautifully arched faces and tiny, well-set ears; and lovely fleece — with black coloration and white-splotched faces as frosting on the cake. By inbreeding to Bear Farm 95 and Baasha, we're producing exactly what we want, and to date we've never produced a lamb with a birth defect.

PEDIGREE 1. WOLF MOON BAAMADEUS (INBRED)

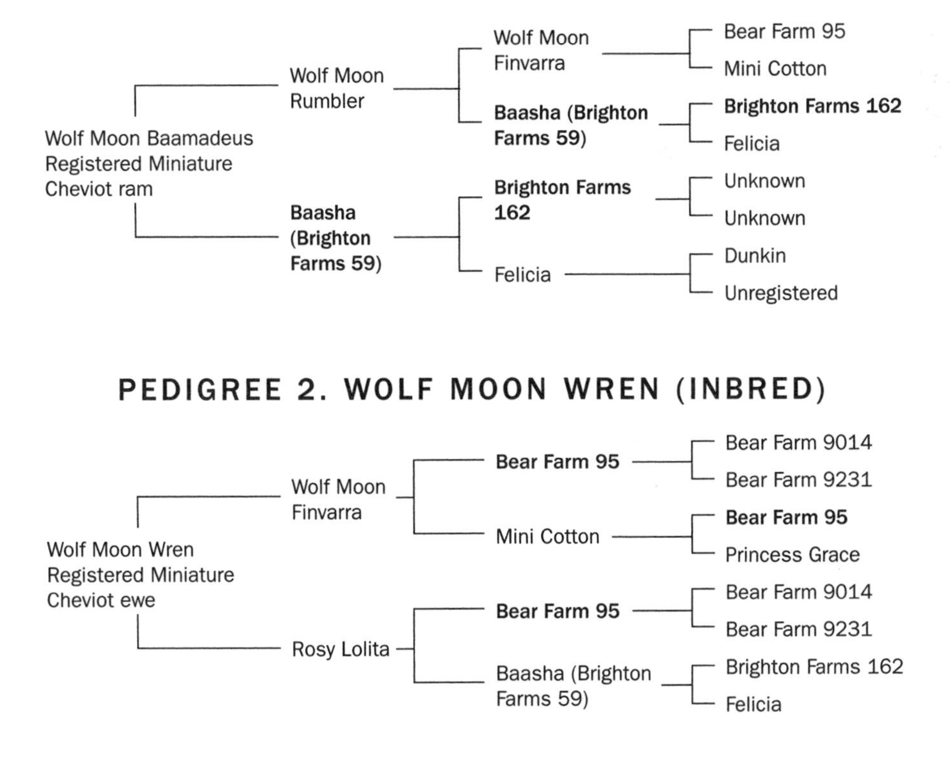

A SCIENTIFIC STUDY OF INBREEDING

Continued inbreeding is sometimes associated with a reduction in reproductive efficiency, including lower fertility and lower birth and growth rates. Dr. Helen Dean at the Wister Institute in Philadelphia bred lab rats, brother to sister, for over 100 generations to prove that inbreeding in and of itself is not harmful. She selected breeding stock carefully throughout the experiment, breeding the best to the best and culling undesirable individuals from the gene pool. At the end of the experiment the rats she produced were larger, lived longer, and produced larger litters than the rats with which she started.

Linebreeding

Most breeders pursue *linebreeding*. In linebreeding, the breeder still mates related animals but the relationship is farther removed, usually by several generations, and through linebreeding he can also introduce genes from other lines into the genetic soup. It's possible to fix desirable traits via linebreeding, especially with many distant crosses to a stellar ancestor, but it takes considerably longer than when inbreeding.

Pedigree 3 illustrates linebreeding through one parent, in this case through the American Quarter Horse Scimitar Star (though Boulder Beauty is inbred, because she is bred to an unrelated stallion, her offspring Slash is considered linebred). Sometimes the common ancestor appears on the top *and* bottom of an individual's pedigree (that is, the pedigrees of both its sire and dam).

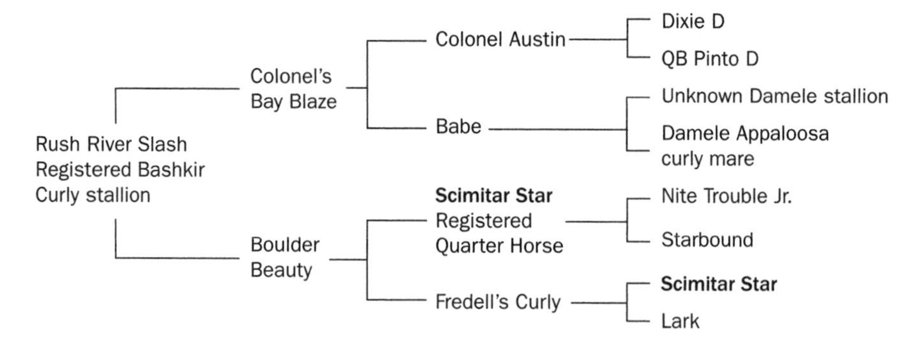

Outcrossing

The terms *outcrossing* and *outbreeding* are often used interchangeably, but for our purposes, outcrossing occurs between individuals of the same breed and outbreeding between animals of two separate breeds.

Outcrossing is defined as the mating of individuals whose four-generation pedigrees indicate that they have no ancestors in common. It's sometimes used to introduce a new trait into an inbred line. More often, however, it's the favored breeding plan for novice breeders wrongly afraid of the stigma of inbreeding and of those who simply want to add a maximum number of individual big-name ancestors to a given animal's pedigree chart.

Outcrossing tends to eliminate the chances of undesirable recessives rearing their ugly heads (provided the unrelated parents don't both carry specific recessives), but it's the least desirable mode of mating for those who hope to breed a degree of uniformity into the livestock they produce. Arabian horse breeders do not call a totally outcrossed pedigree a "fruit cocktail" pedigree ("some of this, some of that") for nothing.

Pedigree 4 is of a young ram we produced by mating a linebred black ewe of our own breeding to a totally unrelated white ram of similar physical type, in an effort to infuse our little flock with new blood. The results were quite an eye-opener: a superlative white ewe lamb and a very correct black-and-white *paint* ram lamb, Wolf Moon Fin Bheara. He is the first paint registered with the American Miniature Cheviot Association, which just goes to show, pleasant surprises happen too!

Outbreeding

Outbreeding occurs when miniature livestock breeders sometimes mate individuals of two separate breeds to create new composite breeds such as Belfair

PEDIGREE 4. WOLF MOON FIN BHEARA (OUTCROSSED)

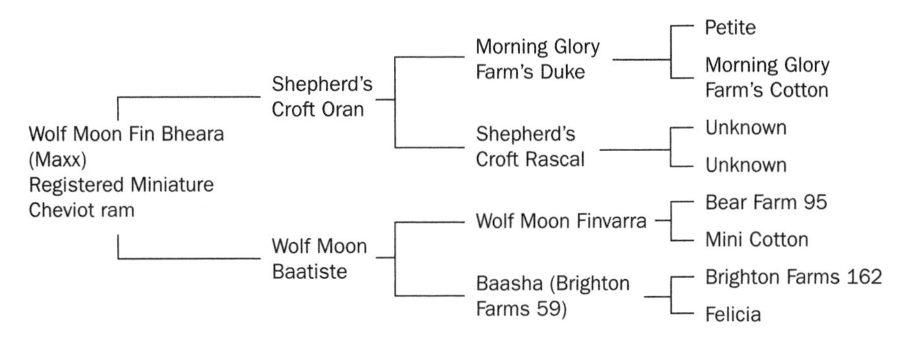

PEDIGREE 5. IMBIR' (OUTBRED)

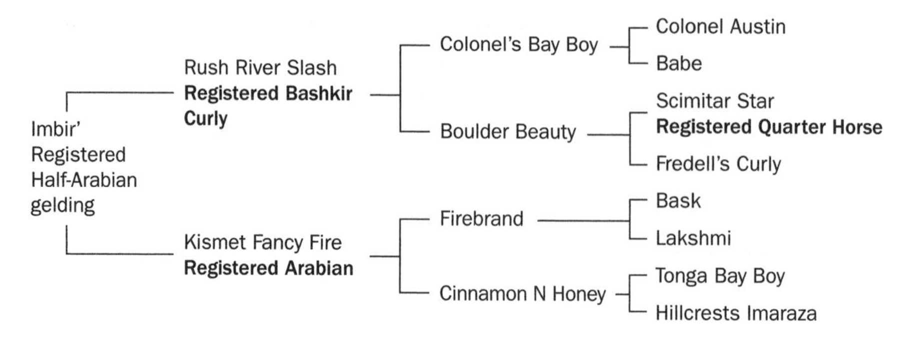

(Dexter bred to Jersey) and Miniature Highland (Dexter to full-size Highland) cattle and Cheviot Doll (Babydoll Southdown to Miniature Cheviot) sheep. It's also done to introduce new qualities (lesser stature, more milkiness, a different type of fleece) and hybrid vigor that often results in faster gains and improved reproductive capacity.

Pedigree 5 belongs to my Half-Arabian gelding Imbir'. We bred his dam, my favorite Arabian mare, to our Bashkir Curly stallion in an attempt to create a Curly horse with the beauty and intelligence of an Arabian. Imbir' inherited the best qualities of both parents — minus the curls!

Other Considerations

There will also be visible factors that you should consider before purchasing your new animals, such as size (to most buyers, smaller is usually better) and conformation (how your animals compare with their breed standard).

Size

Dwarfism aside, a high proportion of miniature livestock breeders do select for size. Small, quality animals always fetch higher prices than their larger peers. Before setting off on a buying spree, decide what size you would like to raise and how much money you're willing to invest to do so.

If raising tiny miniature livestock is your goal, you are far wiser to spend your dollars on a few animals in your chosen height range than to buy larger individuals you plan to breed down, a process that can take many generations to achieve. Conversely, if you're looking for a miniature donkey to drive, a yoke of fancy miniature oxen, or a miniature horse to show at American Miniature Horse Registry sanctioned shows, consider larger individuals. Larger minis cost less, are capable of doing more work, and compete in a less cutthroat environment in the American Miniature Horse Registry's class B division.

Also keep in mind that many veterinarians and breeders believe micro-mini individuals are more prone to birthing problems (dystocia) than their larger kin. Regardless of whether that's true, it's important to mate tiny females to males as small as or smaller than themselves, and this can limit the selection of suitable males.

Standard or Type

Before selecting livestock, be completely familiar with your breed's standard of perfection or know what type of animals you wish to produce. You may wish to produce animals that do not fit your registry's standard of perfection.

I, for example, prefer cob-type miniature horses that resemble British Shetland ponies. If I bought breeding stock that subscribed to the American Miniature Horse Association or American Miniature Horse Registry standards of perfection, I'd never produce the sort of foals I like.

If you show or breed show stock for others, let your breed's standard of perfection be your guide. If showing isn't your main priority, look at a lot of animals of your chosen species and breed. Decide which types and colors tickle your fancy, then buy and raise what you like. Therein lies the joy of raising miniature livestock.

Soundness

Breeding soundness refers to an animal's ability to sire or conceive and give birth to healthy, normal offspring. Soundness means different things to a person buying a yoke of Dexter oxen than it does to one who is buying cows to breed.

Before purchasing performance livestock of any sort — miniature oxen, a miniature donkey or llama to pack with, or a miniature horse to show in driving and jumping — study your species and understand and recognize faults that might prevent your animals from doing the job you have in mind. Better yet, spring for a prepurchase veterinary exam or take along an experienced mentor when choosing your stock. And always ask to see appropriate production records and test results. Don't buy a (breeding) pig in a poke.

Age

Many breeders start with young animals because they cost less and their "cute factor" sells. If you enjoy raising and training young livestock, and you carefully choose youngsters from proven lines, this can work. Youngsters take a while to grow up, however, and they sometimes fail to deliver as expected; the beautiful duckling becomes an ugly swan or the baby from teensy parents grows up too tall. It's a gamble — but it's fun!

Quality animals in their prime naturally fetch the highest prices. If carefully chosen, they are usually the best buy because most unknowns are eliminated. Yet buying healthy, sound animals past their prime (and even downright old ones) can be a fine way for folks on a budget to start or expand a first-class breeding herd. Older animals require additional care, however, and shouldn't be purchased unless you're willing to provide it. Worn teeth and less efficient digestive systems call for special feed and frequent visits by the veterinary dentist; winter may mean blankets or a nice, warm shed. Aged females are more prone to uterine infections and birthing dystocias, but in return, the old ones can put famous names on pedigrees and top-notch youngsters in your barn. Our ewe Baasha is a perfect example. In 6 years she's given us seven strong lambs, including our two best rams, and at 13 years of age she's still going strong.

THE RULE OF THREE

When I was growing up, a sage old horseman told me about the rule of three. Look at a horse when he's three days, three weeks, and three months old, he said, because those are the times you can see what his conformation is going to be. So I've been doing that for almost 50 years — and it works! What's more, it seems to apply to other species as well: cattle, sheep, goats, and most definitely donkeys. So give it a try and see if it works for you. Then contact me with your thoughts; I'd love to hear them.

5

Livestock Guardians

LIVESTOCK OWNERS EVERYWHERE AGREE: It's a heartbreaking, financial disaster when predators raid your flocks and herds and kill your animal friends. Predators are everywhere, from free-roaming suburban dogs to packs of ubiquitous coyotes to mountain lions in up-country meadows, and they all pose a threat to miniature livestock. One of the best ways to protect against predators is to keep a livestock guardian or two.

USDA Statistics Tell the Tale

Consider losses among full-size sheep and goats. The United States Department of Agriculture's National Agricultural Statistics Service (NASS) keeps track of what kills American sheep and goats and periodically publishes findings in a report titled "Sheep and Goats Death Loss." According to the report, predators killed 280,000 sheep and goats during 2004, accounting for slightly more than 37 percent of each species that died of all causes that year! These sheep and goats were killed mainly by coyotes (more than 60 percent of the total) but also by dogs, mountain lions, bears, foxes, eagles, bobcats, and other species (among them wolves, ravens, and black vultures).

Predation is a serious problem, and one you'll have to address in order to raise small livestock anywhere in North America—even in relatively populated areas, where dog predation poses a serious risk.

When NASS surveyed sheep and goat producers to ask how they managed predation in 2004, almost 53 percent indicated they relied at least in part on predator-proof fencing; 33 percent also penned their stock at night; and 55 percent kept livestock guardian animals with their flocks and herds.

Which Guardian Is Best?

The concept of livestock guardian animals goes back a long, long way — about 6,000 years to be precise — to Turkey, Iraq, and Syria, where dogs were first trained to protect sheep and goats. Savvy livestock owners still use guardian dogs to protect their livestock, but some have added guardian donkeys and llamas to the mix.

Donkeys

Donkeys are born with an inherent hatred for anything that (to them) resembles a wolf. Therefore, some donkeys make first-rate herd guardians where coyotes and dogs are troublesome.

Donkeys require no specialized training; they simply dislike dogs and coyotes. They chase while braying, biting, and sometimes pawing or kicking at canid invaders. Donkeys have keen hearing and good eyesight, so dogs and coyotes rarely sneak past a donkey on guard.

They are hardy; long-lived (25- to 30-year life spans are the norm); and with the exception of medicated feed laced with Rumensin (which is poisonous to equines of all kinds), they eat the same sorts of things you probably already have on hand to feed ruminants such as miniature cattle, sheep, goats, and llamas. And you won't break the bank buying a guardian-quality, standard-size or larger donkey. You can usually buy one locally in the $100 to $800 price range, depending on quality, registration status, and size.

Full-size donkeys make good livestock guardians for miniature goats.

CHOOSING A LIVESTOCK GUARDIAN DONKEY

Consider these points when buying a donkey to guard your miniature livestock:

- Get a donkey that's big enough to do the job; this precludes miniature donkeys. Miniature donkeys aren't large enough or strong enough to repel most predators, especially those that work in packs, although they'll usually try and can be seriously injured in the process.
- Choose a sturdy, sound, healthy jenny or gelding. See the donkey section of this book for particulars.
- Consider adoption. Many donkey rescue organizations test donkeys with sheep and goats to evaluate their livestock guardian potential. Another bonus: A donkey adopted through any bona fide equine rescue is returnable if it doesn't work out. Give a donkey a working home; many worthy donkeys need one.

There are disadvantages, however:

1. Gelded (castrated male) donkeys make the best guardians; jennets (females, also called jennies) run a close second. But some jacks (intact males) are aggressive toward humans, most will savage animals they dislike, and many have been known to kill newborn kids and lambs. Also, some jacks try to breed the larger female livestock they're hired to protect, sometimes inflicting serious, even fatal, injuries in the process.
2. Donkeys prefer the company of other donkeys. If you place several guardian donkeys with a flock or herd, or pasture your flock next to a field containing donkeys or other equines, most donkeys will hang out with their own kind instead of watching their charges.
3. Unlike guardian dogs, donkeys haven't been bred for generations to look after other livestock; some simply aren't interested in bonding with animals not of their kind.

If you buy a donkey to guard your herd, ask if you can return him if he's aggressive toward or disinterested in the livestock he's supposed to guard. Chances are good, however, that a carefully chosen donkey will do the job in spades. In one survey, 59 percent of Texas producers who use guardian donkeys rated them as good or fair for deterring coyote predation and another 20 percent deemed them to be excellent or good.

Llamas

In 1990, researchers at Iowa State University polled 145 sheep producers in five western states to determine the effectiveness of llamas for reducing dog and coyote predation. The producers reported losing an average of 21 percent of their ewes and lambs each year prior to adding llama guardians and only 7 percent after guardian llamas joined their flocks. Eighty percent rated their llamas effective or very effective for guarding sheep. In another study conducted

CHOOSING A LIVESTOCK GUARDIAN LLAMA

Many full-size llamas make excellent guardians. Keep these thoughts in mind when choosing one for your flock or herd:

- Llamas are in their element when guarding livestock against less aggressive predators such as foxes or a dog or two; they are *not* effective at dealing with big guns such as mountain lions or bears or with packs of aggressive canids, all of which can easily kill or maim a guardian llama.
- Llamas guard by alarm calling (once you've heard this cry you'll never forget it) and then either attacking the predator or moving their charges to a place of safety. They also can work well in pairs, in which case one may attack while the other herds the animals they guard to safety.
- Choose a sound, healthy, fairly human-friendly female or gelded male to guard your livestock. See the llama section of this book to learn how to choose the best one. If choosing two, it's best to select two llamas of the same sex.
- Again, consider adoption. Many llama rescues also test potential guardian llamas with sheep and goats, and they, too, consider adopted llamas returnable if they don't work out. There are many, many llamas in rescues throughout North America. Visit the Southeast Llama Rescue Web site (see Resources) for more information.

in Utah, 90 percent of sheep producers rated guardian llamas effective or very effective on the job. NASS figures indicate that 14 percent of sheep and goat producers maintained guardian llamas in 2004.

Just like donkeys, llamas naturally dislike dogs and coyotes. Many make excellent guardians; others make poor guardians. Llamas prefer the company of other llamas, so you usually have to keep just one per pen or herd. Intact males are often aggressive and may try to breed small females of other species. A guardian-quality, gelded llama costs roughly $200 to $750; females usually sell for somewhat more. Llamas have certain advantages and disadvantages when compared with donkeys:

- Llamas require the same food (unlike with donkeys, it's safe to feed them Rumensin-medicated products), vaccinations, and foot care as other small ruminants. It's easy to treat them as just another member of the herd or flock.
- Llamas, however, are more aloof than donkeys. While most donkeys crave human interaction, the average llama doesn't. This makes it more difficult to catch and handle them for routine maintenance chores.
- Llamas don't do well in hot, muggy climates, where they're prone to heat exhaustion. Depending on the length of their fleeces, llamas require full or partial shearing at least once a year.
- Llamas don't live as long as donkeys. The average llama lives 15 years.

Some full-size llamas make superlative livestock guardians.

Livestock Guardian Dogs (LGDs)

For thousands of years, stalwart European, Middle Eastern, and Asian guard dogs of dozens of types and breeds have watched over herds of goats and flocks of sheep, protecting them from predation by wolves, bears, jackals, and human thieves.

Some of these dogs eventually made their way to North America. Now, according to current NASS figures, nearly 33 percent of American sheep and goat producers use livestock guardian dogs, representing 20 or more breeds.

How They Work

All of the livestock guardian breeds are large, intelligent, strong-willed, and potentially aggressive and dominant dogs, so they aren't to be casually handled by the faint of heart. Most are aggressive toward other dogs; they sometimes kill household pets that wander in with their charges, and some fight to the death with other guardian dogs of the same sex.

Since most predators strike at night, livestock guardian dogs are most active after dark. Barking is their primary means of warning off intruders, so expect a lot of nighttime barking from your LGDs.

Before buying a livestock guardian dog, visit LGD breeders and livestock owners who keep them. Compare philosophies and training methods. Know what you're getting into before you bring a dog home.

It's best to buy an adult dog that is already familiar with guarding the species you raise, but if buying a puppy is your only option, here are a few pointers to keep in mind:

- To be effective, livestock guardian dogs must be bonded to and stay with the livestock they're expected to guard. During their prime socialization period (from 4 to 14 weeks of age), it's important to minimize human interaction with LGD puppies. However, don't believe those who insist a guardian puppy should never be cuddled or played with. LGD puppies should look forward to interacting with humans, but it's important to play with the puppy on her own turf rather than taking her to the house or for rides in the truck; she needs to understand that watching her livestock pals is her primary job.
- Livestock guardian puppies needn't be trained to watch their charges; the desire to guard has been bred into these dogs for thousands of years. Most begin showing guardian dog behaviors by five or six months of age (scent marking, purposeful barking, and deliberate patrolling), although few become reliable protectors until they achieve mental and physical maturity, usually at about two years of age.

- As they pass through adolescence, most LGD puppies exhibit inappropriate behavior toward their charges, usually in the form of play-chasing. Expect it and work through this phase with reprimands; there's an effective working livestock guardian waiting on the other side.
- The occasional pup comes along who has no interest in guarding. All breed organizations maintain rescue programs for their dogs, so if you get one of these apathetic types, consider placing your nonworking dog with a rescue group.

CHOOSING A LIVESTOCK GUARDIAN DOG

It pays to be picky when choosing a livestock guardian dog since your animals' well-being is involved.

- Crosses between two livestock guardian breeds are often the best of both worlds, but avoid crossbreds having one livestock guardian breed parent and another parent of a non-LGD breed. Some become admirable guards, while others take after the non-guardian parent and may hurt rather than guard their charges.
- Some livestock guardian breeds, especially breeds recognized by the American Kennel Club, are sometimes bred for show-ring conformation and presence ("look at me" alertness) instead of working ability. Always choose a dog or puppy from working stock rather than show-dog bloodlines.
- Whenever possible, choose an adult dog that has already proved itself as a livestock guardian bonded to the species you raise. Watch the classifieds. When people go out of the sheep, goat, or alpaca business, they often advertise guardian dogs in local papers. Species-specific e-mail lists are excellent hunting grounds, too.
- Give preference to a puppy raised among livestock by a working dam. Bring him home when he's seven or eight weeks old and place him with a few friendly, nonaggressive individuals of your herd. Give him a safe place to retreat to where his future charges can't follow him, place his bed and food and water dishes in that spot, and allow him to bond with the species he'll be guarding.
- If you have an adult livestock guardian dog, buy a puppy while that adult is still in its prime. The puppy will learn its job by following the adult's example.

Breeds

The following are descriptions of most of the bona fide livestock guardian dog breeds currently available in North America.

Akbash Dog

Origin: The Akbash region of Turkey
Name in native land: Akbas ("white head"); Coban Kopegi (pronounced cho-bawn co-pey; "shepherd's dog")
Height: 28–32 inches (71–81 cm)
Weight: 90–130 pounds (40.5–58.5 kg)
Coat types: Short to medium-length double coat
Color: White
Life span: 10–11 years

The Akbash is leaner and leggier than the other Turkish guardian dog breeds. It's a highly effective guardian dog, though more aloof toward humans than are some breeds. In Turkey, the Akbash is prized for its intelligence, bravery, independence, and loyalty — qualities that endear it to its North American admirers, too. Although cases of hip dysplasia occur in Akbash dogs, it occurs less frequently than in most other livestock guardian breeds.

Akbash

Anatolian Shepherd

Origin: Semi-arid Anatolian Plateau of central Turkey
Name in native land: Anadolu Coban Kopegi (pronounced cho-bawn co-pey; "Anatolian shepherd's dog")
Height: Dogs 28–35 inches (71–89 cm); bitches 26–28 inches (66–71 cm)
Weight: Dogs 100–160 pounds (45–72 kg); bitches 90–130 pounds (40.5–58.5 kg)
Coat types: Short to medium-length double coat
Color: Usually fawn with a black mask, but any color is acceptable
Life span: 10–15 years

The Anatolian Shepherd is one of the most popular of the livestock guardian breeds. It was developed to be independent and forceful, responsible for guarding its master's flocks without human assistance or direction. Anatolians are known for their strength and courage. Because they're suspicious of and often aggressive toward strangers, they work best in situations where they don't come in constant contact with unaccompanied friendly visitors, though they're in their element against predation by humans. Both the American Kennel Club and the United Kennel Club recognize and register Anatolian Shepherds.

Anatolian Shepherd

Caucasian Ovcharka (Caucasian Mountain Dog)

Origin: The Caucasus mountain region of Armenia, Azerbaijan, Georgia, and Russia
Name in native land: Kavkazskaya Ovcharka in Russian, Nagazi in the Georgian Republic, and Gampr in Armenia
Height: 25–32 inches (63.5–81 cm)
Weight: 100–150 pounds (45–67.5 kg)
Coat types: Short or long; double-coated; abundant ruff and fringing
Color: Usually agouti gray; otherwise any color except red and white like the Saint Bernard, solid black or brown, or solid black and tan
Life span: 10–11 years

Ovcharka is a generic Russian word meaning "shepherd's dog"; also spelled ovtcharka and owtcharka, the word is pronounced *uhf-char-ka*. Caucasian Ovcharkas mature slowly and are somewhat prone to hip and elbow dysplasia. Caucasian Ovcharkas were extensively used as military guard dogs throughout the former Soviet Union. I'm told Ovcharkas are rather like Chow Chows — friendly but watchful.

Caucasian Ovcharka

Great Pyrenees (Pyrenean Mountain Dog)

Origin: The Pyrenees mountain region of southern France and northern Spain
Name in native land: Chien des Pyrénées, Chien de Montagne des Pyrénées
American nickname: Pyr
Height: Dogs 27–32 inches (68.5–81 cm); bitches 25–32 inches (63.5–81 cm)
Weight: Dogs from 100 pounds (45 kg), bitches from 80 pounds (36 kg)
Coat types: Long, coarse outer coat may be straight or slightly wavy; fine undercoat, soft and thick
Color: White, often with gray or tan "badger markings" on the head and occasionally spotting on the body or tail. The color of the nose and the eye rims should be jet black.
Life span: 10–12 years

Great Pyrenees are considered one of the most human-friendly of livestock guardian breeds, making them ideal for use on small acreages with nearby neighbors. Some owners say they patrol out farther than other breeds, making secure fencing a must. Great Pyrenees have double dewclaws on their hind legs. They are somewhat prone to hip dysplasia and, occasionally, skin problems. In North America, they are recognized by both the American Kennel Club and the United Kennel Club.

Great Pyrenees

Kangal Dog

Origin: Kangal District of Sivas Province in central Turkey
Name in native land: Karabash ("black head")
Height: Dogs 30–32 inches (76–81 cm); bitches 28–30 inches (71–76 cm)
Weight: Dogs 110–145 pounds (50–66 kg); bitches 90–120 pounds (41–54 kg)
Coat type: Short, dense double coat
Color: Light dun to gray with black mask and ears
Life span: 11–14 years

Kangal Dogs are recognized by the United Kennel Club. Some authorities lump Kangal Dogs and Anatolian Shepherds together, but they are similar yet different breeds. According to the Turkish Shepherd Dogs Web site FAQs:

Kangal

- The lineal and weight requirements for Kangal Dogs and Anatolians are different.
- Kangal Dog color is restricted. Anatolians are acceptable in all colors, patterns, and markings.
- The Kangal Dog is short-coated. The Anatolian coat may be up to four inches in length.
- Kangal Dogs registered with the UKC originate, or descend from, dogs obtained from the Sivas-Kangal region of Turkey. Anatolians appear to originate from any area of Turkey.
- Their temperaments are different. The Kangal Dog is people oriented and only hostile to traditional predators.

Karakachan

Origin: Bulgaria
Name in native land: Karakachansko Kuche
Height: Dogs 25–29 inches (64–74 cm); bitches 23.5–27.5 inches (60–70 cm)
Weight: Dogs 85–110 pounds (39–50 kg); bitches 75–100 pounds (34–45 kg)
Coat types: Short-haired (hair up to about 2⅓ inches in length), long-haired (anything over 2⅓ inches); ruff; feathering on legs; long hair on tail; double-coated
Color: Most are spotted, but solid-colored dogs are acceptable; colored areas can be black, gray, brown, yellow, or brindle.
Life span: 12–16 years

This breed is critically endangered worldwide; breeders in the United States are working to preserve this extremely ancient, effective livestock guardian breed before it becomes extinct.

Karakachan

Komondor (Hungarian Sheepdog)

Origin: Hungary

Name in native land: Komondor (pronounced KOM-ahn-door; plural is Komondorok; the name Komondor may be derived from the term *komondor kedvu*, meaning "somber" or "angry")

Komondor

Height: 25.5 inches (65 cm) and up

Weight: Dogs up to 125 pounds (57 kg); bitches 10 percent less

Coat type: Very long, nonshedding double coat. Show people divide it into sections and allow felted cords to form; clipping working Komondorok once a year is a better solution.

Color: White

Life span: 10–12 years

Komondorok are prone to hip dysplasia, bloat, and skin diseases. It is one of the strongest and most aggressive of the livestock guardian breeds. The Komondor is one of the most effective breeds in guarding herds against human thieves. Komondorok are recognized by both the American Kennel Club and the United Kennel Club.

Kuvasz

Origin: Hungary

Name in native land: Kuvasz (pronounced KOO-vahz); plural is Kuvaszok. The name most likely comes from the Turkish word *kavas*, meaning "guard" or "soldier," or *kuwasz*, meaning "protector."

Kuvasz

Height: Dogs 28–30 inches (71–76 cm); bitches 26–28 inches (66–71 cm)

Weight: Dogs 100–115 pounds (45–52 kg); bitches 70–90 pounds (32–41 kg)

Coat type: Medium-length, straight or wavy; thick undercoat

Color: White (over dark skin)
Life span: 10–12 years

The Kuvasz is a highly intelligent dog that is fiercely protective and intensely loyal, yet somewhat aloof or independent, particularly with strangers. Kuvaszok are somewhat prone to hip dysplasia, and this breed may drool. Recognized in North America by the American Kennel Club and the United Kennel Club.

Maremma Sheepdog

Origin: The Maremma region of Tuscany in Italy
Name in native land: Maremma (pronounced mair-emma), Abruzzese, Mareemmano-Abruzzese, Cane da Pastore ("sheepdog")
Height: 23.5–28.5 inches (60–72 cm)
Weight: 65–110 pounds (29–50 kg)
Coat type: Long, harsh, double coat; outer coat straight or slightly wavy
Color: White with ivory, light yellow, or pale orange on the ears
Life span: 10–12 years

Descriptions of sheepdogs similar to the Maremma can be found in ancient Roman literature, and depictions can be seen in fifteenth-century paintings. Although Maremmas were developed to guard sheep and goats, cattle ranchers have found that they bond well with cows, and Maremmas are now being used to protect range cattle. A Maremma named Oddball was recently trained to guard the endangered penguin population on Middle Island in Warrnambool, Australia. The breed is recognized in North America by both the American Kennel Club and the United Kennel Club.

Maremma Sheepdog

Polish Tatra Sheepdog

Origin: The Podhale section of Tatra region of the Carpathian Mountains in the south of Poland

Name in native land: Polski Owczarek Podhalanski

Height: Dogs 26–29 inches (66–74 cm); bitches 24–26 inches (61–66 cm)

Weight: 80–150 pounds (36–68 kg)

Coat type: Heavy double-coated; top coat straight or slightly wavy and hard to the touch; profuse, dense undercoat

Color: White over dark skin

Life span: 10–12 years

Tatras are strong, hardy guardian dogs developed to withstand cold, harsh temperatures as well as hot, dry heat. They are courageous; alert; and agile, swift runners. They are also independent, self-thinking, highly intelligent, and able to assess situations without human guidance. These dogs have been known to deter wolves and bears. Occasional cases of hip dysplasia occur. This uncommon breed is recognized by the United Kennel Club.

Polish Tatra

6

Facilities and Fences

CERTAIN PRINCIPLES APPLY to keeping every type of livestock. The material in this chapter applies to housing and fencing for all breeds of miniature livestock. We'll discuss species-specific needs in chapters to come.

Shelters Large and Small

You can certainly build an animal palace for your miniature livestock if you like, but a dry place to sleep in a draft-free shelter meets their basic needs. In fact, in all but the coldest climates, inexpensive structures called "loafing barns," "run-in sheds," or "field shelters" are sufficient.

Other basic shelters for miniature livestock include prefabricated barns; Port-A-Huts (our personal favorite; see Port-A-Hut in Resources); commercial calf hutches; hoop structures (the type designed for pasture-raised hogs); and for miniature kids, lambs, and piglets, Igloo doghouses from the pet store.

You'll probably need enclosed housing, too. Late gestation females, females with newborns, and delicate bottle babies require dry, draft-free housing, especially during the cold winter months. Anyone who dairies, even on a very limited scale, also needs a covered, weather-resistant area separate from their animals' living areas in which to milk.

Field Shelters

Field shelters consist of a roof and three enclosed sides with the remaining side open and facing away from prevailing winds (a southwest exposure helps utilize the winter sun as a source of heat). They're set up in, or adjacent to, pastures, paddocks, or exercise areas, so the occupants can come and go as they will.

Field shelters, also known as run-in sheds, make ideal housing for miniatures.

In southern climates, open-air shelters comprising a roof and sturdy framework provide essential hot-climate airflow, shade, and rain protection. For best results, enclose three sides with roof-high, welded-wire cattle panels. During the colder winter months, it's easy to provide additional protection by covering the fenced sides with removable plywood panels or sturdy plastic tarps.

The rules of field-shelter design are simple: Allow enough floor space for each animal (see Factors to Consider When Choosing a Species on pages 8–11); provide adequate site drainage; and slope the roof away from the shelter's open side so rain and snow cascade off the rear, rather than sliding off in front of the structure. In most climates, packed dirt or clay floors are better than cold, hard concrete. In some locations and with certain species, wooden floors work well, but eventually they'll rot, necessitating periodic replacement.

Bed this type of structure with four to six inches of absorbent material such as straw, poor-quality (but not moldy) hay, wood shavings, sawdust, peanut

THE EASIEST MINI SHELTER

One of the best, least-expensive, and easiest-to-maintain shelters for small miniatures such as sheep, goats, and pigs is a prefabricated calf hutch. Our favorite is the Quonset-style Port-A-Hut (see Resources) made in Iowa and shipped fully assembled to dealers throughout the Midwest. Our sheep and goats love them!

Corrugated steel Quonset huts like the one made by Port-A-Hut make fine housing for pigs or a few sheep and goats.

hulls, ground corncobs, or sand. Continue adding just enough bedding to keep floors dry; then, periodically (in most climates a few times a year is adequate) clean everything out, back to floor level. This system, called "deep litter bedding," is comfortable and warm, and it's easy and simple to maintain.

When building field shelters in northern climates, it's usually best to keep the roof height as low as you can, keeping in mind that the tallest individuals should never crack their polls on the ceiling, even if they throw back their heads. Low-slung roofs hold in body heat. However, squat buildings are harder to clean, especially if someone has to clean them by hand, so you may have to balance your various needs.

Pre-Engineered, Packaged Barns

If you're on a tight budget and need to build a small to midsize barn in a hurry, think pole barns (pre-engineered metal buildings with stout framework poles set in the ground). Pole barns are fast and economical to build; a no-frills, unlined, galvanized steel structure can usually be built in less than a week. Or build a fancier version with skylights, lots of windows, colored siding, and a cupola or two and you've got a structure anyone should be proud of.

Depending on the design, pole barns can be built with posts set directly in the ground (without a concrete slab foundation) or built with the posts fastened to a slab. Siding choices vary, but most pole barns are roofed and sided with galvanized or colored steel.

Steel I-beam packaged barns are another option, as are Quonset-style metal barns. What they have in common with pole barns is that they sell as packaged units designed to reduce construction costs through standardization. You can buy them built on your site or, if you're handy with tools, as do-it-yourself kits. All three come in high-end and economy models with options to meet most any need, and each has its own special virtues.

PACKAGED BARN COMPARISON

Type of barn	Advantages	Disadvantages
Pole barns (wood and steel)	• Short construction time • Easy to finish inside (wood provides natural interior stud wall) • Many choices of roof pitch and color • Can be engineered for high snow and wind codes • Often sold as easy-to-build kits	• Cement costs higher due to close placement of uprights • Wood uprights sometimes rot where contact is made with ground, regardless of treatment • More parts than in I-beam or Quonset structures • Usually limited to a 60-foot (18 m) width • Termite problems are possible in some locales • May sweat, causing condensation
Steel I-beam construction barns	• Short construction time • Virtually maintenance free • Few width limitations allow for large clear span widths • Most come with excellent paint and no-rust warranties • Many choices of roof pitch and colors • Can be engineered for very high snow and wind codes • Usually have high resale value • Noncombustible so it usually costs less to insure all-metal barns	• May sweat, causing condensation (roof insulation relieves this) • Special tools such as high-impact drills are often needed when building larger size buildings
Quonset-style steel barns	• Usually the cheapest type of steel structure • Easy to erect • Easy to disassemble • Virtually maintenance free	• May sweat, causing condensation • Can be difficult to insulate • Cannot place door openings on side of building • Usually no choice of colors • Difficult to finish interior • Sometimes cannot meet high wind or snow codes • Custom designs rarely offered • Often priced without end walls

The right barn is the one that provides for all your needs at minimum cost. Questions to ask yourself when choosing a packaged barn include:

- What is the intended use of the building? Will it loosely house miniature beef cattle or provide stalling for miniature horses or donkeys? Will it incorporate an office or groom's living quarters? How many specialty areas, such as wash stalls and feed or tack rooms, are needed?
- Is your property zoned for the type of construction you want to build?
- How big must it be? Do you need free-span construction to provide for a training arena? Tall doors so you can park your trailer inside in snowy climates?
- Where will you put the barn? The right location may impose some constraints on the size and shape of the building you can erect.
- Are you building a barn that full-size horses may someday use? If so, choose one you can easily finish inside by lining it with plywood at least as high as a tall horse's back because horses have been known to kick through unlined metal barns, with dire results.
- Do the aesthetics of the building matter to you? Will a Plain Jane galvanized steel building do? If not, choose a type you can customize with colored metal siding and roofing material (a plus: prepainted finish colors last 15 to 20 years without refinishing), wooden siding (it has better insulating qualities than metal, where that makes a difference), a solid-deck shingled roof, and fancy windows. And skylights make any interior more appealing.

The Builder's Shortlist

No matter which species you raise or which type of housing tickles your fancy, keep these principles in mind.

Check into applicable zoning laws before erecting any sort of permanent structure, including fencing.

All animal housing must be adequately ventilated. Livestock housed in damp, poorly vented barns are prone to respiratory ailments.

Trees and hedges can provide sufficient shade, but animals raised in rainy or snowy climates need access to weather-resistant, man-made structures, too—no exceptions.

Build to endure. Breeding males are notoriously hard on housing and fences. Don't scrimp; always build their stalls, shelters, pens, and fences out of stout, sturdy materials.

Protect glass windows, electrical wiring, and even lighting fixtures with screens or conduit and keep them well out of curious animals' reach; this is especially true if you keep equines or goats.

Allow your critters to socialize. All livestock species are inherently sociable; they will fret if they can't see others of their own kind or at least some other friendly livestock faces. Fretting equates with stress, and stress leads to lower productivity, illness, and even death. Make sure no animal is totally isolated, particularly for any length of time.

Provide getaways where youngsters or low-ranking herd and flock members can escape bullies. And if your animals don't have access to pasture, they'll all need a safely fenced, roomy exercise area to blow off steam and hang out.

Don't mix horned and hornless livestock in close quarters where jostling and sparring typically occur. Mixing can work in a pasture setting, but whenever possible, horned and hornless animals should be housed and penned separately.

Provide plenty of fresh, clean water kept reasonably cool in the summertime and warm enough to prevent freezing when temperatures dip below 32 degrees (see page 86 for cooling and heating tips). Install running water and electricity to your barn or shelter, or barring that, place the structure within reach of existing utilities.

Provide multiple watering facilities instead of one big one; they're a lot easier to keep clean. And if one water source becomes contaminated before you notice and clean it, there will be other, clean sources for your animals to choose from.

Build or buy sturdy, safe hayracks and grain feeders for your livestock; don't feed directly on the ground. Feeding on the ground equates with excessive parasitism and a lot of wasted feed.

Don't store feed where animals can break in and help themselves. Overeating, especially of grain or rich, legume hay, can kill. Store grain in animal-proof covered containers with snug lids (55-gallon food-grade plastic or metal drums and decommissioned chest freezers work well). And always secure the feed room door with an appropriate closure; there are Houdinis in every livestock species.

Dispose of soiled bedding in a responsible manner; don't let it pile up and attract flies.

Learn from the experts. Visit the housing page on the Maryland Small Ruminant Web site (see Resources), hosted by the University of Maryland Cooperative Extension, where you can access scores of university- and government agency-generated bulletins relating to livestock housing and farm structures, feeders and feed storage, ventilation, bedding, pest and fly control, and manure management.

Touch base with your local county extension agent (see Resources) before building any type of livestock structures or renovating existing structures. He can help you assess your needs and provide material specific to your climate and location, and his expertise is absolutely free.

Finally, buy or borrow *How to Build Animal Housing: 60 Plans for Coops, Hutches, Barns, Sheds, Pens, Nest Boxes, Feeders, Stanchions, and Much More* by Carol Ekarius (see Resources). If you live in the country and keep any sort of livestock, you need this book.

Pens

You'll also need some pens. Pens can range in size from a length of cattle panel to an acre or more. You'll use them for maternity housing; to hold special animals such as pets, old duffers, 4-H projects, and breeding males; for new animals fresh from quarantine that haven't yet been turned out with the main herd; and for thin animals who need more grub to gain some weight. Pens are usually built using planks (for cattle and equines), stout woven wire, or welded-wire livestock panels. No matter how many pens you build, you'll wish you had more.

Livestock Panels Are a Farmer's Best Friend

Livestock panels are prefabricated lengths of sturdy mesh fence welded together out of galvanized one-quarter-inch steel rods; they come in an array of wire spacings and heights.

Livestock panels are ideal for creating pens and corrals.

Cattle panels are 52 inches (132 cm) tall and built using 8-inch (20 cm) stays; horizontal wires are set closer together near the bottom of the panel to prevent smaller livestock from escaping. Cattle panels are usually sold in 16-foot (4.9 m) lengths that can be trimmed to size using heavy bolt cutters.

Sheep panels are similar to cattle panels but manufactured in 34-inch (86 cm) and 40-inch (102 cm) heights, and their horizontal wires are set even closer together. Both cattle and sheep panels are ideal for fabricating pens and corrals.

Utility panels are the toughest of all; they're fabricated using 4 × 4 inch (10 × 10 cm) spacing and are welded out of extra heavy-duty 4- or 6-gauge rods in a full 20-foot (6 m) length. They come in 4- to 6-foot (1.2 to 1.8 m) heights and are ideal for building extra-stout pens.

Premier1 (see Resources) sells panels designed specifically for goat and sheep producers in 36-inch (91 cm) and 40-inch (102 cm) heights and in 4- to 6-foot (1.2 to 1.8 m) lengths. These are perfect for building V-type hay bunks and round-bale hay feeders for miniature livestock.

A bad thing about standard cattle, sheep, and utility panels is that the raw end of each rod is very sharp. To make these panels more user-friendly, smooth each rod end with a rasp to take off its razor edge. (Premier1 panels are pre-smoothed at the factory.)

Feeders

All good livestock feeders are designed to discourage animals from wasting feed. You definitely need them. Feeding grain or hay off the ground contributes to parasitism and disease, not to mention excessive waste. Sheep and goats won't touch hay, grain, or minerals they've peed or pooped in, but they don't mind peeing and pooping in their feed. Goat kids complicate the matter by "nesting" in accessible hayracks and grain or mineral feeders, and they don't vacate their nests when Nature calls. It's your mission to prevent these unsanitary and wasteful practices.

Combination hay and grain feeders save a lot of otherwise dropped and wasted hay.

At feeding time, make sure all of the animals in a herd or flock are fed at the same time; otherwise timid individuals may not get to eat. Here are some things to remember when considering feeders:

- Choose feeders you can move by yourself. Several smaller feeders are more manageable than a single extra-bulky one, and rubber or plastic feeders are lighter than metal models.
- Move feeders in wet weather. During wet spells a mud wallow will form around your feeders if you don't move them every few weeks; this and manure buildup in feeding areas contribute to parasitism and disease.
- For small groups of livestock, consider feeders you can hang on a fence and remove after the animals have eaten.
- Mount grain feeders six inches higher than your tallest sheep and goats' tails and provide booster blocks or rails for their front feet to stand on.
- For feeding small rectangular bales of hay, V-shaped hay feeders with welded-wire sides or vertical or diagonal slats work better than models with horizontal slats or bars.
- Make an effective, inexpensive fence-line hay feeder for small livestock by wiring the bottom of 4 × 4-inch welded-wire mesh panels, trimmed to size, to an existing fence and adding sturdy wire spreader arms at the top.
- To reduce waste when feeding large round bales of hay, limit access to the hay with a feeder ring designed for your species.
- Loose mineral feeders should be placed where they won't be rained on — either in a building or under a canopy. Since most mineral mixes are 10 to 25 percent salt, and salt is highly corrosive, plastic mineral feeders are the durable choice.
- Build an effective gravity-fed loose mineral feeder for small livestock by gluing a Y-type PVC cleanout plug to the bottom of a 3- or 4-foot (0.9 to 1.2 m) length of 4-inch (10 cm) PVC pipe, with the Y facing up. Permanently cap the bottom of the Y and attach a removable cap to the top of the tube. Add minerals through the top, and hang it so it is at chin height on your smallest animals.

Watering Devices

Most animals consume a considerable amount of water each day, and the amount varies depending on weather conditions (they require more water during the hot summer months), mind-set, and whether they're females in the latter stages of pregnancy or in milk. Lactating females have the highest requirements, and to prevent urinary calculi (see page 141 of Health chapter), it's important for male llamas, goats, and sheep to drink a lot, too.

WATERING TUBS FOR CHEAP

If you keep cattle, you probably have empty plastic mineral lick tubs sitting around; these make first-rate watering troughs for all miniatures.

If you don't have cattle, talk to someone who does. Or ask a clerk at your favorite feed store; he may know someone who has a pile of tubs to give away.

When water supplies are contaminated with droppings, algae, dead birds or bugs, leaves, and other debris, animals drink only enough fluid to barely get by. If the water you serve your animals isn't appetizing enough for you to drink, you can bet picky species like llamas, goats, and sheep will tend to shun it, too. Empty scummy or contaminated tubs, tanks, and troughs, and scrub or spray their inner surfaces with a one-part chlorine bleach to ten-parts water solution. Choose a series of small troughs or automatic waterers over one or two mega-model tanks; the littler ones are easier to clean.

Place water-filled tubs and buckets in a shady area during the summer months. This helps inhibit algae growth and keeps the water fresher, which means the animals will drink more, too. When temperatures soar into the 90s or better, freeze ice in plastic milk jugs and submerge one jug in each trough or tub; your animals will appreciate this treat. Refreeze them overnight, and they'll be ready to use again by midmorning.

Another cooling ploy: freeze plain water or electrolyte products in "ice cube trays" made from 8- to 12-ounce plastic food containers such as yogurt cups and cream cheese tubs. Pop out the "cubes" and drop them in buckets of water to cool down the liquid.

Reused cattle lick tubs are ideal water containers for livestock.

During the winter, prevent water supplies from freezing by installing bucket or stock tank heaters. Encase the cords in PVC pipe or garden hose split down the side and taped back together with duct tape; if you don't, animals (especially young equines or goats) may gnaw through the cord and electrocute themselves.

When lambs, kids, or piglets are present, the water should be no deeper than 6 inches (25 cm), lest a youngster leap or fall in and drown. *Never* use 5-gallon recycled plastic food service buckets or other narrow, deep-water containers in pens where neonatal livestock is housed.

Fences

Some folks think that because you keep miniature livestock you don't need full-size fences. Not true. You have to think of what you are keeping out as well as what you are keeping in. Although it may (or may not) be easier to keep them confined, smaller animals are more vulnerable to predation than their larger kin. While livestock guardian animals should be part of every miniature livestock breeder's menagerie (and we've devoted a whole chapter to that subject), it's still important to fence out predatory creatures.

The type of fences you build on your farm will likely depend on their purpose (for example, permanent, perimeter fencing; cross fencing; or temporary fences to provide for controlled grazing) and the species of livestock you raise. Most types of fences, for instance, work with cattle, but barbed wire is unsuitable for horses, because they can sustain serious injuries when they run into it or catch their legs in it; nor is barbed wire appropriate for sheep, whose thick wool renders the barbs ineffective. Some species, such as pigs, require super-stout fencing, as do breeding males of virtually every species.

Permanent Fences

You'll want to erect sturdy, permanent fences around the perimeter of your property. These establish your property line; they also prevent your livestock

THIS FARMER'S TAKE

Fencing is a costly investment both in time and money, so it pays to build what you want and to build it correctly from the start. And if you think you'll raise another species in the future, plan ahead. We installed high-tensile wire fences when we moved to our Ozarks farm; then a few months later we got miniature sheep. Six years later we tore it out and replaced it with woven-wire field fence, because the sheep walked right through or under the high-tensile wire. An expensive lesson? You bet!

A WORD ABOUT GATES

Avoid putting gates in the middle of a straight line of fence. Instead, put them in corners where adjacent fence lines help funnel livestock through the gates more easily.

from escaping, should gates or barn doors be left ajar, and keep other people's wandering livestock from joining your own. Permanent fences should be well constructed of high-quality materials so they'll last a long time with minimal repairs.

Be absolutely certain you know the exact location of your property lines before installing any type of permanent fence. And check with adjacent neighbors to see if they're interested in pooling resources to build a better fence than what you can afford by yourself. In many states adjacent landowners are required by law to foot half of the bill for erecting shared fences. (Forcing the issue, however, may not be worthwhile if it leads to strained relationships between rural neighbors.)

Temporary Fences

Movable fences are considered temporary fences. They are, by nature, easy to put up and take down, so planning where to put them isn't as important an issue as it is when planning permanent fences. They are usually used to break larger pastures into paddocks to provide for controlled grazing conditions. They should never be used for perimeter fencing, especially along roadways or in situations where animals that breach them can damage adjacent properties.

Fencing Materials

The most common types of fences encountered on today's small farms are board, barbed wire, woven wire, cable, and electric.

Board

Board fences, also referred to as post-and-plank or post-and-rail fences, are popular on many farms because they're attractive, highly visible, and relatively safe. This group includes fences constructed of treated or painted 1- to 2-inch (2.5 to 5 cm) thick, 4- to 6-inch (10 to 15 cm) wide wooden planks nailed or screwed to the inside of wooden posts; split wooden rails that slide into holes in wooden posts; PVC plastic boards and posts; and wooden boards coated with vinyl. Board fences are usually supported by wooden posts set 8 to 10 feet (2.4 to 3 m) apart. They can be built to any height, though heights of 4 to 5 feet (1.2 to 1.5 m) are most common. The primary disadvantage to board fencing is its cost, but with proper upkeep, these fences last 20 to 25 years.

Wooden planks are usually made of rough-cut oak or treated pine. Treated pine has a more finished look, it accepts paint, and the treatment it undergoes resists rotting and discourages animals from chewing — at least until the treatment wears off. Keep in mind that pine isn't as strong as oak, so thicker boards are needed.

Rough-cut oak lends a rustic appearance, has high-tensile strength if animals lean or scratch their rumps on it, and most animals prefer not to chew oak planks. Oak boards warp when freshly cut, however, and some spots on some boards weather and rot faster than others (where the tree was weakened by natural processes). Rough-cut oak can be stained but not painted.

Solid PVC plastic fences cost more than wooden planks and rails, but they don't require painting (they're the same color throughout the material), which means they cost considerably less to maintain. Vinyl-coated plastic fencing is crafted of boards dipped in vinyl, so unlike solid PVC products, they can warp. White fences built from either type of vinyl fencing require washing with mildew-removing products at intervals, especially in the humid southern states.

EVALUATING EXISTING FENCES

Will the fencing already in place on your farm work for miniature species? Evaluate old fences with a critical eye: Are posts rotting and falling over? Do they move easily if you push on them? Are wires rusted through and breaking? Are boards so warped or rotted that they won't stay up? You might be able to use that fence for a time by erecting temporary electric fencing on the inside, but it could ultimately save you time and money to rip it out and start over.

If the fence is in good repair but not particularly suited for the species you want to raise, again, a few strands of electric wire might bring it up to snuff. Board fences can be predator-proofed (and also goat-proofed, if you raise miniature goats) by installing field fence along the inside.

Barbed Wire

Barbed wire is made of two or more strands of smooth, galvanized steel wire twisted together with two or four sharp barbs spaced every 4 to 5 inches (10 to 13 cm). It's sold in 80-rod rolls (1,320 feet, or one-half mile [402 m]) in a variety of sizes and barb patterns. It can be erected on wood or steel T-posts, and it's sometimes reinforced by installing twisted wire or plastic stays between each set of fence posts. Fences usually consist of three to seven strands of wire stretched between posts set 15 to 25 feet (5 to 8 m) apart.

Barbed wire, invented in the mid-1850s, was the first wire technology capable of restraining cattle. It was (and is) widely used on vast Western ranches where wood was (and is) scarce and the cost of alternate fencing prohibitive. It's still the only legal fence in many states.

Barbed wire for agricultural fencing is available in two styles: soft (mild) steel and high-tensile. High-tensile wire is made with thinner but stronger steel. Its greater strength makes fences last longer because cattle can't stretch and loosen it. It also supports longer spans, but because of its springy nature it's hard to handle and somewhat dangerous for inexperienced fencers to install. Soft wire is much easier to work, but it's less durable.

Barbed wire should never be used to fence an area that contains or may contain equines (full-size or miniature) now or in the future. Equines have trouble seeing barbed wire and may run into it at full speed, causing horrific injuries; they also tend to catch their legs in it when pawing at equines on the opposite side of the fence.

Barbed wire shouldn't be used to fence any type of livestock in small areas such as pens or paddocks, where animals could jostle each other. Although a barbed wire fence is fairly cheap to build, vet bills incurred by using it usually offset any money saved.

FENCE LAWS

In states where free-range laws are still in effect, a landowner must fence out the neighbor's animals; but in most places it's the livestock owner's responsibility to contain his animals within adequate fencing.

Fence laws define who is responsible for constructing and maintaining a fence, who has liability when animals get out and cause property damage, and what constitutes a "legal fence." They are often quite explicit.

TYPES OF FENCE POSTS

Steel posts come in U bar, studded Y, punched channel, and studded T types, but they're all called T-posts. Although they lack the eye appeal of wood posts, T-posts are fireproof, long-wearing, comparatively lightweight, and relatively easy to drive. They also ground the fence against lightning when the earth is wet. They do, however, tend to bend if larger livestock lean against them. Unbent T-posts last 25 to 30 years.

Wooden fence posts come in treated and untreated varieties. Treated posts last 20 to 30 years; the longevity of untreated posts depends on the type of tree they're made from. The strength of wood posts increases as its top diameter grows larger; a 4-inch (10 cm) post is twice as strong as a 3-inch (8 cm) post, a 5-inch (13 cm) post is twice as strong as a 4-inch post, and so on. Corner and gate posts should have a top diameter of at least 8 inches (20 cm), brace posts 5 inches (13 cm) or more, and line posts can be anything 2½ inches (6 cm) or greater, but the bigger the posts the stronger and more durable the fence.

Step-in fiberglass and plastic posts are used to string up temporary fences, particularly fences made of electric string or tape. If you use them, buy good ones; low-end step-in posts don't last.

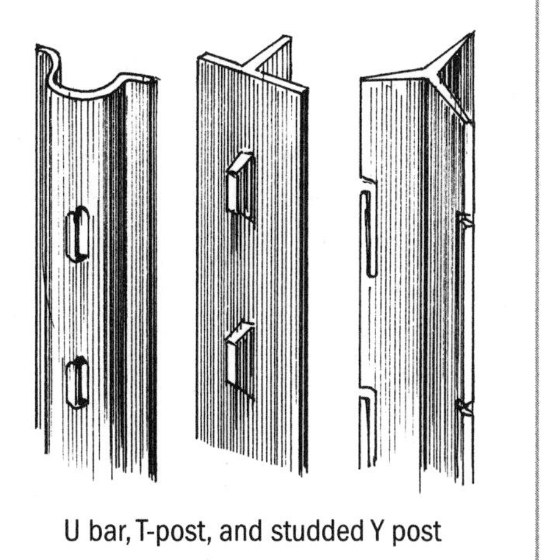

U bar, T-post, and studded Y post

Woven Wire

Woven wire, also called field fence, is the fence of choice for many farms, except in high snowfall areas where deep drifts tend to pull it down. It's made of horizontal lines of smooth wire held apart by vertical wires called "stays." Horizontal wires are usually spaced closer near the bottom of the fence. The vertical stays in standard woven wire are 6 or 12 inches (15 or 30 cm) apart.

Correctly installed, woven wire is the most secure form of affordable sheep, goat, and llama fencing, making it ideal for perimeter or boundary fences. Four-foot-high (1.2 m) woven wire will contain most animals. Installing one or two strands of barbed or electric wire above woven wire helps keep predators at bay.

One drawback is that some animals, especially goats and miniature horses, lean in to wire-mesh fences and then stroll along from post to post, scrubbing their sides. It's a great way for them to shed winter hair or shoo a vexing fly, but it's very, very hard on the fence. To prevent this, install a strand of offset electric wire at shoulder height on the inside of woven-wire fences.

High-tensile woven wire costs more than regular woven-wire fencing; however, it's rust-resistant, sags less, and is lighter in weight.

Woven wire comes with galvanized (zinc) or aluminum coatings. Both are further classified as I, II, or III wire; the higher the number, the thicker the coating and the more durable the fence. Class I galvanized woven wire generally shows signs of rusting in 8 to 10 years, while Class III fencing begins to rust in 15 to 20 years. Aluminum-coated wire resists corrosion three to five times longer than galvanized wire with the same thickness of coating. Since a major part of fencing cost is installation, it's best to use the longest-lasting wire you can afford.

When buying woven wire, read the tag; the numbers printed on it will tell you how it's made. For instance, 10-47-6-9 fencing has 10 horizontal wires; it's 47 inches tall; there is a 6-inch spacing between stay wires; and the fence is made of 9-gauge wire. Woven wire is sold in 20-rod (330-foot [100.5 m]) rolls and is generally supported by wood or steel posts erected at 14- to 16-foot (4 to 5 m) intervals.

Next to livestock panels, woven wire works best for most species.

Cable

Cable fences are strong and attractive, but because of their cost they're usually used only to fence arenas and corrals. They're generally made of ⅜-inch smooth steel cable made of seven heavy wires twisted together. Cable is stretched between sturdy anchor posts with heavy-duty springs installed between post and cable to absorb shock on the cable caused by animals running in to or pressing against them. Along the length of the fence, cables usually pass through holes in wooden or heavy steel posts.

Electric

Standard electric fencing wire is sold in aluminum, regular steel, and high-tensile steel varieties. For permanent fencing, high tensile is the way to go.

High-tensile smooth fencing wire is installed using wooden or T-posts and plastic insulators. It comes in 11- to 14-gauge wire and has a breaking strength of roughly 1,800 pounds (816 kg). High-tensile fencing is durable and relatively easy to install, and it can be stretched tightly without breaking. Strong corner

ELECTRIC FENCE SAFETY

In most cases, being shocked by an electric fence isn't fun, but the pain quickly passes. There are, however, exceptions to that rule. To prevent injuring yourself when dealing with electric fencing and electric fence chargers:

- Never electrify barbed wire fencing; the chance of an animal or person becoming entrapped in it is much too great, and multiple shocks over a long period of time can kill.
- Take care not to touch electrified fence with your neck or head. Don't crawl under live electric fences. Step over low ones, turn off the charger, or find a way around.
- Don't tamper with or try to repair a fence charger; take it to an authorized service agency or replace it.
- Never charge a battery on a battery-type fence charger while the charger is connected to the fence.
- Never attach electric fence wire to a utility pole. High-voltage current leaking down a wet pole can be very dangerous.

and end braces are needed, however, along with tensioners and strainers to keep it bowstring-taut.

Where high-tensile fences aren't needed, standard steel or aluminum wire works well. Aluminum wire is the better of the two; it's rustproof, so easy to work with, and it conducts electricity better than steel wire.

Electric fence failures happen when the fence chargers (also called "fencers," "controllers," or "energizers") that power them aren't up to the job. Chargers are distinguished by voltage (4,000 volts and up will hold most livestock) and the number of joules they put out (a joule is the amount of energy released with each pulse). One joule will power 6 miles (9.7 km) of single wire fence; a 4½ joule fencer will energize 20 to 60 acres, depending on the length of the fence and the number of wires that are used in its construction.

When building an electric fence, choose an adequate charger. The box will tell you how many miles of fence it charges, but that's the greatest distance for one strand of fencing operating under tip-top conditions. Think big. The more powerful your charger, the fewer problems you'll have.

There are two types of chargers on the market: high- and low-impedance models. High-impedance chargers put out a relatively high voltage with low amperage. When these short out — and even a weed or blade of grass can do it — they don't work. Low-impedance chargers put out a lower-voltage, higher amperage charge resulting in a short, intense pulse and more energy that isn't as easily shorted out.

According to University of Maryland sheep and goat expert Susan Schoenian, an estimated 80 percent of the electric fences in the United States are improperly grounded. Follow the instructions in your charger's product manual; they vary between brands and models. Here are some other important tips when installing electric fencing:

- Use quality insulators. Sunlight degrades plastic; choose high-quality insulators, preferably a brand treated to resist damage done by ultraviolet (UV) light.
- Don't skimp on wire. The larger the wire, the more electricity it can carry. And don't space wires too closely. To get the most from your charger, place them at least 5 to 7 inches (13 to 18 cm) apart, even near the ground.
- Train your animals to respect electric fences. Electric fences are psychological rather than physical barriers. Place untrained animals in a small area, then entice them from the sidelines with a pail of tasty grain till they get zapped.

- Buy a voltmeter and use it. A good one costs in the neighborhood of $50 to $100, and it will help you keep your fences nice and hot. Check voltage every day. If the fence runs low on voltage or shorts out, you need to know it and correct the problem right away.

Polywire, polytape, and rope-style electric fencing that is used with step-in plastic or fiberglass posts make fine interior fences and can be moved around with ease. Wide, flat tape offers high visibility—especially important when your fences hold horses. However, flat tape whips around in the wind more than rope-style temporary fencing. To minimize whip, twist the tape once or twice between fence posts rather than installing it flat.

Net-style portable electric fencing is used to subdivide pastures for rotational grazing; to keep coyotes and stray dogs out and livestock guardian dogs in; to erect small pastures for special needs animals; to create lanes for moving livestock without assistance; and to fence steep, rocky, or otherwise uneven land. There are two basic types: brands with built-in posts and those without. Most roll up onto easy-to-use reels, making moving them a breeze. However, horned sheep and goats have been known to tangle their horns in the mesh, go into shock, and die. The same thing can happen to the heads and necks of curious llamas, lambs, and kids.

Strands of electric fencing used to contain miniature species must be set close together to keep livestock in and predators out.

7

Feeding

WHILE THE NUTRITIONAL NEEDS of the species covered in this book are dissimilar in certain respects, the principles are enough alike to discuss them here. Species specifics are addressed in section 2.

You might assume that you would feed miniature livestock smaller portions of the same things you would feed their larger peers, but that's only partially correct. Miniatures, as a rule, convert feed so much more efficiently than full-size animals that certain differences apply.

Take concentrates, another term for grain and cereal feeds. Most miniature species don't need them unless they're still growing, in the last trimester of pregnancy, or nursing young. High-quality pasture and hay are by far the most important feedstuffs for raising most types of miniature livestock; good hay, lots of clean drinking water, and species-specific minerals are all that most nonlactating minis really need.

COUNTY EXTENSION AGENTS ARE A FARMER'S BEST FRIENDS

The best, most reliable way to learn how to feed livestock using locally available feed is to discuss your needs with your county extension agent. What is available (and works) for a llama farmer in New Hampshire won't work for a Pygmy goat owner in Arizona or a Miniature Zebu breeder in southern Florida. To locate county extension offices in your area, visit the USDA National Institute of Food and Agriculture Web site (see Resources).

In fact, the protein levels in some types of hay can be too high for most classes of miniature species. Again, talk with your county extension agent or a successful breeder of your chosen miniatures to work out feeding plans specific to your animals' needs.

Certain rules of feeding apply across the board:

- Change feeds gradually, allowing your animals' digestive process time to adjust. Abruptly switching from one feed to another can trigger colic or laminitis in mono-gastric species (horses, donkeys, mules) and bloat or enterotoxemia in ruminants (cows, goats, sheep, llamas).
- Store feed where animals can't break in and overeat.
- Feed at the same times every day. Messing with routine plays havoc with animals' digestive systems.
- Don't feed musty, moldy, or spoiled feed to your livestock. This includes feeding moldy hay to cattle; many cattle keepers think this is okay. You might get by with it once or twice, but the law of averages is against you.
- Buy the best feed (especially hay) you can afford. Although it may cost more than other choices, you'll reduce your vet bills and increase the animals' performance, seeing higher fertility; multiple births in species that give birth to more than one young; and faster, greater weight gains in meat animals.
- Get the biggest bang for your buck by feeding hay and grain in feeders instead of off the ground; feeding on the ground results in excess waste and increased internal parasitism. Choose sturdy, safe feeders that your stock can't easily demolish or get hurt on.
- If possible, provide your livestock access to pasture and browse (leaves, twigs, and other high-growing vegetation). Goats, pigs, donkeys, and some breeds of cattle and sheep derive a good deal of nutrition from browse.
- Always provide plenty of clean water. If it isn't clean enough that you'd drink it, picky species such as sheep, goats, and llamas probably won't drink it either.

All about Hay

Hay is a generic term for grass or legume plants that have been cut, dried, and stored for use as animal feed, particularly for grazing species such as cattle, equines, llamas, goats, and sheep. Pigs will eat hay, but they don't digest it very efficiently, so they do need concentrates in their diet. Next to good pasture, high-quality hay is the ideal feed.

FEED VALUES

This chart, adapted from many sources, compares common livestock feed components. Total Digestible Nutrients (TDN) is a commonly used method of expressing the energy value of feed. Calcium and phosphorous contents are given because balancing these two minerals in a 2:1 ratio helps prevent urinary calculi in male livestock (we'll talk more about this in chapter 8).

Feed	Protein (%)	TDN (%)	Calcium (%)	Phosphorus (%)
HAY				
Alfalfa, early bloom	18.0	60	1.41	0.22
Alfalfa, full bloom	15.0	55	1.25	0.22
Alfalfa, mature	12.9	54	1.13	0.18
Bahia grass	8.2	51	0.50	0.22
Bermuda grass, early	12.0	60	0.53	0.21
Bluestem, common	5.4	45	n/a	n/a
Fescue, tall, early bloom	8.0	62	0.44	0.28
Kudzu, early	14.3	55	2.35	0.35
Lespedeza, bloom	14.6	52	1.21	0.27
Oat, with head	9.3	61	n/a	n/a
Orchard grass	9.0	42	0.25	0.25
Peanut, no nuts	10.8	55	1.23	0.15
Red clover	14–18	57–67	1.5	0.2–0.25
Sudan grass, early	15.6	58	0.77	0.36
Sudan grass, late	9.7	57	0.43	0.30
Timothy	8.0	45	0.4	0.4
Wheat straw	3.6	44	0.18	0.05

Grasses used for making hay include timothy, Bermuda, Bahia, Sudan, brome, fescue, big bluestem, prairie hay, bluegrass, Dallis grass, and orchard grass. Farmers in most regions put up mediocre hay composed of wild, native grasses called different names in different localities, such as upland hay (swamp grasses including rushes and reed canarygrass) in northern Minnesota and Wisconsin and bottomland hay in the South (plants like tall Johnson grass that thrive along southern waterways).

Feed	Protein (%)	TDN (%)	Calcium (%)	Phosphorus (%)
BROWSE				
Acorns	4.8	47	n/a	n/a
Hackberry, mature	14.0	41	4.00	0.13
Honeysuckle buds and leaves	16.0	72	n/a	n/a
Honeysuckle leaves, late	10.0	69	n/a	n/a
Sumac, early	13.7	77	n/a	0.20
ENERGY FEEDS				
Barley	13.5	84	0.05	0.38
Corn	10.6	89	0.03	0.29
Corn, cob, ground	6.6	74	n/a	n/a
Milo	11.4	88	0.04	0.32
Molasses, beet	8.5	79	n/a	n/a
Oat	13.3	77	0.07	0.38
Wheat	16.0	88	0.09	0.39
Wheat bran	17.1	70	0.13	1.38
PROTEIN FEEDS				
Brewers grains	29.4	70	0.33	0.55
Cottonseed meal	45.2	76	0.18	1.21
Linseed meal	38.3	78	0.43	0.89
Peanut meal	52.3	77	0.29	0.68
Soybean meal	49.9	88	0.34	0.70

There are cereal grain hays as well, such as barley, oat, wheat, or rye, all of which are also types of grasses. These are cut while green and growing, before seed heads mature as grain.

Legumes commonly used for hay making are alfalfa, clover of several varieties (red, crimson, alsike, and ladino to name a few), and lespedeza. Others sometimes encountered include kudzu, vetch, birdsfoot trefoil, soybean, peanut, and cowpea.

Many hay producers put up bales combining one or more grasses with a legume variety; these are referred to as mixed hay. And not every variety grows in every locale. Bermuda and bahai grasses, for example, grow well in the South, whereas timothy and orchard grasses do best farther north.

Discuss hay with your county extension agent and local farmers who aren't invested in selling hay. Sellers aren't always upfront about what's in the hay they sell, so ask an impartial party to show you the good stuff; then you'll recognize it on sight. Finger a typical sprig to memorize its feel in your hand. Take a deep sniff. Good hay not only has a clean, sweet smell, but each variety also has its own distinctive aroma. Only then should you set about buying hay.

Choosing Quality Hay

Since high-quality and poor-quality hay often sell for the same price, it pays to learn how to evaluate hay and buy the best.

HAY BUYER'S CHECKLIST

- ☐ Buy by the ton, not the bale. It's hard to visually tell the difference between a 50-pound bale and one that weighs 60 pounds. If they're of comparable quality and you pay the same price per bale for both, the 50-pound bales are no bargain.
- ☐ Open several bales so you can evaluate the hay inside. Don't worry about slight discoloration on the outside, especially in stacked hay.
- ☐ Refuse hay that is dark and brittle, excessively sun-bleached, or smells moldy, musty, dusty, or fermented.
- ☐ Choose hay that has soft, flexible stems. Hay with tough, thick stems is less palatable than softer hay. As grass matures, its stems become tougher and more fibrous, and protein and energy levels decrease. The presence of seed heads and thick stems is the hallmark of hay cut at a mature stage of growth, therefore making it a lower grade of hay. Leaves contain most of the plant's digestible energy and protein, so leafy hay with few or no seed heads is the most nutritious.

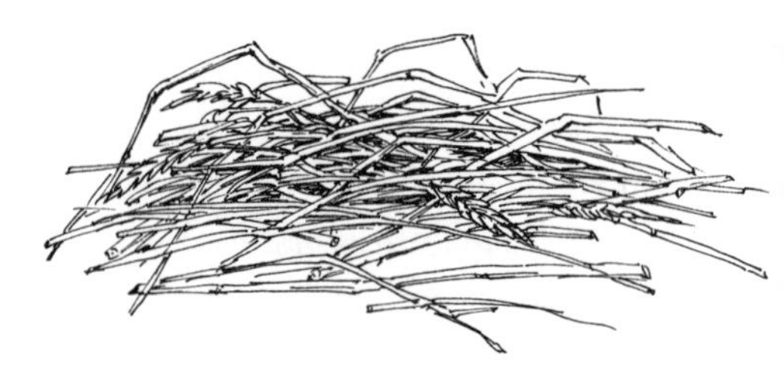

Avoid coarse, unpalatable hay with seed heads. It was harvested after its prime.

Color

Green hay usually contains a high level of protein and vitamins. Yellow or beige-colored hay may be sun-bleached or it might have been rained on prior to baling. Rain leaches nutrients from hay and decreases its quality. Dark brown coloration indicates that the hay overheated because it was baled when it was too moist or it was rained on after baling.

Mold

Hay should have a fresh, clean aroma. If it smells musty, it was probably baled wet or it was stored improperly and it's moldy. Hay should also not be dusty. Dusty hay causes respiratory problems in many animals; in some cases, the dust particles are actually mold spores. To distinguish between dust and mold, shake out a flake of hay. If the "dust" appears as a grayish white color, it's mold. Also, if the flakes are hard or stick together in clumps, the bale is moldy.

- ☐ Avoid hay containing significant amounts of weeds, dirt, trash, or debris.
- ☐ Examine hay for signs of insect infestation. Be careful to check for dried blister beetles in alfalfa. Blister beetles have long (¾- to 1¼-inch), narrow bodies, broad heads, and antennae that are about one-third the length of their entire body. They are most abundant in arid regions of the United States but occur elsewhere as well. Cantharidin, the poisonous compound found in blister beetles, is comparable to cyanide and strychnine in toxicity, and it remains toxic in dead beetles. Livestock can be poisoned by ingesting beetles in cured hay. Blister beetles are particularly toxic to equines.
- ☐ Reject bales that are uncommonly heavy for their size or that feel warm to the touch. Unusually heavy bales are usually moldy, and warm bales may spontaneously combust.
- ☐ When possible, purchase enough good hay to last a season. To preserve its nutritional value, however, don't buy any more than what you'll feed within a year of harvest.
- ☐ Store hay in a dry, sheltered area out of the rain, snow, and sun, or cover outdoor stacks to protect them from the elements.

WHICH HAY?

While there are many regional-specific hays, one or more of these standards probably grows wherever you live.

Hay	When to cut	Positives	Negatives
Alfalfa	Bud to 10 percent bloom	Nutritious and tasty; high yields; heat and drought resistant	Too high in protein for many species; freezes out in northern climates; requires well-drained soil
Red clover	20 percent bloom	High protein; does well in heavy soils	Difficult to put up at correct moisture content
Timothy	Early bloom	High yields beginning first year; winter hardy; performs well in a wide range of soils	Low drought resistance
Orchard grass	Boot (grass heads are still enclosed and no seed heads are showing) or early bloom when baled with alfalfa	High palatability and yield; reaches early bloom at the same time as bud stage of alfalfa grown in mixtures; establishes quickly	Gets tough after early bloom, so must be harvested at optimal time
Bromegrass	Heads emerged	High yield, drought and heat tolerant; winter hardy; grows well in a wide range of soils	Unpalatable when overmature
Tall fescue	Boot	Adapts to a wide range of growing conditions; winter hardy	Often endophyte-infected; should not be fed to late-gestation females, particularly equines
Reed canarygrass	Boot	Particularly well adapted to wet soils; flood, heat, and drought tolerant; winter hardy; adapts to frequent harvest	Older varieties unpalatable, especially when overmature; potential for nitrate poisoning in drought conditions
Bermuda grass	Boot	Drought tolerant; high yields; pest and disease resistant; tolerant of close grazing	Not winter hardy in the North; requires frequent harvest

Stage of Maturity

The quality of hay is determined in large part by its stage of maturity at time of harvest. The optimal stage varies by plant species:

- **Alfalfa.** Alfalfa should be harvested in the bud stage (when it has buds at the tip of its stems) or slightly later, in early bloom (when it has some purple flower petals and its stems are somewhat heavier than they were in bud stage). Alfalfa cut in full bloom or later may have seed pods, there will be fewer leaves, and its stems will be coarse and woody. Many leaves and soft, pliable stems are excellent indicators of quality alfalfa hay.

Prime alfalfa is harvested in bud stage or early bloom.

- **Red clover.** Red clover should be harvested when 20 percent of its blooms are present.
- **Grasses.** Orchard grass, tall fescue, Bermuda grass, and reed canarygrass should be harvested in the boot (when grass heads are still enclosed by the sheath of the uppermost leaf and no seed heads are showing) to early heading (when seed heads are just beginning to emerge) stages. Timothy and brome should be cut in early bloom (when the plant's tiny flowers are beginning to form in the uppermost seed head) and when fully headed. Early-cut grass hay is greener than late-cut; plump brown seeds are indicators of less nutritious, fully matured grass hay.
- **Mixed hay.** Grass and legume mixes should be cut when their legume component reaches its ideal stage for harvest.

Junk

Make sure the hay you feed is free from trash, insects, and weeds (I'd be suspicious of any weeds you couldn't identify). Sticks and chunks of mud are heavy nuisances when buying hay by the ton, and baled-up briars and stickers injure mouths and tongues. Blister beetles, sometimes found in western-grown alfalfa hay, are extremely toxic to equines. And toxic weeds can be hard to distinguish from nontoxic species once the plant has dried and been baled into hay.

CHECK IT OUT

Hay producers use electronic sensing devices that measure the moisture content of hay by means of a probe inserted into windrows and bales. If you buy a lot of hay and you're worried about the quality throughout a stack, you can use a moisture tester to sample bales without having to open them. Reliable testers aren't cheap — they typically cost around $200 or more — but the device can quickly pay for itself, given the rising cost of quality hay.

Thick-stemmed weeds, like milkweed, that aren't completely dried prior to baling can cause moldy areas within the bale.

Occasionally a snake or something larger is caught and killed in the baling process. If you find a dried-out critter in baled hay, throw the hay away. These finds are fairly common, and livestock that eat hay exposed to spoiled carcasses are at risk for botulism.

Tested Hay

To be certain that your animals are receiving adequate nutrients, purchase tested hay or have large batches of homegrown hay tested. To find a feed-testing service in your neck of the woods, talk to your county extension agent. He or she will also loan you the tools you need to gather core samples and show you how to use them; later, your agent can show you how to read and evaluate the results.

Bales — Large or Small?

Small square bales (traditional rectangular-shaped, string or wire-tied bales weighing 40 to 100 pounds each) are most everyone's choice for easy feeding. Putting them up, however, takes far more labor, time, and money than baling large round or large square bales, so they're increasingly harder to find and buy every year. The solution for some is feeding large square bales (rectangular bales generally 3 × 3 × 8 feet in size, weighing 600 to 700 pounds [272–317 kg]) or, more commonly, large round bales.

Large square bales are easy to use, breaking off in flakes like their smaller counterparts. Their only failing is that you'll need a tractor with a loader spike to move them. (Perhaps it's not a failing when you consider that having the tractor lift the hay is protecting your back; however, the tractor and loader spike do cost money to purchase and operate.)

Large square, small square, and round bales of hay all work for feeding miniature livestock.

Large round bales, ranging in weight from 500 to 1,800 pounds (227–816 kg), are the most economical type for farmers to bale and for livestock owners to buy. They have several drawbacks, however.

Large round bales are designed to be fed whole, in the field, with or without a hay ring (a cage designed to prevent livestock from lying in or pulling down hay and wasting it). It's possible to remove individual feedings by gradually unwinding the bale and forking them off, but it's hard work. And once opened, big bales exposed to the elements start to spoil, so you need to have a lot of animals that can polish off a bale in two or three days. Alternately, they work well if fed under cover.

HOW MUCH IS A BALE OF HAY?

Most hay dealers sell hay by the ton instead of by the bale. Purchasing hay by the ton allows the buyer to know precisely how much hay he is getting for his money, provided the hay is cured properly and accurately weighed. To convert the price per ton to price per bale:

1. Determine the average weight (in lbs) of the bales you are purchasing.
2. Divide the price per ton by 2,000 and multiply the result by the average weight of the bales to determine the price per bale: Price per ton ÷ 2000 × average bale weight (lbs) = price per bale.

Storing Hay

Protect valuable hay from the elements by covering it and placing it in a building that doesn't leak and has good ventilation.

If possible, don't store hay in or near a building that houses livestock, even though it's tempting to use overhead space as a hayloft. If you can't store your hay in a separate location, the next best thing is to keep only small amounts of hay in the barn at one time to help reduce the risk of fire. Hay is highly flammable; you don't want it overhead if your barn catches on fire. Also, improperly cured, high-moisture hay creates an environment where bacteria and mold fungi survive and create heat by consuming the hay. In worst-case scenarios, the heat generated can cause hay to spontaneously combust.

If the storage area is open on one or more sides, or it's a shed with only a roof, cover the hay with a tarp to keep out weather and light. Sunlight bleaches hay, causing it to lose as much as 20 percent of its nutritional value, especially protein and vitamin A.

Store hay where it's accessible. Choose a building in an elevated, well-drained area so the hay doesn't soak up moisture from wet soil or standing water. Storage should be near a road or driveway that large trucks can navigate, with room to back up and turn around. Doorways must be wide and tall enough for the hay to be easily unloaded and not far from where it needs to be stacked.

Don't stack hay higher than you can safely navigate and move (this is especially important when stacking large round bales).

FINDING HAY TO BUY

Virtually all state county extension services publish lists of hay dealers within their state and possibly other states as well. Hay dealers have a reputation to uphold and generally market good hay.

Feed stores usually sell hay by the bale. Although their overhead makes it pricier than most, convenience may be more important than cost.

Watch for hay sellers' ads on feed store bulletin boards or the classified ads, and post "hay wanted" ads of your own.

During drought years, hay can be hard to find. When that happens, consider chopped, bagged forage products like U.S. Alfalfa and Chaffhaye. They're convenient to store and feed, and lack of waste neatly compensates for greater cost.

Don't store hay near machinery (trucks, tractors, mowers) or near any type of heat source, and don't let anyone smoke in your barn. By the same token, don't store fire accelerants like gasoline, kerosene, oil, and aerosol cans near stored hay.

If you stack hay on bare ground it will wick moisture up through the bottom tiers, ruining the hay. The floor of a hay storage building can be earthen, but top it with a 4- to 6-inch (10–15 cm) coarse rock base to minimize bottom spoilage. Other materials that successfully break contact with wet soil and provide some air space are telephone poles, wooden pallets, and tires. Concrete floors also draw moisture, so treat them the way you would a dirt floor.

Don't stack new hay in front of older bales. Pull the old bales to the front and feed those first. Stack the bottom layer of small square bales on their sides, with the strings facing sideways instead of up. The uneven surface allows better air circulation. Stack the second layer with the strings facing up, all pointing in the same direction. Then, stack the third layer perpendicular to the second layer, so that if the second layer of bales is pointing east and west, the third layer of bales should point north and south. This will lock the stack in place and make it more stable. Be sure to leave some space between rows to promote airflow and allow moisture to escape.

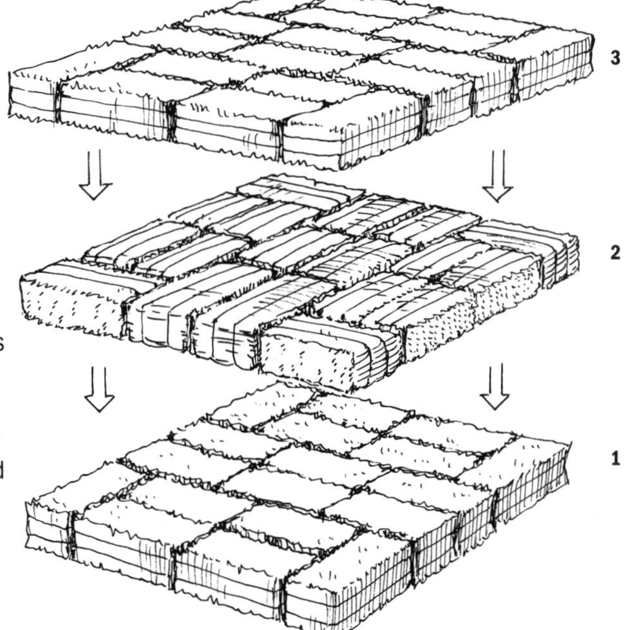

Proper way to stack square bales of hay: stack the first layer on its side, with strings facing sideways; stack the second layer with strings facing up; stack the third layer the same way as the first.

Outdoors, cover stacks of square bales (large or small) with tarps, securing the tarps in place with strong tie-downs. A sloped top created by pyramiding the final layers will shed snow and rain better than a flat top. When covering hay with tarps, make sure they're *good* tarps — strong, sturdy ones free of holes or rips that might let moisture through. If you have to use more than one tarp to cover the stack, overlap them by 3 to 5 feet (0.9–1.5 m) to prevent water from seeping in where the tarps meet. Monitor tarps on a regular basis to make sure

SMALL-SCALE HAYMAKING

If your winter hay needs aren't great and you have an unused corner of an outbuilding for storage, consider growing and putting up premium hay for a miniature milk cow or a few head of miniature sheep or goats.

The trick is to make fantastic hay so that every single mouthful packs a punch. Touch base with your county extension agent to determine which forage works best for your climate and your critters, then prepare, fertilize, and plant a small hay plot (one-quarter to one acre in size) accordingly. Just like the big haymakers, be set to harvest at the optimal time during a window of warm, sunny days.

You'll need a mower or a scythe. A scythe will do an admirable job and it's fun to use one, but you'll need to learn how to operate it first. Visit the Scythe Supply Web site or read *The Scythe Book* (see Resources) to see if the old way is the way for you. Otherwise, the scythe's twenty-first–century counterparts — heavy-duty string trimmers and brush cutters — will save your back and do the work in a fraction of the time.

Your hay should be turned at least twice. Our forefathers used wooden peg-toothed haying rakes, and if you're handy with tools you could make one, but a wide-toothed garden rake will do in a pinch.

Check mown hay frequently, and when it's fully dry but still green and sweet-smelling, hook a cart to your lawn tractor, grab your pitchfork, and bring in the harvest.

Store your hay in a well-ventilated area out of direct sunlight. Don't fork it directly on the floor; place it atop old tires or wooden pallets to prevent spoilage from contact with the ground. Pack it down and pile it high. It's best to leave new hay uncovered for a few weeks until it finishes curing, and then top it with tarps or old bed sheets to preserve its cleanliness and quality.

they're securely tied down; loose, flapping tarps wear out quickly and allow moisture to reach stored hay.

Large bales stored outside invariably sustain some spoilage, depending on a combination of factors including moisture of the hay at baling time, amount of rain and snowfall during the storage period, drainage of the soil on which the bales are stored, amount of space between the bales, type of hay (grass, legume, or grass-legume), and the skill of the operator making the bales.

It's best to place round bales in such a way that there is about 1 foot (30 cm) of air space on all sides for good ventilation. Round bales also store well when flat ends are butted end-to-end in long rows. Orient these rows north and south so prevailing winds won't create snow drifts and so both sides of the row receive sunlight for drying. Don't arrange them in a row with the twine sides touching, creating a water-absorbing valley between bales. If bales must be stored side by side, leave at least 24 inches (60 cm) between bales. If more than one line of bales is needed, space adjacent lines at least 3 feet (0.9 m) apart so water from one row doesn't run off onto another. Never store bales under trees; storing hay in damp, shaded environments prevents the sun from drying the hay after a rain and encourages the bottom of the bales to remain wet.

Choose densely packed large round bales. These sag less, resulting in less contact with the ground. Opt for big bales put up in plastic or net wrap or that use plastic twine; these factors reduce bale sag and help maintain bale shape. Plastic twine resists weathering, insects, and rodents better than natural fiber twines. Twine should be snug and spaced 6 to 10 inches (15–25 cm) apart. Densely packed bales are also more stable when stacked, though stacking does increase losses because moisture gets trapped in the hay that is not exposed to sun and wind.

Whenever possible, store large round bales under cover. The outer 4-inch (10 cm) layer of a round bale that is 6 feet (1.8 m) in diameter contains about 25 percent of the total bale volume. Studies indicate that outdoor storage losses range between 5 and 35 percent, depending on the amount of precipitation, the storage site location, and the original condition of the bale. Storing the hay indoors can reduce storage losses by approximately two-thirds.

What about Fescue?

No discussion of hay would be complete without mentioning fescue, the Dr. Jekyll and Mr. Hyde of forage crops.

Tall fescue (*Festuca arundinacea* or *Schedonorus phoenix* [scop.] Holub) is one of America's most widely used forage crops. It's hardy, productive, insect and nematode resistant, tolerates poor soil and climatic conditions very well, and has a long growing season.

Fescue has been around a long time. Tall fescue came to this country from Europe in the late 1800s and was officially discovered in Kentucky in 1931; it was marketed in 1943 as Kentucky 31 Fescue. It was an immediate hit with farmers and spread quickly throughout the midwestern and southern United States. Unfortunately, it also has a downside. Roughly 80 percent of the tall fescue grown in North America (about 40 million acres at last count) is infected with a fungal endophyte — so called because it is in (*endo*) the plant (*phyte*) — that produces ergotlike alkaloids that are toxic to animals. The seed head is the most toxic portion of the plant. Toxicity is especially severe when pregnant livestock graze tall fescue during the last trimester of gestation.

Grazing species react slightly differently to fescue toxosis, but it's a serious thing in every species.

Tall fescue

Fescue and Horses and Donkeys

Mares and jennies grazing endophyte-infected tall fescue experience reproductive problems such as late-term abortions and stillborn foals, prolonged gestational periods, dystocia, thickened or retained placentas, laminitis (founder), poor conception rates, and agalactia (diminished ability to produce milk).

Foals born alive are frequently larger than normal and have poor suckling reflexes, bad coordination, and lowered body temperatures. Foals may also have poor immunity due to lack of colostrum production by their dams.

Because equines are especially sensitive to the toxins in endophyte-infected tall fescue, even the lowest levels of endophyte can produce equine fescue toxicosis.

Fescue and Cattle

Multiple studies show that consumption of endophyte-infected tall fescue decreases the feed intake of cattle and therefore lowers their weight gains. They also produce less milk, have higher internal body temperatures and respiration rates, retain a rough hair coat, demonstrate unthrifty appearance, salivate excessively, and experience multiple reproductive problems.

Tall fescue fertilized with high levels of nitrogen can lead to bovine fat necrosis. As hard masses appear in the fatty tissues surrounding the intestines, this condition causes digestive problems and can interfere with calving.

Fescue and Sheep, Goats, and Llamas

Ewes and does suffering from fescue toxicosis kid up to 10 days late. When they do go into labor their cervixes may not dilate, contractions are often weak, and offspring may be too large to be delivered without human intervention. Unusually thick umbilical cords are hard to break, and placentas are so thick that the young can't get out unless someone tears them open. Female llamas are prone to the same sort of ills.

Weight gains are often reduced by 50 percent (the same rate as in cattle). Also like cattle, goats and sheep are prone to fescue foot, a type of dry gangrene in which parts of the hooves and tail rot and fall off.

Fescue — Yea or Nay?

So should you feed fescue to your livestock? Probably not if you have a choice — but many people don't. Fescue is so well established in the United States, especially the Midwest and parts of the South, that unless it's purposely killed out, most grasslands and hayfields contain at least some fescue. Eradicating ultra-hardy fescue isn't easy, and replacing it with new, "improved" endophyte-free fescues is not the answer in many cases; the endophyte itself contributes to fescue's hardiness, so endophyte-free varieties often fail to thrive.

If you live where fescue prevails, assume it's in your pastures and in most of the grass hay you buy. The best ploy is to remove late-term pregnant females to a dry lot and feed hay known to be fescue-free. Some hay dealers and feed stores sell guaranteed fescue-free grass hay for just this reason. Judiciously feeding baled or chopped and bagged alfalfa works, too.

THE BEAUTY OF COMMERCIAL FEED

When your livestock need concentrates, consider feeding nutritionally balanced commercial feeds formulated for your animals' species and class. Commercial mills employ livestock nutritionists who design feeds with precisely the right nutritional ratios each group requires, eliminating guesswork on your part. Bagged feeds also eliminate the need to keep barrels of various grains on hand in order to mix your own at home.

When you feed commercial feeds, however, heed the information on the bag. Don't add "extras" to the mix and risk upsetting its nutritional balance.

BREEDER'S STORY

Lauren Simermeyer, DVM, and P. Colleen Simermeyer
Stubby Acres Farm, New Jersey

COLLEEN SIMERMEYER and her daughter, Lauren, raise Babydoll Southdown sheep at their Stubby Acres Farm in southern New Jersey, so named for their love of short-legged Babydoll sheep and corgis. Lauren, who is a veterinarian, and Colleen specialize in rare, spotted genetics not encountered in the average flock of Babydolls and in conformationally correct, healthy, happy sheep.

We asked Lauren to comment on their involvement with this popular breed and this is what she told us.

"I was exposed to sheep eight years ago through training and testing my herding corgis. I immediately realized they were wonderful creatures and I wanted to have some of my own, but I had a lot of schooling to accomplish before I could add sheep to my life. So my mother and I didn't obtain our sheep until 2007.

"I researched several breeds and fell in love with Babydolls. My mother and I contacted one of the largest breeders in the East at the time, Catherine Doty, and I visited her flock back in 2001. Her sheep were small, fuzzy, and gentle; this is exactly what we wanted. We later came to realize that Babydolls are generally too laid-back to make good herding sheep, but we were already sold.

"We are attempting to breed for more spotting patterns and unusual colors. We plan to maintain solid blacks and whites as well but would like to place an emphasis on spotted patterning, as this is unique. Some breeders do not feel that spotted sheep are true Babydolls, but we found several documented sources dating as far back as the 1840s that state that Southdowns were occasionally piebald and had spotting patterns. Our intention is to breed selectively to accentuate this particular trait, while maintaining essential Babydoll characteristics like smiling wooly faces; stocky, short legs; and good overall conformation. Every sheep needs to be able to stand solidly on four legs and be appropriately muscled with a good topline.

"Many people show their Babydolls, but this is not an area we have ventured into just yet. Many also shear them for their wool. I have been told that Babydoll wool is very nice to work with when breeders select for better-quality fleece. We are taking classes to learn how to spin and knit. It will be very rewarding to use our own wool to make blankets and scarves.

"There is a market for breeds that make nice additions to the family. These sheep are very endearing, and their gentle natures make them wonderful pets. We easily sell all available females. As with all breeds of sheep, males are a different story, but anyone considering breeding needs to be aware that wethers make nice pets, too. While this breed is generally known as a pet breed, some do go to market, and there are several breeders who breed them for this purpose.

"There is always room for good breeders. It is extremely important to select for healthy animals with good temperaments and favorable Babydoll characteristics. Also, as in any animal-breeding community, new blood offers new ideas and experiences that will help the breed grow as a whole.

"The biggest piece of advice I could give new breeders would be to 'do your homework.' Keeping sheep is no small venture. They have many needs that you should be prepared to provide.

"Babydoll sheep are typically healthy sheep, but it is very important that you have a veterinarian to work with you and your flock. You should look into potential veterinarians before you purchase sheep — this way you have one to call in case of an emergency.

"Babydolls have a couple of health needs specific to their breed. They need much more frequent hoof trimming than other breeds — a minimum of once a month unless you have rough terrain that wears down their hooves. And they can have a large amount of wool on their faces. This is adorable but can cause wool blindness. Regular trimming of the wool surrounding the eyes may be necessary. Some breeders are breeding for cleaner faces because of this.

"To maintain a healthy flock, you must deworm your sheep. It is essential that you find a veterinarian in your area, as each area of the country has its own deworming needs, not to mention the unique needs of each individual farm.

"Many people are not familiar with and often neglect mineral and baking soda supplementation. Sheep need these, and sheep in some areas of the country need more supplementation than sheep in other areas.

"Babydolls tend to gain weight easily and keep it on. Most books on shepherding advise feeding large amounts of food to sheep. If this advice is followed, your Babydoll flock will be obese, and this will increase health concerns.

"Most Babydolls lamb with ease, and rams are fertile when very young, so it's not a good idea to buy a pair of lambs and keep them together. This could result in your ewe lambing too young. Buy from a good, reputable breeder; it's your best bet for a good start in sheep."

Choosing Concentrates

It can't be said often enough: Most minis don't need grain. If you elect to use it, add it gradually and back off if your animals gain too much weight. Should you choose to create your own mix based on locally available grains, get advice; your county extension agent or a nutritionist at your state agricultural university will be happy to help you formulate a plan.

Buying Grain

If you feed individual grains or homemade mixes, it's important to learn to evaluate the grain you buy. You'll probably order it in bulk at your feed store, so ask to see samples before your bags are filled. There are several factors to consider. They are, in order of relative importance: plumpness, foreign material, color, and a few other considerations.

Plumpness

Grain is used as an energy source in livestock rations. The energy-supplying portion of a grain kernel, called the *endosperm,* is made up of high-energy carbohydrates (starch and sugars) and proteins. A plump kernel has a well-developed endosperm. Plumpness also indicates a lower percentage of hull or bran (or both), and thus a lower percentage of crude fiber, resulting in a higher feeding value for livestock. Kernel plumpness is particularly important when evaluating oats or barley, as these grains are naturally high in fiber due to their hulls.

Choose grain with plump, full kernels.

Foreign Matter

This is any undesirable substance found in grain, such as stones, chaff, bits of sticks or wire, and straw. They add significantly to its cost because you're paying the same price for inedible junk as for grain. Also, for your animals' health and safety, you may have to clean the grain before it's fed.

Insects, living or dead, are sometimes found in grain. Their presence indicates that the grain was stored at too high a temperature or moisture content. Insect feeding activities can rapidly deteriorate grain, and some dead insects are toxic.

Grain sometimes contains unwanted weed or other crop seeds. Some seeds and bulblets (such as wild onion or garlic) affect the flavor of feeds and should be removed. Weed seeds, unless they're toxic or present in large amounts, don't greatly affect the feeding value of grain; however, some can pass through an animal's digestive system and, in doing so, inadvertently seed your fields with weeds.

Ergots are also found in grain. These are black, hornlike fungal bodies that are poisonous to stock when fed in large quantities or if fed in small quantities over a long period of time. They mainly occur on rye (its most common host), wheat, barley, milo, and millet. Ergot affects oats only rarely.

Color

Weathered seeds are bleached or darkened (seeds darken due to the growth of decay molds). If molds aren't present, weathering causes only minor deterioration in quality.

Grain that is stored at high temperatures or exposed to high temperatures during processing may be dark brown and have a burnt taste or odor. Heated grain can be less palatable to animals and therefore problematic.

Oats naturally occur in a wide range of colors; their colors are genetic in origin and have no effect on grain quality.

Mold

Never feed moldy grain of any kind. Moldy grains, particularly corn, are capable of producing dangerous amounts of poisonous fungal mycotoxins called "aflatoxins." Aflatoxins are caused by mold fungus, usually *Aspergillus flavis.*

Aflatoxin contamination is greater in corn that has been produced under stress conditions; drought, heat, insect, and fertilizer stress are all conducive to high levels of aflatoxins.

Aflatoxin levels are regulated by the Food and Drug Administration (FDA) at 20 ppb (parts per billion) in food and feed. Aflatoxin can also appear in the milk of lactating animals fed aflatoxin-contaminated feed.

Mycotoxins, including aflatoxins, are known to cause serious health problems in animals, including equine leukoencephalomalacia in horses and porcine edema in swine. Reduced weight gain, capillary fragility, reduced fertility, suppressed disease resistance, and even death have been attributed to mycotoxins. No animal is known to be resistant, but in general, older animals are more tolerant than younger livestock. Some mycotoxins have also been associated with human health problems.

Because aflatoxin contamination is fairly common, many livestock producers eschew feeding corn.

Cracked Grain

Cracked kernels have no major impact on grain quality. One disadvantage, however, to cracked grain is that it doesn't store for long periods of time. Oils in the cracked grain break down, causing rancid taste and odor. Cracked grain is also more susceptible to insect infestation.

Treated Seed

Seed grain is often treated with chemicals to prevent diseases like seed and seedling rot. Most seed treatments are red colored for easy recognition. Should you find colored kernels in your grain, discard the entire bag; some of these substances are exceedingly toxic.

Mineral Supplements and Salt

Responsible livestock owners provide their animals with loose or lick-type mineral supplements appropriate to their locale, the species and type of animal consuming the product, and the sort of other feedstuffs fed. As is true of every other aspect of livestock nutrition, mineral supplementation is a very complex issue and a one-size-fits-all approach won't work. Discuss your needs and your feeding program with your county extension agent, a nutritionist associated with your state university's agriculture college, or with a knowledgeable representative from the company that manufactures your bagged feed.

Some mineral supplements are fed in conjunction with salt and others aren't, so read the label on mineral supplement packaging to be certain.

Mineral supplements and salt come in blocks, tub licks, and loose forms. Blocks work well for rough-tongued species such as cattle but are next to useless for equines, goats, and sheep. Most species relish tub licks because the majority of them are formulated using plenty of molasses. However, all species fare best with loose minerals, which should be served up in mineral feeders designed for that purpose. Llamas, because of their unique tongue structure, must have loose minerals in lieu of blocks or licks.

8

Health

IT'S INCREASINGLY DIFFICULT to find large-animal vets who make farm calls. Thus, livestock owners tend to become their own vets. This is, however, not without risk. The most workable solution: Find a vet who's qualified to treat your species (reread chapter 3 for useful tips on how to find a good one) and establish a relationship, but also learn to address minor problems and routine veterinary procedures yourself.

Calling the Shots

The first thing to learn about keeping your animals healthy is how to give their shots yourself. You need this skill to vaccinate your stock and to treat them when they're ill. If you've never given a shot before, it sounds scary; but if you understand the basics and practice before giving your first shot, it is quite manageable.

PRACTICE MAKES PERFECT

When I learned to give intramuscular injections, I practiced by injecting water into a thick-skinned naval orange. This works, but if you're a roasted chicken connoisseur, try this instead: Buy a whole, fresh (not frozen), dressed chicken at the supermarket. Load a syringe with marinade, then gently lift the chicken's skin and practice your subcutaneous injection technique; for intramuscular practice, inject marinade into the bird's thighs and breast. When you're finished, cook the chicken. Enjoy!

Pick the Right Equipment

Choose the smallest syringe and largest needle you can use to perform the task at hand. Large syringes are cumbersome to use, especially in women's smaller hands, and it's better to use a large-bore needle to quickly inject a substance than to slowly force viscous fluids through a smaller needle.

Opt for disposable syringes and discard them after use. Don't try to sterilize and reuse them, as boiling compromises their integrity. Keep a variety of sizes on hand.

There are two types of disposable syringes: smooth top and luer-lock; the needle slides onto the first and screws onto the second. When giving intramuscular shots, the smooth top is best if you prefer to insert the needle into muscle before attaching the syringe — a ploy sometimes used when injecting fractious animals. The luer-lock, however, is more secure and is the type we use and recommend.

Assemble a collection of needles. Subcutaneous injections (SQ; under the skin) should be given using 18- or 20-gauge needles that are ½ to ¾ of an inch long. Give intramuscular injections (IM; into muscle) using 18- or 20-gauge needles that are 1 to 1½ inches long. Transfer needles (for pulling vaccine or medicine out of the bottle) that are long and large-gauge are easiest to use. Some antibiotics are syrupy-thick, and the carriers used in their production make these injections really sting; for these, choose 16-or 18-gauge needles so you can inject the fluid quickly before the animal objects.

Start with enough needles to do the job. You'll need a new needle for each animal, plus a transfer needle to stick through the rubber cap on each product you're using (e.g., if you're vaccinating 10 sheep using CD/T toxoid and Case-Bac, you'll need 12 needles). Using a new needle each time is less painful for the animal, and it eliminates the possibility of transmitting disease via contaminated needles.

Preparing for the Injection

Don't combine vaccines or drugs. Mixing could destroy the effectiveness and value of the individual products.

Reread the label on each product and follow directions. The label on health products includes the dosage to be given, the timing of administration, and the route of administration. The most common routes of administration are intramuscular (IM), subcutaneous (SQ), intravenous (IV), intranasal (IN), and topical. The label also states the best method of administration. Generally, bacterins (killed products) can be given subcutaneously. Modified live virus products are

PREVENTION IS THE KEY TO GOOD HEALTH

It's always easier to prevent disease and health problems than to treat sick animals. Here are some tips for ensuring your herd is in the best of health:

- Check each of your animals every day (twice is better). Are any sick, injured, or off-color? Take care of problems right away; don't wait to see if they get better by themselves. If you don't know what's wrong with an animal or you're not positive you know how to treat what ails her, get help.
- Monitor pregnant females extra closely and be there when they give birth. Keep a well-stocked birthing kit on hand (see page 192) and know how to use it.
- House your livestock in clean, safe surroundings. Check fences for weaknesses on an ongoing basis, removing or repairing accidents waiting to happen. Don't overstock your facilities. Crowding leads to stress and overgrazing.
- Quarantine incoming animals, whether they're new purchases, long-held animals returning from shows, or females visiting for breeding purposes. Make no exceptions.
- Provide a hospital area for sick animals (use your quarantine pen if it isn't already occupied); don't leave them with the rest of the flock or herd.
- Feed wisely; so many health concerns are feed-related. Make changes gradually and don't overfeed. Make certain every animal is getting the amount of feed he or she needs; feed shy or aggressive animals separately, if necessary.
- Exclude mice, rats, raccoons, cats, dogs, and poultry from feed storage areas to prevent them from soiling feed.
- Have dead animals, especially aborted fetuses, necropsied to determine the cause of death; ask your vet for details. Burn, bury, or compost the birthing tissues (don't let your dogs or barn cats eat them).
- Trim hooves or toenails on an ongoing basis to maintain soundness and prevent deformities.
- Discuss a vaccination program with your vet. Unless you're raising your animals organically, at least vaccinate for tetanus and diseases endemic to your species and region.

usually given intramuscularly, because this allows the virus to reproduce and reach the lymphatic system more easily. Whenever possible (if allowed on the label), use the subcutaneous route. The label will also list warnings, indications for use, withdrawal times if any, and proper methods of storage and disposal (some products can be stored at room temperature; others must be refrigerated), and expiration date.

Restrain the animal using a stout halter and lead, head gate, grooming or milking stand, or restraint chute, or recruit a helper strong enough to hold the animal reasonably still. Proper restraint reduces the potential of hurting the animal or yourself.

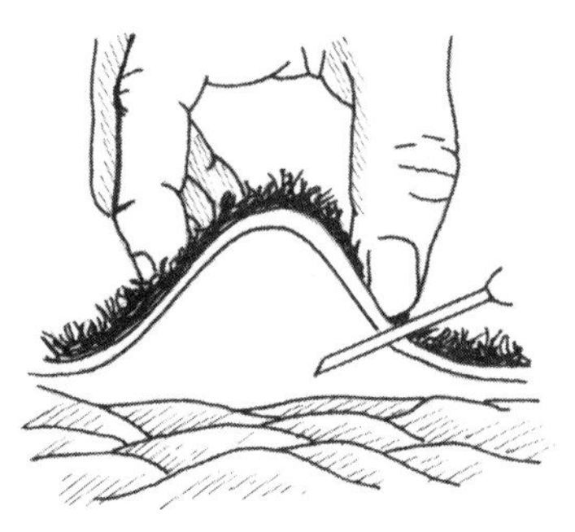

Subcutaneous shots are given just under the skin (see step 1 on page 122).

How to Prepare the Needle (for SQ and IM injections)

1. Insert a new, sterile transfer needle through the rubber stopper cap of each product. Never poke a used needle through the cap to draw vaccine or drugs.
2. To draw a shot, attach a syringe to the transfer needle and draw the vaccine or drug into it.
3. When you're drawing fluid from the bottle, first inject half the amount of air into the bottle (e.g., if you're drawing 8 cc of fluid, inject 4 cc of air) to avoid the considerable hassle of drawing fluid from a vacuum, then pull a tiny bit more of your required fluid amount into the syringe.
4. Detach the transfer needle, and attach the needle you'll use to inject the pharmaceutical into your animal. Press out the excess fluid to remove any bubbles created as you drew the vaccine or drug.

Giving the Injection

Injection sites vary from species to species, but overall you want to inject the vaccine or drug where it will work well without injuring the animal or, in some cases, damaging expensive cuts of meat. To minimize the risk of infection and the incidence of injection site lesions, avoid injecting into or under wet or dirty skin.

Never inject more than 10 cc into one site. When making multiple injections, make sure your injection sites are at least 5 inches (13 cm) apart and be careful not to reuse injection sites. When injecting more than one product, don't inject them in close proximity to one another.

Sheep, goat, and llama vaccines and drugs are usually given subcutaneously. Occasionally nodules develop on the injection site a few days later, and these can be mistaken for caseous lymphadenitis in sheep and goats. To minimize this concern, many producers inject all pharmaceuticals between the front legs where they're less likely to be noticed and where caseous lymphadenitis abscesses don't normally occur. Other preferred sites for subcutaneous injections are in the neck, over the ribs, and into the "armpit." With llamas, the best place is between where the shoulder ends and the neck begins.

PROPER INJECTION SITES

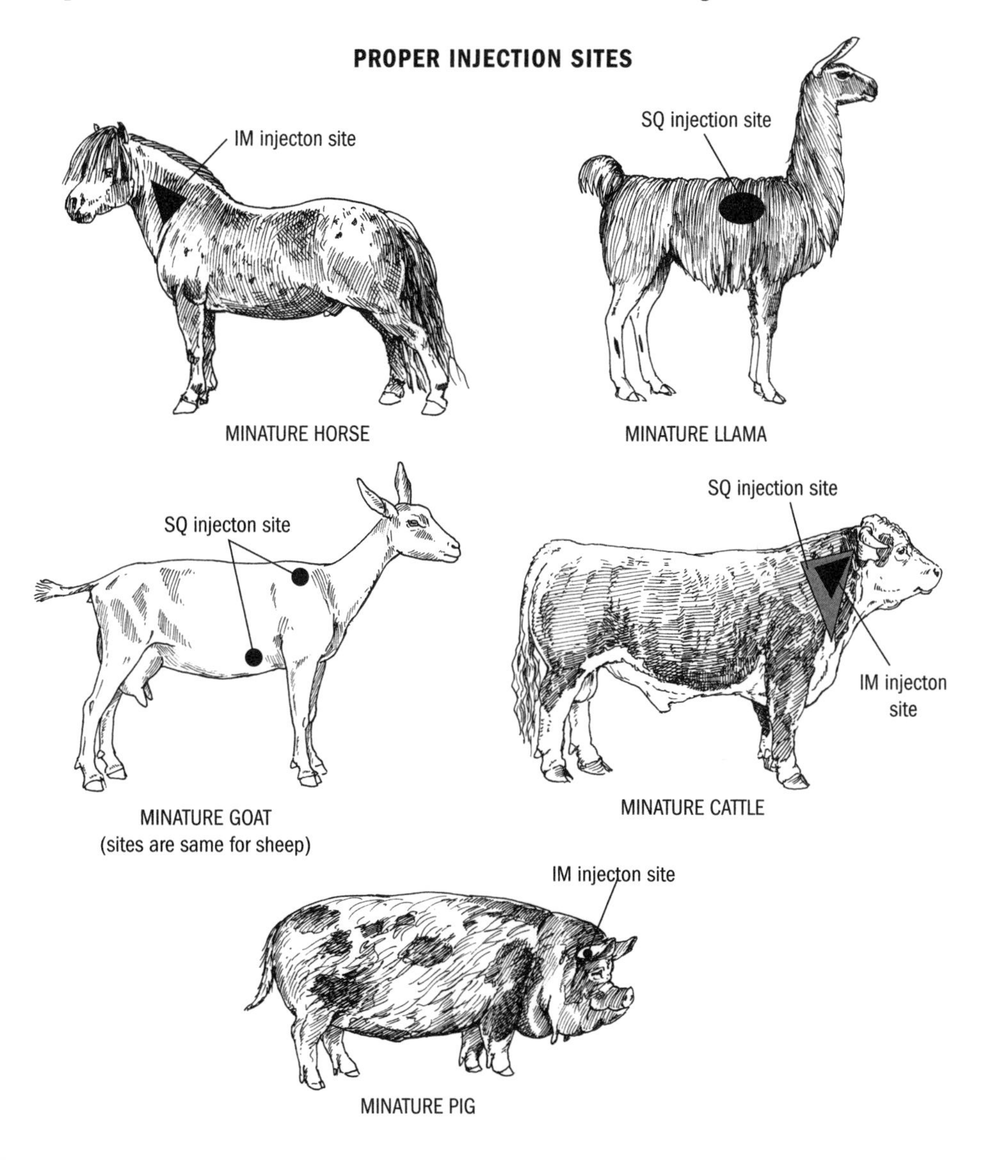

How to Give a Subcutaneous (SQ) Shot

1. Tent the skin to allow the product to be injected under the skin and not into the muscle.
2. Pull the skin away from the animal's body, then insert the needle into the fold of skin sideways parallel to the animal's body — never at a right angle. Check to make certain you haven't poked it out the other side of the fold.
3. Slowly depress the plunger until all product has been dispensed, and then withdraw the needle.
4. Rub the injection site to help distribute the drug or vaccine.

For intramuscular (IM) injections, cattle and equines are injected into the thick muscles in the "triangle" of the animal's neck. Llamas are generally injected into the upper thigh just below the tail or into the stifle area of the hind leg. Pigs are injected high on the neck behind the base of the ear (piglets can be injected into the top of the hip or the thigh).

How to Give an Intramuscular (IM) Injection

1. Quickly but smoothly insert the needle deep into muscle mass. Be sure to inject straight in, not at an angle.
2. Aspirate (pull back on) the plunger ¼ inch to see if you hit a vein, which is something you want to avoid. If blood rushes into the syringe, pull the needle out, taking great care not to inject any drug or vaccine as you do so, and try another injection site.

Absorption rates vary for subcutaneous and intramuscular injections, and some substances behave differently when given by different routes. If your vet prescribes one route, consult with him before switching to a different one.

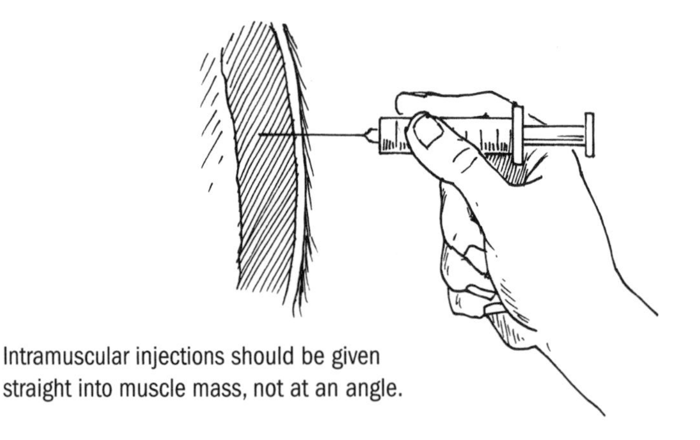

Intramuscular injections should be given straight into muscle mass, not at an angle.

BENDING OVER TO DO IT RIGHT

When injecting minis, it's often easier to stand on the side opposite the injection site and bend over the animal than to stand on the side where you plan to give the injection. If you're right-handed, stand on the left side and bend over the animal's back to the right-side injection site. The handler stands on your side. To give two injections, change sides.

Don't Forget the Epi!

Epinephrine, also called adrenaline or epi, is a naturally occurring hormone and neurotransmitter manufactured by the adrenal glands. It's widely used to counteract the effects of anaphylactic shock.

Any time you give an injection, no matter which product or the amount injected, be prepared to immediately administer epinephrine to counteract an unexpected anaphylactic reaction. Signs of anaphylactic shock are glassy eyes, increased salivation, sudden-onset labored breathing, disorientation, trembling, staggering, or collapse. If the animal goes into anaphylaxis, you won't have time to race to the house to grab the epinephrine. You might not even have time to fill a syringe; you have to be ready to inject epinephrine immediately, on the spot.

Keep a dose of epinephrine drawn up in a syringe in your refrigerator. Kept in an airtight container, it will keep as long as the expiration date on the epinephrine bottle. Take a loaded syringe of epi with you every time you give a shot. The standard dosage is 1 cc per 100 pounds; don't overdose, as it causes the heart to race. Previously available over the counter, epinephrine is now a prescription drug, so you have to get it through a vet.

COMMON CONVERSIONS

1 milliliter (1 ml) = 15 drops = 1 cubic centimeter (1 cc)
1 teaspoon (1 tsp) = 1 gram (1 gm) = 5 cubic centimeters (5 cc)
1 tablespoon (1 tbsp) = ½ ounce (½ oz) = 15 cubic centimeters (15 cc)
2 tablespoons (2 tbsp) = 1 ounce (1 oz) = 30 cubic centimeters (30 cc)
1 pint (1 pt) = 16 ounces (16 oz) = 480 cubic centimeters (480 cc)

Administering Liquids and Pastes

Liquid medicines and some dewormers are given orally as drenches. Drenches can be administered using catheter-tip syringes, but the most efficient way to drench is with a dose syringe.

To drench an animal, restrain him and slightly elevate his head — just enough so gravity can help — using one hand under his chin. Insert the nozzle of the syringe between the animal's back teeth and his cheek (this way he's less likely to aspirate part of the drench) and *slowly* depress the plunger, giving the animal ample time to swallow.

When giving semisolid drenches such as paste-type dewormers or gelled medications, deposit the substance as far back on the animal's tongue as you can reach. In either case, keep the animal's nose slightly elevated until he visibly swallows. And be careful; if you stick your fingers between the animal's back teeth you're likely to be cut (in some species, such as sheep and goats, those molars are *sharp*) or accidentally bitten.

BUILD A BETTER FIRST-AID KIT

We keep our farm-based first-aid kit in two 5-gallon plastic food service buckets fitted with snug lids. On the top and both sides we've affixed big Red Cross symbols using red duct tape, so the buckets are easy to spot when we need them. We keep the buckets in the house in a walk-in closet; they're returned to their place immediately after each use — no exceptions.

One bucket houses emergency equipment that we use for the goats, sheep, and horses. It contains lead ropes and a half dozen halters ranging from our smallest small-ruminant halter (actually an alpaca halter, but it neatly fits full-size goat kids and our miniature sheep) to a huge one that fits our Thoroughbred mare; the horse halters are hand-tied rope versions to conserve space. When we need something in an emergency, we carry the bucket to the site and simply dump everything on the ground. The bucket also contains a fencing tool and a small length of aluminum electric fence wire for making impromptu fence repairs if an animal must be extracted from a fence.

The second bucket is organized using resealable plastic bags in several sizes. Using a felt-tipped marker, we've labeled each bag according to the basic uses of its contents:

- **Wound cleanup and bandaging materials:** gauze sponges, Telfa pads, three rolls of Vetrap self-stick disposable bandage, a roll of 2½-inch-

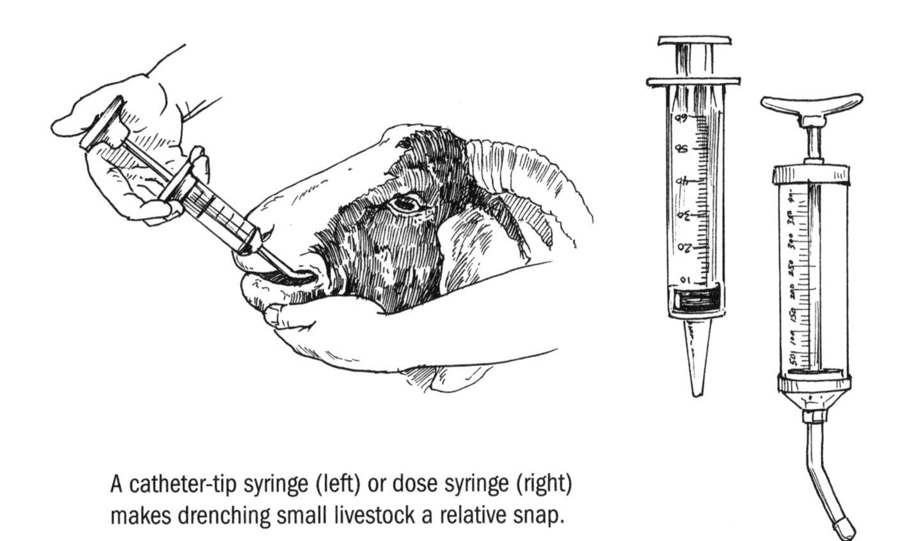

A catheter-tip syringe (left) or dose syringe (right) makes drenching small livestock a relative snap.

wide sterile gauze bandage, 1- and 2-inch-wide rolls of adhesive tape, a partial roll of duct tape with ¾ to an inch of tape left on it, two heavy-duty sanitary napkins (they can't be beat for applying pressure wraps to staunch bleeding), a sandwich-size bag of cornstarch (it works better than commercial blood stopper powder), a small bottle of Betadine Scrub, another of regular Betadine, and a 12-ounce bottle of generic saline solution.

- **Hardware:** blunt-tipped bandage scissors, a hemostat (we prefer it to tweezers), a flashlight (the flat kind you can hold between your teeth), a stethoscope, and a digital thermometer in a hard-shell case.
- **Basic medicines:** wound treatments of choice (Neosporin antiseptic ointment, emu oil, and Schreiner's Herbal Solution), topical antibiotic eye ointment, and a full tube of Probios probiotic paste.

We also store over-the-counter and prescription drugs we'd need in an emergency in a separate, easy-to-grab-when-we-need-it plastic basket in our pharmaceutical refrigerator (a dorm-size model from the used-a-bit store).

A separate, scaled-down first-aid kit is kept stowed behind the seat of the truck.

Wound Care

Animals have a knack for injuring themselves, even in seemingly safe surroundings, so you'll need to learn how to treat minor wounds.

Keep a supply of saline solution on hand for cleaning cuts, bites, and abrasions. If you're out of solution, flush wounds using lots of cool water from a hose. After flushing, apply a mild disinfectant like dilute Betadine solution to kill bacteria left on the wound. Be gentle; don't scrub.

What do you put on this nicely cleaned wound? In many cases, nothing; clean, open wounds heal better (and faster) than injuries coated with gunk. When we do dress a wound, we use holistic liquid dressings such as Schreiner's Herbal Solution (see Resources) and emu oil.

Don't attempt to treat a serious wound (including all puncture wounds) yourself. Call your vet if an injury is extensive or bleeding profusely, contaminated by any sort of debris, on or near a tendon or joint, or if it's already infected.

Taking Vital Signs

It's important to know how to take your animals' temperatures and assess their respiration and heart rates.

Temperature

To take an animal's temperature, you'll need a rectal thermometer. Veterinary models are best, but digital rectal thermometers designed for humans work, too.

Traditional veterinary thermometers are made of glass and have a ring on the end, to which you can attach a cord. Add an alligator clamp to one end of the cord and knot the other end to the thermometer; that way you can apply the clamp to your patient's hair or tail before inserting the thermometer, which prevents it from falling to the ground and breaking or from being sucked too far into the

ANTIBIOTICS — YEA OR NAY?

The overuse of antibiotics is a real and rapidly expanding problem as disease-causing organisms become increasingly antibiotic resistant. For most farms, however, strictly avoiding antibiotics isn't feasible.

If your vet says to use them, follow directions to the letter. Use precisely the recommended dosages and complete the series as directed. Because antibiotics destroy good bacteria as well as bad, follow antibiotic treatment with oral probiotics to restore the patient's digestive system to good health.

rectum to be easily retrieved. Glass thermometers must be shaken down after every use: hold the thermometer firmly and shake it in a slinging motion to force the mercury back down into the bulb. Because a digital thermometer is faster, beeps when done, and needn't be shaken down, digitals are better for working with livestock.

Restrain the animal in the same manner as when giving shots. Insert the business end of a lubricated thermometer a short way into the animal's rectum. Use KY Jelly or lubricant designed for veterinary purposes; plain old saliva works in a pinch. Hold a glass thermometer in place for at least two minutes; hold a digital model until it beeps.

After recording the reading, shake down the mercury in glass models, clean the thermometer (glass and digital) with an alcohol wipe, and return it to its case. Always store thermometers at room temperature.

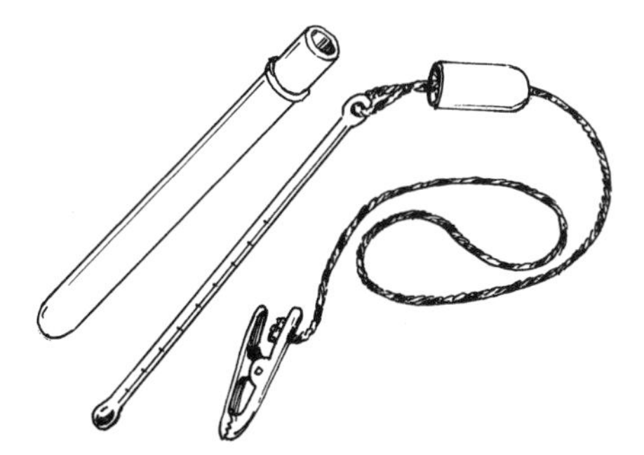

Veterinary thermometers have a ring on the end to accommodate a cord and alligator clip.

ADULT TEMPERATURE BY SPECIES

(temperatures in young may be slightly higher)

Species	Temperature (°F)
Mini cattle	100.4 to 103.1
Mini donkeys	99.5 to 101.3
Mini goats	102.2 to 104.9
Mini horses	99.5 to 101.3
Mini llamas	99.5 to 101.5
Mini mules	99.5 to 101.3
Mini pigs	100.4 to 104.0
Mini sheep	102.2 to 104.9

Heart Rate

The exact places and means of checking heartbeat vary slightly from species to species, but behind the left elbow and under the jaw are generally good spots. The easiest way to check heart rate is with a stethoscope; count the number of pulses in 15 seconds and multiply that number by 4 to get the number of beats per minute (bpm).

HEART RATE BY SPECIES

Species	Beats per minute (bpm)
Mini cattle	40 to 70
Mini donkeys	38 to 40
Mini goats	60 to 90
Mini horses	38 to 40
Mini llamas	60 to 90
Mini mules	38 to 40
Mini pigs	60 to 80
Mini sheep	70 to 80

Respiration

Watch the animal's rib cage as it moves and count the number of breaths the animal takes in 15 seconds, then multiply that number by 4.

External conditions can affect readings. Body temperatures rise slightly as the day progresses and may be higher on hot, sultry days. Extreme heat and fear or anger elevates pulse and respiration. Slightly elevated readings are sometimes the norm.

RESPIRATION RATE BY SPECIES

Species	Breaths per Minute
Mini beef cattle	10 to 30
Mini dairy cattle	18 to 28
Mini donkeys	12 to 16
Mini goats	12 to 20
Mini horses	8 to 15
Mini llamas	15 to 30
Mini mules	10 to 15
Mini pigs	12 to 30
Mini sheep	12 to 20

Identifying, Preventing, and Treating the Most Common Illnesses

It's impossible to discuss every disease encountered by the species covered in this book; these are but a few to be aware of. There are vaccines approved for the ones marked with an asterisk (*); ask your vet for details.

Bloat

Bloat occurs when ruminants gorge on grain (perhaps through raiding an unlocked feed room), legume hay (when they aren't accustomed to eating it), or tender, high-moisture spring grass. Gas becomes trapped in the rumen and expands until it presses so hard against the animal's diaphragm that the animal will suffocate without immediate treatment. Equines also suffer from a similar, serious form of bloat caused by the same factors and referred to as tympanitic, or wind, colic.

Symptoms. Bulging, taut sides; the animal looks back at or kicks at her abdomen, grunts, cries out in pain, grinds her teeth; labored breathing; horses may want to roll.

Treatment. This is an emergency situation; call your vet.

Prevention. Store grain and legume hay where animals can't overindulge; feed grass hay in the morning before turning livestock onto lush, spring pasture.

*Bovine Viral Diarrhea Virus**

Bovine viral diarrhea virus (also called BVD or BVDV) is a widespread problem for cattle raisers, but it also affects other species such as sheep, goats, and especially llamas. The BVD virus has the ability to replicate into many different variants. If the virus finds itself in a stressful environment or situation, another variant forms. These variants lead to changes in the virus's disease-causing capabilities, making the disease an especially difficult one to diagnose and treat. Anyone involved in raising miniature cattle or llamas should discuss BVD with his or her vet.

*Brucellosis**

Brucellosis is a serious, federally reportable disease. Called Bang's disease in cattle, *B. ovis* in sheep, and undulant fever in humans, brucellosis is caused by bacteria from the genus *Brucella*. Brucellosis can be passed to humans who handle aborted fetuses or consume unpasteurized milk from infected livestock. Brucellosis in animals cannot be treated; any animals that test positive must be destroyed, and all other animals on the same farm must be tested for the disease.

Symptoms. Triggers spontaneous abortions, retained placentas, intermittent fevers, and sometimes manifests in males as orchitis (inflamed testicles) and in females as mastitis.

Prevention. An effective brucellosis vaccine is available for cattle, and it's sometimes used to vaccinate other species, especially equines; however, before using it off-label (meaning in a way other than what's prescribed on the product label), check with your vet. There is a blood test for brucellosis (and it's required for shipping into and out of several states), but it sometimes gives false readings; any animal that tests positive should be retested right away.

*Caseous Lymphadenitis (CL or CLA)**

Caseous lymphadenitis (also called cheesy gland) is caused by the bacterium *Corynebacterium pseudotuberculosis*. Although primarily an infection of sheep and goats, CL occasionally occurs in equines, cattle, camelids (including llamas), and pigs. The same bacterium causes ulcerative lymphangitis (an infection of the lower limbs), chronic abscesses in the chest region, and contagious acne in horses.

CL is contagious and incurable. Transmission is by way of pus from ruptured abscesses.

Symptoms. Thick-walled, cool-to-the-touch external abscesses containing odorless, greenish white, cheesy-textured pus. CL abscesses form on lymph nodes and lymphoid tissue, particularly on the neck, chest, and flanks, but also internally on the spinal cord and in the lungs, liver, abdominal cavity, kidneys, spleen, and brain.

Individuals with internal abscesses may waste away, depending on which organs are involved. However, don't automatically panic when a goat or sheep develops a lump; few of them are caseous lymphadenitis. Sheep and goats are notorious for developing injection site nodules that mimic CL abscesses to a tee. Abscesses can also form when any of hundreds of organisms breach the skin through puncture wounds, splinters, and cuts or abrasions. The only way to be certain an abscess is CL is to have the contents cultured. Isolate animals with ripening (progressively softer-centered) lumps and proceed as indicated below.

Treatment. Any animal with a ripening abscess (whether or not you know for sure it is CL) should be quarantined and the abscess drained and treated according to a veterinarian's instructions. Don't allow pus to contaminate your property. Because CL is transmissible to humans, it's important to wear protective clothing. When the procedure is completed, sterilize or burn the clothing along with anything else contaminated by pus. The animal should remain quarantined until the abscess has healed.

Prevention. Buy only from CL-free flocks and herds. Blood tests are available; have your animals tested, then vaccinate. Colorado Serum's Case-Bac is an effective vaccine for healthy sheep; to date, there are no species-specific vaccines for llamas or goats. To effectively suppress future abscesses, CL-positive sheep and goats can be vaccinated with an autogenous vaccine custom-made from bacteria of one of your infected animals.

Choke

An animal chokes when it has an object (nearly always a bolus of poorly chewed food) stuck in its throat. It's a serious matter but not the extreme medical emergency it is when humans choke, because the blockage is in the animal's esophagus instead of its windpipe and it can still breathe. Choke generally occurs when a famished or greedy animal gobbles its food; animals are particularly likely to choke on pelleted concentrates, unsoaked beet pulp, dry hay cubes, and bulky treats like whole apples or large chunks of carrot.

Symptoms. Most animals panic while experiencing choke. More stoic species like sheep, llamas, and donkeys may stand with their heads down, neither eating nor drinking. Copious amounts of slimy, green matter sometimes streams from the nose.

Treatment. This is an emergency situation; call your vet.

Prevention. Switch to nonpelleted grains or eliminate concentrates altogether. Soak dry feeds such as beet pulp and hay cubes before feeding. Slice or break treats into manageable chunks. Place obstructions such as fist-size rocks in feeders so animals have to shove them out of their way or eat around them and thus stop gobbling their food.

Colic

Colic is defined as severe abdominal pain; it's a clinical sign or a symptom rather than a disease. All livestock species are prone to colic. There are several types of colic and many causes; all are serious and require prompt veterinary intervention. Colic is a major cause of premature death among equines.

Symptoms. Whereas most colicky horses show dramatic signs of abdominal pain (pawing, looking back or nipping at their sides, stomping, and rolling), species more stoic than horses (such as sheep, llamas, and donkeys) are likely to draw back into themselves, refuse food, and suffer in relative silence. Always suspect colic when a generally outgoing animal seems depressed, stops eating, and stops making manure. Other symptoms include increased respiration, excessive sweating, and sometimes (but not always) lack of gut sounds.

Treatment. If colic lasts longer than 30 minutes, call your vet. In the meantime (and until the vet arrives), allow the animal to rest but prevent horses from rolling as this can contribute to twisted gut. If your vet doesn't live nearby, discuss colic treatment strategies before you need him. He may prescribe Banamine (flunixin meglumine), a powerful prescription painkiller in paste or injectable form, to have on hand to administer while he's in transit.

Prevention. It's better to prevent colic than attempt to cure it. Unfortunately, different things cause different types of colic. The most common type, spasmodic colic, is caused by gas buildup in the digestive tract. Things apt to trigger gas-related colic are eating spoiled feed or drinking stagnant water; overeating rich grass, legume hay, or grain; and sudden changes to the diet or feeding routine. Stress can be a contributing factor, too. Animals that ingest sand and dirt while grazing or eating feed directly off the ground may suffer from sand impaction colic. Old animals unable to properly chew their food, as well as animals with heavy intestinal parasite loads, sometimes suffer impactions, too. Other contributing factors include feeding excess concentrates to animals that don't need it and not providing enough drinking water — as is apt to occur when water sources freeze during the winter months.

Cushing's Disease

Cushing's disease or Cushing's syndrome (also called hyperadrenocorticism) is a disease of the endocrine system caused by an abnormality of the pituitary gland, which then causes the body to produce excessive amounts of cortisol — the body's natural steroid hormone. Cushing's is primarily seen in equines and, less commonly, goats. Elderly equines are more prone to Cushing's than are younger ones, but it's not strictly a disease of old age. There is currently no method of prevention.

Symptoms. Increased thirst, appetite, and/or urination; a thick, wavy summer hair coat; a potbellied appearance with loss of topline muscling; chronic laminitis; depression; lethargy; and a compromised immune system leading to increased susceptibility to respiratory disease, abscessed hooves, skin infections, and periodontal disease.
Treatment. There is no cure for Cushing's, but several effective drugs are used in its treatment.

*Encephalitis (Equine)**

There are at least six viruses that cause encephalitis (sleeping sickness) in horses, donkeys, and mules: eastern equine encephalitis (EEE), western equine encephalitis (WEE), Venezuelan equine encephalitis (VEE), West Nile virus (WNV), the neurological form of equine herpes virus (EHV-1), and rabies. EEE, WEE, VEE, and WNV are transmitted by mosquitoes, which pick up the respective viruses from birds and can pass them on to horses, humans, and other vertebrate species. There are effective vaccines available for all of these diseases, so if you live where mosquitoes are endemic, vaccinate.
Symptoms. Depression, lack of muscular coordination, weakness, circling, head tilt, paralysis, muscle tremors, and convulsions.
Treatment. Treatment is generally unsatisfactory; most equines that contract encephalitis die.
Prevention. Vaccinate.

Encephalitis (Ontario)

Ontario encephalitis is caused by a coronavirus called haemagglutinating encephalomyelitis virus (HEV). The virus is widespread in the pig populations of North America but is generally of little concern because clinical disease is rare. This is because most sows have been infected and are immune. They pass their immunity to their piglets in colostrum, and this protects the young animals through the vulnerable period. Although the virus can infect susceptible pigs at any age, it only causes clinical disease in newborn piglets born to sows who are not immune to the virus.
Symptoms. There are two different syndromes. Both manifest around four days of age and affect whole litters. Piglets huddle together, they vomit bright green-yellow vomit, and they are constipated. One syndrome causes piglets to lose their ability to suck or swallow, become very thirsty, and stand with their head over water but be unable to drink. These piglets rapidly waste away and die. The other syndrome causes them to froth and champ at the mouth, tremble, and

develop bluing of their extremities and bloated abdomens. They have a stilted gait, which rapidly progresses to partial paralysis. They lie down, go into convulsions, and die within two to four days.
Treatment. There is no cure.
Prevention. None.

*Enterotoxemia**

Five types of enterotoxemia (A, B, C, D, and E) have been identified in livestock; various types affect cattle, equines, sheep, goats, llamas, and pigs. Also known as entero, overeating disease, and pulpy kidney, the disease is caused by common bacteria found in manure, soil, and the rumens and guts of perfectly healthy animals. When for one reason or another (overeating on grain or milk, abrupt changes in quantity or type of feed, drastic weather changes) these bacteria quickly proliferate, they produce toxins that, in some types, can kill their host in hours.
Symptoms. Bloating, rocking-horse stance, teeth grinding, crying out in pain, seizures, foaming at the mouth, coma, and death.
Treatment. Treatment is generally ineffective.
Prevention. Vaccinate with species-specific vaccine.

*Hoof Rot**

Hoof rot (sometimes called foot rot) is a disease of equines, sheep, goats, cattle, and pigs caused by an interaction of two bacteria, *Bacteroides nodosus* and *Fusobacterium necrophorum*. *Fusobacterium necrophorum* is commonly present in manure and soil wherever livestock is kept; it's when *F. necrophorum* forms a synergistic partnership with *B. nodosus* that hoof rot occurs. *F. necrophorum* can live in soil for only two to three weeks; it can, however, live in an infected hoof for many months.
Symptoms. Animals with hoof rot are excruciatingly lame; they may hold up an infected hoof and hop on three legs; if one or both forefeet are infected, some species kneel to feed. Trimming infected hooves exposes a putrid-smelling pasty substance lodged between the horny outer surface of the hoof and its softer inner tissues. Infected hooves exude a horrible stench.
Treatment. The infection is spread from infected hooves to the soil to the hooves of healthy livestock, so isolate all infected animals and make no exceptions. To expose disease-causing bacteria to oxygen, trim infected hooves back to the affected areas and remove as much rot as possible. Treat according to your veterinarian's directions; treatments include the use of topicals (Coppertox, merthiolate, mercurochrome, bleach solution), foot baths and soaks (zinc sulfate, copper sulfate, formaldehyde), and antibiotics.

Prevention. Don't buy livestock from sale barns or from infected herds. Make certain commercial transporters who carry your animals disinfect their trailers after every trip. Disinfect your shoes after visiting infected facilities. Because hoof rot flourishes in manure and muck, keep barnyards and holding areas as dry as you can. Quarantine infected animals and set up a foot bath (ask your vet which chemicals to use and at what strength to mix them) for noninfected stock where they must walk through it at least twice a day. Vaccinate.

Hypocalcemia (Milk Fever)

Hypocalcemia is caused by a drop in blood calcium a few weeks prior to and immediately after birthing. It is easily confused with pregnancy toxemia. Both are life-threatening situations; call your vet.

Symptoms. Affected animals lose interest in eating, experience muscle tremors, and become progressively weaker until they lie down and won't get up again; mild bloat; subnormal temperatures as the condition progresses.

Treatment. Proceed according to your vet's recommendations. The usual treatments include orally dosing with energy boosters such as glucose or NutriDrench and using oral or injectable calcium substances such as calcium gluconate or CMPK (a fluid calcium, magnesium, phosphorus, and potassium product).

Johne's Disease

Johne's (pronounced YO-nees), also called paratuberculosis, is a contagious, fatal, slow-developing disease of cattle, sheep, goats, and, to a lesser degree, llamas; it's most commonly seen among dairy cattle. Johne's is caused by *Mycobacterium paratuberculosis*, a close relative of the bacterium that causes tuberculosis in humans, cattle, and birds. According to Johne's Information Center (see Resources) statistics, 7.8 percent of America's beef herds and 22 percent of dairy herds are infected with *M. paratuberculosis*. Johne's disease typically enters flocks and herds when an infected but healthy-looking animal is added to the mix. The infection then spreads to other animals in the group. Young animals are also infected by nursing from infected dams.

Symptoms. Progressive weight loss; weakness.

Treatment. None.

Prevention. Adults should be tested and the flock or herd divided into infected and disease-free herds, then maintained separately from one another. The offspring of infected females must be removed from their dams before they nurse and grafted onto disease-free foster mothers or bottle raised using milk replacer or milk known to be free of *M. paratuberculosis*.

Hyperlipemia

Hyperlipemia is a potentially fatal condition triggered by starvation or a stressful event that causes an enormous mobilization of fat from the tissues to the liver. The liver is completely overcome by the overload; it fills with fat and fails, usually resulting in death. Late-gestation female llamas, as well as donkeys, miniature horses, and small-breed ponies, are very prone to developing hyperlipemia if they stop eating for more than a day or two. Jennies and mares (especially if they're pregnant or lactating) are more likely to develop hyperlipemia than geldings, stallions, or jacks.

Symptoms. Initially, anorexia, lethargy, weakness, and depression; followed by jaundice, ventral edema, head pressing and circling, and other indicators of liver and kidney failure.

Treatment. This is an emergency situation; if an animal, especially a late-gestation female, stops eating and you suspect hyperlipemia, call your vet.

Laminitis (Founder)

Laminitis is inflammation of the sensitive laminae (layers) of the hooves; chronic laminitis is often referred to as founder. Equines and cattle are most prone to laminitis but goats, sheep, and pigs are also affected. Predisposing causes include overeating or sudden access to concentrates, high-grain and low-roughage diets, or high-protein diets. Laminitis can also develop as a complication of acute infections such as mastitis, metritis, or pneumonia, especially after giving birth.

Symptoms. Heat in the hoof wall; increased digital pulses in the pastern; lying out flat, reluctant to rise; hesitant gait ("walking on eggshells"); standing in a sawhorse stance with the front feet stretched out in front to alleviate pressure on the toes.

Treatment. Treatment varies by species; call your vet.

Prevention. Feed a proper diet; make certain livestock can't gain access to feed rooms to overeat.

*Leptospirosis**

Leptospirosis is a worldwide fatal disease of humans, wildlife, dogs, and domestic livestock — particularly cattle, horses, sheep, and pigs. It's caused by more than 17 species of bacteria known as *Leptospira*. These bacteria localize in the kidneys or reproductive organs and are shed in the urine, sometimes in large numbers and over a period of months or years. These organisms survive in surface waters such as swamps, streams, and rivers for extended periods, so disease is often waterborne. They also survive in mud and moist soil. Floods often trigger new outbreaks.

Symptoms. Fever, kidney failure, infertility, abortion, anemia, production of abnormal milk (thick, yellow, and blood-tinged, with thick clots and a high somatic cell count), and uveitis (severe eye infection often leading to blindness).

Treatment. Intensive antibiotic therapy depending on species.

Prevention. Vaccination in species for which a vaccine exists; fencing livestock away from potentially contaminated streams and ponds.

Listeriosis

Listeriosis is an uncommon but serious disease caused by a bacterium called *Listeria monocytogenes*. The bacterium is found in soil, plant litter, water, and even in healthy animals' guts. Listeriosis is a type of encephalitis (inflammation of the brain). Problems arise when dramatic changes in feed or weather conditions occur, causing bacteria in the gut to multiply. Parasitism and advanced pregnancy can trigger bacteria proliferation, too.

Symptoms. Disorientation; depression; stargazing, staggering, weaving, circling; one-sided facial paralysis, drooling; rigid neck with head pulled back toward flank. Symptoms resemble polioencephalomalacia, rabies, and tetanus. Cattle, goats, sheep, llamas, pigs, and very, very rarely equines are affected.

Treatment. Treat according to your vet's recommendations.

Prevention. Avoid drastic changes in type and amount of feed, and never feed moldy hay or grain.

Mastitis

Mastitis is inflammation of the udder and can be caused by a number of bacterial and staph agents. It can be triggered by substandard hygiene when milking and by delayed milking in dairy animals, udder injuries, stress, and milk buildup after weaning.

Symptoms. Swollen, hot, hard udder; extreme pain; lameness; loss of appetite; fever; decreased milk production; clumps, strings, or blood in the milk; watery-looking milk. Gangrene mastitis: bruised-looking, extremely swollen and painful udder turning blue as infection takes hold.

Treatment. Both intramammary infusions of antibiotics and systemic antibiotics are generally used. It's best to have milk samples cultured to determine which bacteria are involved in order to know which medications to use.

Prevention. Practice good milking sanitation. Reduce lactating females' grain rations for several successive days before weaning and switch them from legume to grass hay, then eliminate all grain postweaning until their udders have dried up.

Metritis

Metritis is a fatal inflammation of the uterus caused by bacterial infection. It's fairly common in cattle, equines, sheep, goats, llamas, and pigs during the immediate post-birthing period, especially after difficult or assisted deliveries. After giving birth, it is normal for a female's uterus to contract and squeeze mucus, fluid, and afterbirth out through the vagina. Discharge can continue for several days up to a week or more, depending on the species. If the female is eating well and has a normal temperature with no mastitis, ignore the discharge. However, abnormal discharges at any other time indicate the presence of infection and require treatment. Simple metritis is not to be confused with contagious equine metritis (CEM), an acute, highly contagious venereal disease of horses.

Symptoms. Post-birthing: fever, lethargy. Any other time: white or brown discharge not associated with recently giving birth; inability to conceive.

Treatment. Varies by species; consult your vet.

Prevention. Assisted deliveries should be followed with antibiotic treatment; consult your vet for specific treatments.

Pinkeye

Pinkeye, also known as infectious keratoconjunctivitis, is a bacterial eye infection usually spread by flies and dust. Because it is contagious and can cause ulceration of the cornea and permanent blindness, it should be aggressively treated.

Symptoms. Watery eye, cloudy cornea, light sensitivity.

Treatment. Isolate infected animals in shade or a darkened building. Numerous over-the-counter products are effective against pinkeye; ask your vet for recommendations. Systemic and topical antibiotics are also used.

*Pneumonia**

Pneumonia, a serious and potentially fatal inflammation or infection of the lungs, is caused by any of a host of bacteria, fungi, and viruses that often gain a toehold due to environmental factors such as stress; aspiration of milk, vomit, or drench material; damage caused by lungworm infestation; drafts or being hauled in goat totes or stock-type trailers in freezing weather; and dusty feed, bedding, or surroundings. Some forms are contagious.

Symptoms. Loss of appetite; depression; rapid respiration and labored breathing; standing with forelegs braced wide and neck extended; thick, yellow nasal discharge; congestion, coughing, an audible "rattle" in the chest.

Treatment. Call your vet.

Prevention. Remove environmental factors. Effective cattle vaccines are readily available and sometimes used off-label for other species; discuss their use with your vet.

Polioencephalomalacia

Polioencephalomalacia (also called PEM, cerebrocortical necrosis, or goat polio) isn't related to the viral disease called polio (poliomyelitis) in humans. Polioencephalomalacia is a neurological disease caused by a thiamine (B_1) deficiency that culminates in brain swelling and the death of brain tissue. Goats, cattle, sheep, and llamas can be affected.

Symptoms. Disorientation; depression; stargazing, staggering, weaving, circling, tremors; diarrhea; apparent blindness; convulsions; death.

Treatment. If treatment of thiamine injections begins early enough, affected animals begin improving in as little as a few hours.

Prevention. Thiamine deficiencies can be triggered by eating moldy hay or grain; overdosing with amprollium (Corid) when treating for coccidiosis (and never use it to treat goats); ingesting certain toxic plants; reacting to dewormers; and sudden changes in diet, including weaning. Overuse of antibiotics contributes to thiamin deficiencies, too.

Pregnancy Toxemia

Pregnancy toxemia is a fairly common, serious metabolic condition that afflicts cows (particularly beef cattle), does, ewes, and, more rarely, llamas during their final few weeks of pregnancy and the first week or two after giving birth. It's mainly a condition of obese or thin females and those carrying multiple offspring.

Symptoms. Affected animals become depressed, stop eating, and often separate themselves from the main flock or herd; they are reluctant to walk and spend considerable time lying down. Some will have the odor of acetone (comparable to aromatic solvents such as pain thinner) on their breath. As the condition worsens, the animal develops weakness in the hindquarters and muscle tremors and may collapse.

Treatment. This is an emergency; call your vet. Affected animals that refuse food must be treated aggressively, as a decrease in energy intake causes the disease to progress rapidly. Tempt the animal with favorite foods; dose with glucose given orally or IV; or administer propylene glycol as an oral drench. Cows should be lifted by a hip hoist two or three times a day for 15 to 20 minutes; smaller animals should be encouraged to rise and, if unable to, turned several times per day.

Prevention. Prevention involves maintaining pregnant females in moderate flesh, neither fat nor thin; identifying females with twins and triplets and feeding accordingly; and according to the *Merck Veterinary Manual* (see Resources), including niacin in the diet of goats and sheep at the rate of 1 gram per day during late gestation.

Rain Rot

Rain rot, also referred to as rain scald or streptothricosis, is caused by a fungus-like actinomycetes (an organism that behaves like both bacteria and fungus) called *Dermatophilus congolensis*. Some authorities claim it's present in soil; others say it isn't. Whichever is true, several conditions must be present for it to take hold and proliferate: the organism has to be present on the animal's skin; the skin must be kept moist; and the skin must be damaged in some way (a cut, scrape, or fly bite) in order for the organism to invade the epidermis layer of the skin. Cattle, sheep, equines, and goats are commonly infected; pigs and llamas less frequently.

Symptoms. In most cases rain rot manifests as small, crusty scabs or slightly raised, matted tufts of hair. The crusts and tufts are easily (though painfully) lifted to reveal pus and raw, pink skin. Lesions generally appear in clumps across the infected animal's neck, back, rump, and legs; the underbellies and udders of dairy goats and cattle are often affected, too.

Treatment. Bathe the animal using antifungal or antibacterial shampoo. Soak crusty areas and remove crusts and tufts of hair. Allow the areas to dry, then apply Betadine or Nolvasan and continue applying it once a day.

Prevention. Provide adequate shelter for animals to take cover when raining or snowing. Using rain-resistant blankets or sheeting will also help protect against rain rot.

Ringworm

Contrary to popular belief, ringworm (also known as dermatophytosis) is not a worm but a fungal infection. Called "club fungus" by those who show livestock, it is commonly picked up at public venues such as shows, fairs, and sale barns.

Symptoms. Ringworm lesions form round, dime- to quarter-size crusty patches that, when removed, leave scaly red skin and hair that lifts out in clumps. An infected animal may have two or three small patches or be covered with lesions from nose to tail. Ringworm is highly contagious, and it easily spreads from one species to another, humans included. It's transmitted through both direct (animal to animal) and indirect (for example, wall surface to animal) contact.

Treatment. Although self-limiting (ringworm usually resolves itself in six weeks to three months), it must be treated quickly and aggressively lest it spread. The best way is by bathing the infected animal with an iodine-based shampoo formulated specifically for fungus problems. Wear disposable gloves. Also, disinfect anything an infected animal touches.
Prevention. Don't take your animals to, or buy animals from, any place where ringworm is endemic.

Stomach Ulcers

Stomach ulcers are surprisingly common in livestock, especially horses and llamas. And they can be fatal, so it's important to recognize and prevent this disease. Most ulcers are triggered by stress. Newly weaned foals, crias, females whose young have recently been weaned, and animals transported into new surroundings away from familiar faces are particularly prone to stress-induced ulcers. Diets high in concentrates can contribute to ulcer formation, too.
Symptoms. Teeth-grinding (an indication of pain), pained facial expression, depression, lying down in abnormal positions, horses "cat stretching" with legs camped out in front and behind them, kicking at belly, excessive rolling, not eating, black stools indicating internal bleeding.
Treatment. Remove the offending stress and consult your vet for treatment options.

*Strangles**

Strangles is one of the oldest known equine diseases. It's a highly contagious and serious infection of equines caused by the bacterium *Streptococcus equi*. Strangles is most common in animals less than five years of age and is especially common in groups of weanling foals or yearlings. Foals under five months of age are usually protected by colostrum-derived passive immunity from their mothers. Recovered animals shed *S. equi* from their noses and in their saliva for up to six weeks following infection.
Symptoms. High fever, poor appetite, a soft cough, and depression. Thin, watery nasal discharge becomes thick and yellow and the upper respiratory lymph nodes, particularly the ones between the jawbones, become enlarged. Abscesses form on the lymph nodes; these usually rupture and drain copious amounts of nasty, yellow pus. About 15 percent of sufferers develop complications, often of a serious nature; the rest recover with only supportive care.
Treatment. Isolate the animal in comfortable quarters and provide good-quality feed. Your vet may prescribe phenylbutazone (bute) to reduce fever, pain, and

swelling. Encourage swollen lymph nodes to rupture and drain externally by applying hot packs to the swollen areas.
Prevention. Quarantine all incoming equines. A vaccination against strangles is readily available. It doesn't fully protect against the disease; however, it lessens its incidence and severity; ask your vet for details.

*Tetanus**

Tetanus occurs when wounds become infected by the bacterium *Clostridium tetani*. These bacteria thrive in anaerobic (airless) conditions, such as those found in deep puncture wounds, fresh umbilical cords, and wounds caused by recent castration. Unless treated very early and aggressively, tetanus is nearly always fatal.
Symptoms. Early on: stiff gaits, mild bloat, anxiety. Later: a rigid rocking-horse stance, drooling, inability to open the mouth (hence tetanus's common name: "lockjaw"), head drawn hard to one side, tail and ear rigidity, and seizures.
Treatment. If you suspect tetanus, call your vet.
Prevention. Vaccinate.

Tuberculosis

Bovine tuberculosis (TB) is a serious, reportable disease that can be transmitted from livestock to humans and other animals. It's caused by the bacterium *Mycobacterium bovis*. Host animals typically show no symptoms of infection until slaughter, at which time lesions may be found on any of their organs, intestines, and lymph nodes. No other TB organism has as great a host range as bovine TB: it can infect all warm-blooded vertebrates. Currently 49 states and territories are tuberculosis free (meaning there have been no confirmed cases for five years). Nevertheless, testing of cattle, sheep, and goats is required for shipping between some states. Consult your vet for further information.

Urinary Calculi

Also called urolithiasis or UC, urinary calculi are mineral salt crystals ("stones") that form in the urinary tract and block the urethras of male animals (both sexes can develop urinary calculi, but stones pass easily through relatively larger, straighter, and shorter female urethra), particularly cattle, goats, and sheep. The condition is an emergency that requires immediate medical attention; it won't correct itself, and if left untreated the afflicted animal's bladder will burst and the animal will die.

Symptoms. Anxiety, restlessness, pawing the ground, teeth-grinding, crying out in pain; straining to urinate; rocking-horse or hunched-over stance; impaired flow of urine (dribbling).
Treatment. Call your vet.
Prevention. Feed male animals, intact or castrated, a balanced 2:1 calcium: phosphorus ration. To accomplish this, feed high-quality grass hay instead of alfalfa and offer very little (if any) grain. Adding minute quantities of ammonium chloride to the diet may prevent some types of calculi from occurring, and always provide a copious supply of clean, palatable drinking water.

White Muscle Disease

White muscle disease, also called nutritional muscular dystrophy, is caused by a serious deficiency of the trace mineral selenium. Most of the land east of the Mississippi and much of the Pacific Northwest is selenium deficient; these are the areas where white muscle disease is most likely to occur. It affects all livestock species.
Symptoms. Neonates and young stock: weakness, inability to stand or suckle, tremors, stiff joints, neurological problems. Adults: infertility, abortion, difficult births, retained placentas; stiffness, weakness, lethargy.
Treatment. Injections of Bo-Se or Mu-Se (prescription selenium and vitamin E supplements) often dramatically reverse symptoms, especially in neonates.
Prevention. All livestock raised in selenium-deficient areas should be fed selenium-fortified feeds, have free access to selenium-added minerals, or be given selenium–vitamin E shots under a vet's direction. To prevent birthing problems and protect unborn young, females should be injected with selenium–vitamin E three to six weeks prior to their expected birthing date.

Dealing with Internal Parasites

No chapter about livestock health care would be complete without a discussion of internal parasites, also known as nematodes and better known as "worms." Because of space constraints, we'll cover deworming in general instead of discussing each of the many species and varieties that might afflict your flock or herd. For up-to-date information about deworming issues in your species and locale, always consult your veterinarian or county extension agent.

Parasitism causes a host of problems ranging from weight loss, dull coat, poor appetite, mild colic, and an itchy backside to more serious ills such as persistent diarrhea, anemia, susceptibility to infections, nonhealing sores, coughing, and significant or recurrent colic. Internal parasites can ultimately cause pneumonia, emaciation, debilitating diarrhea and colic, and gut emergencies such as torsion, telescoping bowel, and gut perforations. Some of these emergencies result in death. Internal parasites are not to be ignored.

Test Your Herd for Parasites

Your vet is your number one source of information about internal parasites. She can work with you to create a deworming regimen ideal for your farm. To discover how many parasites your animals are carrying and which are involved, she'll run EPG (eggs per gram) fecal tests on some or all of your animals. To do this, she'll need fresh fecal material from each animal you want tested. The key

DEWORMERS

- **Ivermectin** is effective against large strongyles, small strongyles, pinworms, ascarids, hairworms, large-mouth stomach worms, bots, lungworms, and intestinal threadworms, as well as external parasites such as mange mites, biting and sucking lice, nose bots, grubs, and horn flies. It's a very safe product. In some regions parasites are resistant to ivermectin.
- **Moxidectin,** marketed as Cydectin, controls large strongyles, small strongyles, pinworms, ascarids, hairworms, large-mouth stomach worms, bots, and kidney worms. It's effective against external parasites such as grubs, mites, lice, and horn flies. It shouldn't be used for deworming young stock, debilitated animals, or animals suspected of having large parasite burdens.
- **Febendazole** works against large strongyles, small strongyles, pinworms, and ascarids. It's an extremely safe product; however, worm resistance to this deworming agent is a problem in many locales.
- **Piperazine** is a very mild agent somewhat effective against large strongyles, small strongyles, large-mouth stomach worms, and pinworms.
- **Oxibendazole** is very safe. It's effective against large strongyles, small strongyles, pinworms, and threadworms.
- **Praziquantel** controls tapeworms and roundworms.
- **Pyrantel Pamoate and Pyrantel Tartrate** control large strongyles, small strongyles, pinworms, and ascarids; they are also effective against tapeworms in equines when double-dosed.

word is *fresh*. Stand by with a labeled plastic sandwich bag turned inside out on your hand and when the animal delivers, pick up a sample, preferably from the top of the heap so it didn't touch the ground, turn the bag right side out around the material, and seal it up.

Parasites are rarely visible in droppings (most are too small to see), so after processing the manure, your vet will view it under a microscope to identify what types of parasite eggs are present. Then she'll count the number of eggs of each parasite species found in one gram of prepared sample. By identifying species and counting the numbers of eggs, she can recommend the perfect deworming agents for your needs.

Fecal exams are also a cost-effective follow-up to deworming to determine whether the dewormer worked, so have your vet run another fecal egg count about two weeks after deworming.

Develop the Best Deworming Program

In the past, experts told livestock producers to deworm all of their stock at the same time and to rotate dewormers to reduce drug resistance. Since few weighed their stock before deworming, many under-dosed their animals. In this manner, parasites were exposed to all of the available anthelmintics (chemical dewormers), often in doses too light to be fully effective. Weak parasites died but the strong survived. Now dewormer-resistant "super worms" have evolved to the point that most anthelmintics aren't effective, and the drug companies aren't developing new products. Therefore it's important to deworm correctly. To get the most from your deworming program, heed these tips:

- Read dewormer packaging and follow the instructions.
- Don't rotate dewormers every time you deworm your stock. If an anthelmintic is working, use it for at least a year or until it loses its effectiveness.
- Use fecal testing to find out which deworming chemicals the internal parasites on your farm are resistant to, and choose products from another class.
- Weigh each animal and dose accordingly. Never under-dose. If you don't have a scale, use a weight tape appropriate to your species to calculate your animals' ballpark weights.
- Deworm all females immediately after they give birth, when changing estrogen levels cause arrested larvae to molt and proliferate.
- Run fecal counts on all new animals and deworm them while they're in quarantine; have fecals run again two weeks later and deworm again based on those results.

- Deworm livestock 48 hours before moving them to clean pasture.
- Keep accurate records.

Don't deworm unless your animals need it. Rely on fecal testing to identify individuals with heavy parasite loads or learn the FAMACHA system for evaluating parasitism in small ruminants (see Southern Consortium for Small Ruminant Parasite Control in Resources).

Prevent Parasitism as Best You Can

These practices can help cut back on the numbers of internal parasites on your farm:

- Feed hay and grain from feeders instead of directly off the ground.
- Provide a clean water supply free of manure contamination.
- Remove manure from stalls, field shelters, pens, paddocks, and pastures on a weekly basis; don't let it pile up. Spread it on cropland or other ungrazed areas, compost it, or sell it as fresh garden enhancer, but don't stack it where your animals can reach it.
- Mow and chain-harrow pastures to break up manure deposits and expose parasite eggs and larvae to the elements.
- Rotate pastures with one species following another (for example, sheep and goats following cattle following equines) to interrupt the life cycles of species-specific parasites.
- Avoid overstocking to prevent overgrazing and reduce fecal contamination.
- Set up a deworming schedule and stick to it.

Factors such as local climate, the season, soil conditions, and the number, age, and type of animals using a facility all need to be considered when formulating a parasite control program. Consult your vet or county extension agent for management tips particular to your region.

Dealing with External Parasites

External parasites such as flies, ticks, lice, and mites feed on body tissue such as blood, skin, and hair, and they sometimes transmit diseases from sick to healthy animals. External parasites spoil hair coats and reduce weight gains and milk production while making life unpleasant for their hosts.

Flies

Several classes of bothersome flies can make summertime on the farm a trying time for you and your livestock alike. Among the worst offenders are blackflies, heel flies and cattle grubs, horn flies, horse- and deerflies, midges, and stable flies.

Blackflies

Blackflies, also called buffalo gnats, are smaller cousins of the horsefly. More than 1,000 species are found worldwide, in boreal to tropical climates.

Both sexes feed on nectar, but the female also drinks blood. Her bite is painfully out of proportion to her size. Like horseflies, she slashes and sucks pooled blood, injecting an anticoagulant that triggers mild to severe allergic reactions. The swelling and itch that follow can last two weeks or more. Hordes of blackflies pose a serious threat. Mega-bitten hosts sometimes die from acute toxemia or anaphylactic shock.

Most blackflies are daytime feeders, and they rarely venture indoors. Biting varieties target animals' ears. Not much repels them; even DEET-based repellents are only minimally effective.

Heel Flies and Cattle Grubs

The heel fly (*Hypoderma* sp.) is a large fly that resembles a bee in size and coloration. Heel flies deposit eggs on the hairs of cattle, usually on their hind legs or bellies. The larvae hatch and bore through their host's skin, usually at a hair follicle. These grubs spend about eight months migrating through the tissues of the animal and end up in the loin area of the back where they form warbles (cysts with breathing holes). Warbles were once common on American bison and have been found on sheep, goats, and horses. Some cattle grubs have even been removed from humans.

The carcasses of warble-bearing cattle are worth less money because the flesh where the grubs were located is greenish yellow, jellylike, and unfit for consumption. Warble holes reduce hide values, too.

Sprays, dips, feed additives, and pour-ons provide adequate cattle grub control. Toxic reactions to deworming agents are common.

Horn Flies

Horn flies are roughly one-half to three-quarters the size of the common housefly. They are more slender, with a brownish gray to black body with a slight yellowish cast, a set of parallel stripes just behind the head, brownish red antennae, and two wings with a smoky tinge. They also have painfully effective piercing-sucking mouthparts.

Although primarily attracted to cattle, horn flies also harass sheep, goats, and equines, but they don't bite human beings. Adults spend most of their life on a specific host, congregating on the back and shoulders or on the underside on hot days. Persistent feeding causes irritation and bleeding sores.

Both sexes bite. They feed up to 40 times per day, and when not feeding they tend to rest around the horn region of their host. Females require blood meals for egg production and can lay several hundred eggs during their life span. The entire life cycle from egg to adult is completed in two to four weeks, and three to four generations may hatch in a single summer.

Pesticide dips, sprays, dusts, pour-ons, and ear tags are all effective against horn flies.

Horse- and Deerflies

Horseflies (*Tabanus* spp.) are stout-bodied flies up to 1½ inches in length. Deerflies (*Chrysops* spp.) are smaller; most are the size of common houseflies. Only female horse- and deerflies feed on blood. Females harvest their meals the way most biting flies do, by slashing a host's skin with their sharp mouthpieces, then lapping pooled blood. Most varieties prefer to feed on equines, cattle, and deer. Of the two, deerflies are more likely to bite humans.

Deer- and horseflies are day feeders. They're bothersome on warm, sunny days and are attracted to moving objects, warmth, and carbon dioxide emissions. They are voracious but flighty feeders, often skipping from host to host to complete a meal; thus they're often implicated in the spread of disease. Under sustained attack from these flies, animals cease grazing and huddle together for protection, sometimes resulting in considerable weight loss.

Permethrin-based insecticides offer short-term relief for livestock, but other chemical repellents seldom work.

Midges

Biting midges, also called sand flies, sand gnats, punkies, and no-see-ums, are among the world's tiniest biting flies. Most are dark gray or black with spotted wings. Only females suck blood. Bites are largely painless, but tissues swell and itch intensely within 8 to 12 hours.

Most midges feed at dawn and twilight, from early spring through midsummer. A few species are daytime biters, especially on damp, cloudy days. Biters frequent salt marshes, sandy barrens, riverbanks, and lakes. Chemicals won't repel them.

Midges are highly attracted to dogs and livestock, particularly to their ears and lower legs. Keep animals indoors during prime feeding time.

Stable Flies

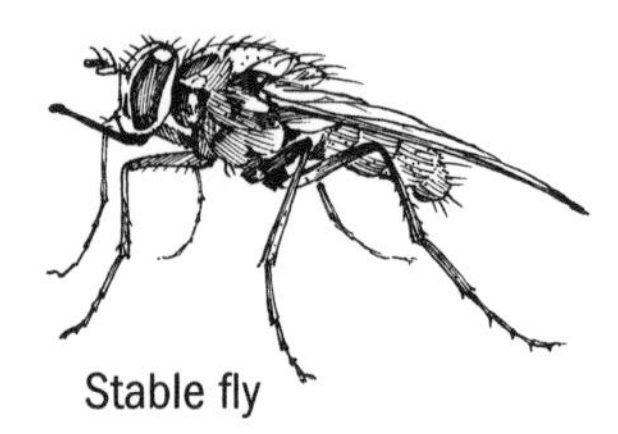
Stable fly

At first glance stable flies look like houseflies, but a distinguishing feature, visible to the naked eye, is the stiletto-like proboscis of the stable fly extending beyond its head, which is used to pierce the skin and draw blood. The look-alike housefly cannot bite because it has sponging mouthparts. Both male and female stable flies feed on blood and are persistent feeders that cause significant irritation to their host.

Stable flies feed on the blood of practically any warm-blooded animal, including humans. Peaks of feeding activity occur during the early morning and again in the late afternoon. Stable flies prefer to feed outdoors and rarely come indoors. They also prefer to feed on the lower parts of their hosts, such as their legs. Females deposit their eggs in a variety of decaying animal and plant wastes, but they are rarely found in fresh manure. Their entire life cycle from egg to adult is generally completed in three to six weeks.

Keep livestock indoors during the day and on pasture at night when stable flies are bothersome. Many types of residual and knock-down (spray and immediate kill) insecticides are effective against stable flies.

Ticks

Ticks transmit more disease worldwide than any other insect pest. In the United States alone, ticks are responsible for the spread of nine serious human diseases. Livestock diseases spread by ticks include Lyme disease, East Coast fever, ehrlichiosis, and tick paralysis.

There are two kinds of ticks: hard ticks (Ixodidae family) and soft ticks (Argasidae family). Ticks neither jump nor fly, and their bites cause little, if any, initial discomfort, although many become itchy a day or so later. Both sexes attach and suck blood, but only the female engorges to many times her unfed size.

Hard Ticks

Hard ticks have hard plates on their backs, and their mouthpieces are visible from above. When they bite, they secrete cementlike saliva that glues the feeding tick in place. They can go several months without feeding.

Hard ticks are attracted to heat, vibration, shadow, and carbon dioxide emissions. They locate a host through a process called "questing" whereby they perch on vegetation, waiting until something happens by. When the tick senses any of the things it is attracted to, it extends its front legs and snags the passing host as it brushes past.

In its six- to eight-day life as an adult, engorged female hard ticks are capable of expanding from 2 to 600 times their unfed weight.

Adult Rocky Mountain wood ticks (*Dermacentor andersoni*) feed on large mammals, especially deer, humans, dogs, and livestock. This tick is well known as a vector of the Rocky Mountain spotted fever, the Colorado tick fever virus, and the bacteria that causes tularemia. It is also responsible for tick paralysis in humans, livestock, and wild mammals.

The blacklegged tick (*Ixodes scapularis*), also known as the deer tick or bear tick, is a major vector for Lyme disease—a serious disease affecting humans, dogs, and horses, and it's suspected that other livestock species can get it as well.

Soft Ticks

Soft ticks have soft, leathery bodies. Their mouthpieces aren't visible from above. A few species seek hosts by questing, but most are nest dwellers preferring established burrows and nests (including dog beds). Females engorge 5 to 10 times in a few hours, and they look like inflated balloons.

The spinous ear tick (*Otobius megnini*) is a common pest of livestock and horses throughout the western United States. Heavy infestations result in intense irritation, rubbing, and hair loss in livestock.

Tick Removal

If you find a tick attached to one of your animals, take it off. The longer a tick is attached, the more likely it will transmit any disease it's carrying.

To remove a tick, use your gloved fingers, needle-nose tweezers, a hemostat, or a commercial tick remover to grasp the tick's mouthpiece as close to the host as possible, and then pull slowly and steadily straight back. Be careful; squeezing the tick will cause it to inject additional toxins. If all or part of its mouthpiece is left imbedded in the animal's skin, it will eventually wither and fall out on its own, but watch closely over the next few days to make certain the bite doesn't become infected.

Dispatch the tick by dropping it into a container of alcohol (soapy water will do in a pinch). If the tick is engorged, place it on a hard surface and step on it.

Many tick species mate while the female is feeding. After removing an engorged female, look closely for tinier males attached in the same location.

Controlling Ticks

The best way to control ticks is by altering their habitat, so mow your pastures and keep barn and outbuilding areas weed and litter free. Also consider adopting a flock of free-roaming guinea fowl. Folks who keep guineas swear by their tick-gobbling efficiency. Free-range chickens are tick-eaters, too.

Lice

There are more than 500 species of lice worldwide, including two that infest humans. Other species infest cattle, swine, horses, sheep, goats, dogs, and rabbits, but not cats or fowl. They are generally species-specific, but the same species infests sheep and goats.

Lice spend their entire lives on their host. Both immature and adult stages suck blood and feed on skin. Louse-infested animals have dull, matted coats and display excessive scratching and grooming behavior resulting in raw patches on the skin or loss of hair. Weight loss may occur when infested animals don't eat. Milk production is reduced by about 25 percent. A louse-infested animal is generally listless, and in severe cases the loss of blood to sucking lice can lead to serious anemia.

There are two types of lice: sucking lice that pierce their host's skin and suck blood and biting lice that have chewing mouthparts and feed on particles of hair and scabs. Lice are wingless; adults are usually 1/16 to 1/8 inch long (a few, for example the hog louse, are nearly 1/4 inch long) and range from pale yellowish to blue-black or brown. Lice are generally spread through direct contact, often when infested animals join an existing herd.

Populations vary seasonally. Lice proliferate during autumn and reach peak numbers in late winter or early spring. Summer infestations are rare. Wintertime infestations are usually the most severe. Treatment, generally with an over-the-counter residual pesticide designed for livestock, is needed whenever an animal scratches and rubs to excess. It is difficult to control lice because pesticides kill lice but not their eggs. Since eggs of most species hatch 8 to 12 days after pesticide application, you must treat again 2 or 3 weeks following the first application.

Mites

Mange mites can be a problem in cattle, sheep, goats, equines, and pigs; five types afflict cattle alone, three of which are serious enough to be reportable diseases.

Mange mites feed on the skin surface or burrow into it, making minute, winding tunnels from 1/10 to 1 inch long. Fluid discharged at the mouth of each tunnel dries and forms scabs. Mites also secrete a toxin that causes intense itching. Infested animals rub and scratch themselves raw. Infestations are highly contagious; if one individual has mites, contact your vet for advice and treat the entire flock or herd.

Miniature donkeys are placed in one of two classes. Class A donkeys are 36 inches and under; class B donkeys are 36.1–38 inches tall.

MINIATURE DONKEYS

MINIATURE HORSES

left page, clockwise from top left: American Miniature Horse (max. height 34 inches for AMHA; 38 inches for AMHR), Falabella (average height 28–34 inches), British Shetland Pony (max. height 46 inches), American Shetland Pony (average height 42 inches)

this page, clockwise from top: American Shetland Pony, British Shetland Pony, American Miniature Horse, British Shetland Pony

CONTINUED

MORE MINIATURE HORSES

top: American Miniature Horse (max. height 34 inches for AMHA, 38 inches for AMHR)
middle: Falabella (average height 28–34 inches)
bottom, left and right: American Miniature Horse

Miniature mules are placed in one of two classes. Class A mules (shown here) are less than 38 inches tall. Class B mules are 38–48 inches tall.

MINIATURE MULES

MINIATURE GOATS

left and right pages: All are crossbreeds with Nigerian Dwarf heritage

CONTINUED

left page, top: Crossbreed with Nigerian Dwarf heritage
left page, bottom: Crossbreed with Pygmy heritage
this page, top: Both are crossbreds with Nigerian Dwarf heritage
this page, middle: Crossbreed with Pygmy heritage
this page, bottom: Both are crossbreeds with Nigerian Dwarf heritage

MINIATURE PIGS

left page, top: Vietnamese Potbelly pig (at 3 years, average height is 16 inches; average weight is 100–125 pounds)

left page, bottom: Kunekune (average height 24–30 inches, weight 120–240 pounds)

this page, clockwise from top: Kunekune, Ossabaw Island Hog (average weight 100–250 pounds), Vietnamese Potbelly pig.

MINIATURE COWS

left page, top: Dexter cow (average height of cows is 38–42 inches; bulls are proportionally larger)

left page, middle, left to right: Dexter bull, Miniature Zebu (max. height 42 inches; cows average 300–500 pounds, bulls average 400–600 pounds)

left page, bottom, left to right: Lowline cow (cows average 38–46 inches, bulls 40–48 inches; cows average 700–1,100 pounds, bulls 900–1,500 pounds), Dexter cow

this page, clockwise from top: Pineywoods guinea cow (average weight 300–500 pounds) Dexter cow, Florida Cracker full-size cow (cows average 600–800 pounds, bulls average 800–1,200 pounds)

MINIATURE SHEEP

this page, clockwise from top: Classic Cheviot (average height 18–24 inches), Classic Cheviot, Classic Cheviot, Shetland (ewes average 75–100 pounds; rams 90–125 pounds)

right page, clockwise from top: Soay (average height 22 inches, weight 45–90 pounds), Classic Cheviot (average height 18–24 inches), Shetland, Babydoll Southdowns (max. height 24 inches)

Miniature llamas are 3 feet tall at the withers, on average, and 120–200 pounds.

MINIATURE LLAMAS

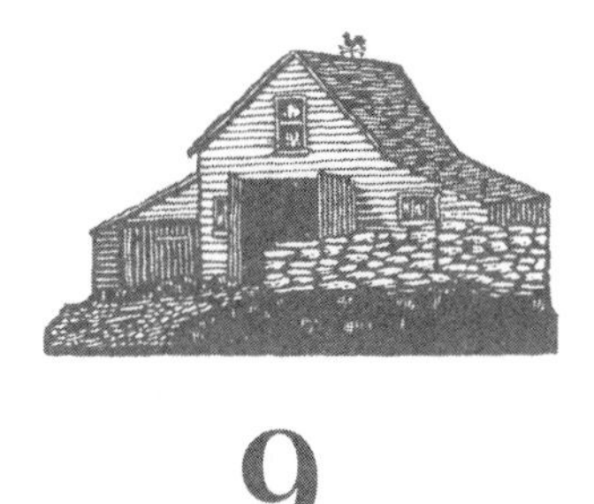

9

Identification

IT'S IMPORTANT TO PROPERLY IDENTIFY your animals and keep accurate records, especially if you raise miniature livestock as a business. In this chapter we'll look at some common means of identification.

Permanent Identification

Purebred livestock must be permanently identified before most registries will admit them to their stud, herd, or flock books. This is especially true of breeds in which individuals tend to look alike. Tattooing and ear tags are the norm in smaller stock, freeze branding is a logical option for larger species, and all species can be microchipped.

Tattoos

Tattoos can be used to identify all types of livestock. They are sometimes put in out-of-the-way places like the inside of the upper lips of equines or the tail webs of Mini-LaMancha goats, but they generally go inside an animal's ear.

Tattoo pliers come in 3⁄10 inch, 5⁄16 inch, and 3⁄8 inch digit sizes; if you have a five-digit herd prefix (get this from your registry if you're tattooing to their standards), make sure you choose a set that accepts five digits (some don't). Better pliers automatically release. Most tattoo kits come with a roll-on applicator of black ink. Throw it away and buy green paste ink. Green is better than black because it shows up well in dark ears and paste won't drip like roll-on tends to do. You'll also need disposable gloves, a soft toothbrush to scrub ink into newly applied tattoo piercings, and alcohol and cotton balls or paper towels to clean the site before tattooing.

Step-by-Step Guide to Tatooing an Animal

1. Assemble your tools before you begin. Sterilize the tattoo digits by immersing them in alcohol before placing them in the pliers.
2. Test the tattoo on paper (index cards work well) to make certain you've inserted the digits right side up and facing the right direction.
3. Place the animal in a handling chute or a fitting or milking stand, or halter and tie him where you can easily access his ears. A little feed will help distract most animals, at least until you clamp the pliers on his ear.
4. Have a helper steady the animal's head. If you aren't using a chute or stand, ask a second helper to crowd the animal against a wall to help secure it in place.
5. Swab the inside of the animal's ear with alcohol and pat it dry.
6. Flatten the ear as best you can. Position the pliers in the ear, making certain the digits are inside the ear and you aren't holding the pliers upside down. Avoid tattooing into blood vessels, ridges of cartilage, or scar tissue. Press quickly but firmly and then release. If the needles stick to the ear or pierce through it, peel the ear away and try again, applying less pressure this time.

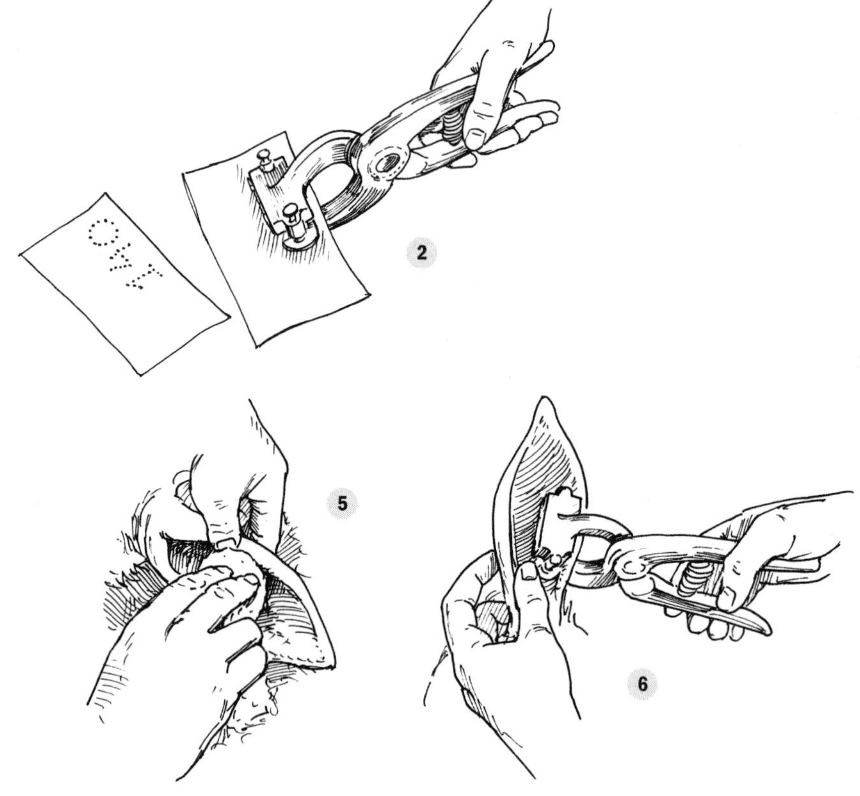

7. If some of the punctures bleed, pinch a paper towel or cotton ball over the bleeder until it stops. Don't apply ink until the bleeding stops because blood flow washes ink out of the holes.
8. Squirt a blob of ink onto a soft toothbrush or your gloved finger (tattoo ink is hard to get off), making certain you get pigment, not just oil, out of the tube. Rub ink into the piercings with your finger or a soft-bristled toothbrush, taking care to get ink into every hole. When the holes have been filled, leave the excess ink on the ear (scrubbing it off may dilute ink in the piercings as well).
9. Give the animal a shot of tetanus antitoxin as a precaution.
10. Remove and disinfect the digits before moving on to another animal. When you've finished all of the tattooing, disinfect the equipment before storing it.

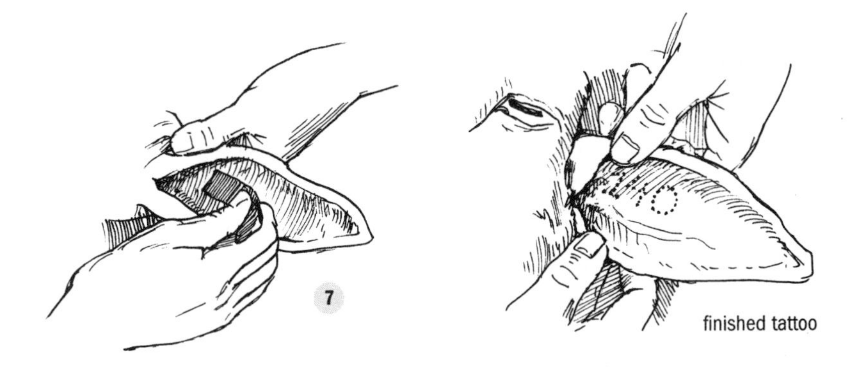

TATTOOING TIPS

- Tattoo at least a month before a show or production sale to allow tattoos time to heal and be fully legible.
- Tattoo needles must make holes that are large enough to accept an adequate amount of ink or the tattoo will be difficult to read. Brand-new, ultra-sharp digits don't do that; lightly filing the tips of new digits often helps.
- Use small-digit pliers when tattooing young animals, as the tattoo will get bigger as the ear grows.
- If a tattoo fades, you'll have to retattoo that ear. Check with your registry regarding their policies on reapplication of tattoos.
- When it's difficult to read the tattoo in a dark-pigmented ear, scrub the ear with alcohol to remove grease and dirt, then shine a flashlight behind the outside of the ear.

Ear Tags

Some producers like ear tags; others don't. Their main failing is that ear tag numbers fade with age and the tags themselves are easily lost (it's for this reason many producers tattoo *and* tag).

The beauty of ear tags, however, is that they're easy to read from a distance and inexpensive to buy and apply. They come in metal and plastic in a wide range of types and sizes.

The United States Department of Agriculture's mandatory scrapie eradication program for sheep and goats is overseen on the state level, so if you keep small ruminants, you may not have much choice in the type of ear tags you use. However, when you can, choose your ear tags wisely. Double-sided, button-type plastic tags and metal tags tend to cause infections, so avoid using them. Choose small, two-piece plastic models that swivel — they're far less likely to snag on fencing or brush and tear the ear.

Plastic tags may be purchased preprinted by companies like Premier1 (see Resources) or as blanks. When writing on blank tags, use a marking pen designed specifically for writing on ear tags. Mark the tags the night before tagging to allow them to dry overnight.

Some colors are easier to read than others. When visibility is an issue, in descending order choose yellow, white, orange, light green, black, pink, purple, gray, brown, red, medium or dark green, then blue.

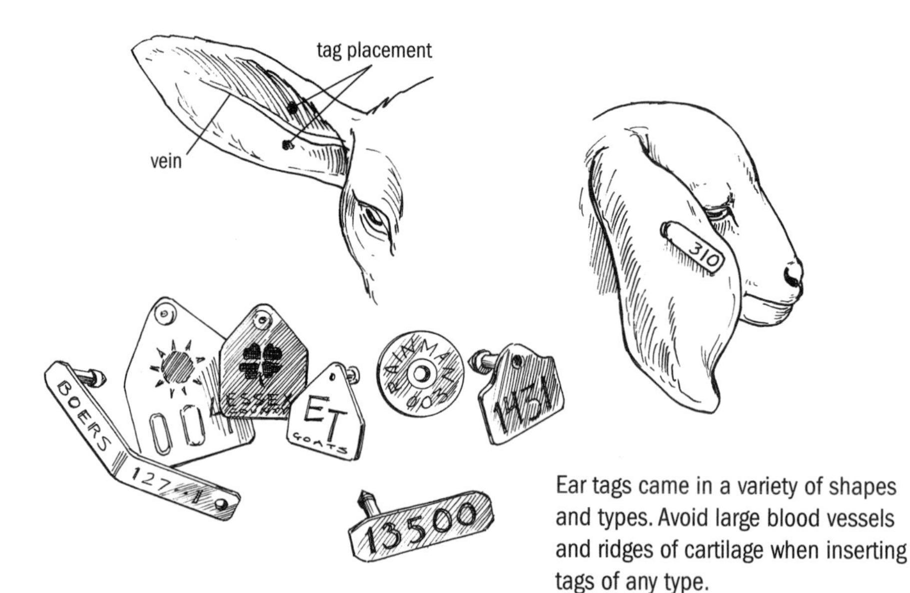

Ear tags came in a variety of shapes and types. Avoid large blood vessels and ridges of cartilage when inserting tags of any type.

How to Install an Ear Tag

Restrain the animal using the same protocol used for tattooing.

1. Clean the ear with alcohol and pat it dry.
2. Place the tag in your tag applicator, and apply antiseptic salve to the male (pointed) tip of the ear tag.
3. Place the side with the hole (the female side) of the tag inside the ear and the pointed part (the male side) outside. When setting the tag, avoid blood vessels, ridges of cartilage, and scar tissue. With sheep and goats, insert it 1 to 2 inches from the skull, where ear tissue is thicker and more difficult to tear.
4. Press quickly and firmly, and it's in.

Use colors and location to reap information at a glance. For instance, tag males in one ear, females in the other. Colors can indicate such things as sex, sire, year of birth, and commercial versus purebred animals.

Premier1 also prints individual tags to order. Name-bearing ear tags are a very handy thing when farm sitters or others unfamiliar with your animals have to look after your stock.

Freeze Branding

Freeze branding is the process of using a special branding iron to freeze a mark on the hide of an animal. The extreme cold of the iron kills pigmentation cells in the animal's skin. Therefore, a freeze-branded animal will have white hair where the freeze branding iron touched her skin. The white hairs produced by a freeze brand are not very visible on a light-colored animal, however, so in these cases the iron is applied to the skin for a longer interval to intentionally halt future hair growth. This freeze brand looks much like a traditional hot brand, but it's usually clearer and more easily read and always more pain-free to apply.

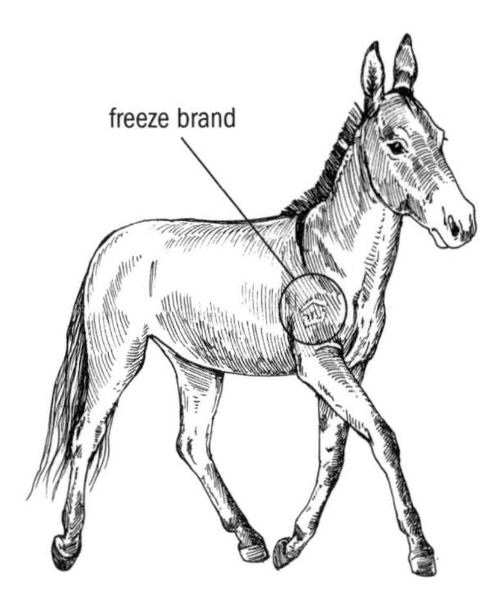

Freeze brands are widely used on large species such as equines and cattle.

The process itself is fairly straightforward, but you should ask an experienced person to help the first time. Some vets also freeze brand for a fee. Keep in mind that if you have to experiment, your first few attempts might not be perfect.

Two types of coolants are commonly used: a combination of dry ice and 95 to 99 percent alcohol, and liquid nitrogen. Both the dry ice/alcohol and the liquid nitrogen methods work well, but liquid nitrogen is much colder, and many experts feel it's easier to use; so for our purposes let's describe freeze branding using liquid nitrogen.

Freeze Branding Basics

First, assemble your tools. You will need:

- A freeze branding iron
- Liquid nitrogen (buy this from artificial insemination companies or welding suppliers) and a container to hold it
- A stiff grooming brush or curry comb
- Clippers with a surgical blade
- 95 to 99 percent isopropyl alcohol solution
- A squeeze chute for cattle; a means of mild sedation for equines
- Gloves and safety goggles. Heavy gloves should be worn to insulate your hands when handling a cooled iron because the handle will become cold enough to cause frostbite. Safety goggles protect your eyes from liquid nitrogen splash.

Cool the iron in liquid nitrogen for 20 minutes before using it on the first animal. The liquid nitrogen should cover the head of the iron by at least one inch. A frost line on the handle of the iron indicates it's ready. Recool the iron for two minutes between uses.

Restrain the animal in a squeeze chute or sedate her and wait until she is relaxed. Clip the hair from the area to be branded. The hip, between the hook and pin bones, is the preferred location for cattle; the shoulder or hip is good for horses (be sure to brand at the same location every time and in the location recorded with your state brand inspector's office). Brush the clipped area to remove dirt and loose hair. Saturate the branding area with alcohol, working it all the way to the skin.

Immediately apply the freeze branding iron to the animal. The iron must move directly from the coolant to the animal. Apply firm pressure, and don't let the iron slip. The area will quickly become numb, and the animal will experi-

CONTACT TIME FOR FREEZE BRANDING WITH LIQUID NITROGEN

Species	Contact time (seconds)*
Bull or other especially thick-skinned bovine	35
Cow	25 to 30
Calf	21 to 24
Horse	8 to 12
Long yearling colt, filly, or gelding (don't freeze brand equines under 18 months of age)	6 to 12

*Shorter times are for dark-colored animals; longer times are for light-colored animals

ence minimal discomfort. Rocking the iron gently from top to bottom and from side to side makes a better brand. Have an assistant keep track of time so that the branding iron is in place for the correct length of time (see table above), then remove the iron from the animal and immediately place it in the liquid nitrogen for recooling.

The brand will show up in indented form on the animal immediately after branding. Within minutes, the area will swell, creating a raised version of the brand. The swelling will subside in a few hours. In two to four weeks the area will begin to peel. Unpigmented hair will grow back in its place.

Designing Your Freeze Brand

When designing a brand, keep it simple and use open letters and symbols. For example, the letters *C* and *X* are far more open (have fewer enclosed areas) than do the letters *A* and *B*. The more open a symbol or letter, the greater the likelihood the brand will show up clearly and be easily readable. The simpler the brand, the lesser the likelihood that it will blotch and be unsightly and hard to read. When designing your freeze brand, become a keen observer of other people's freeze brands to see what did and did not work for them.

Keep in mind that in most western states and in certain other parts of the United States, you can't simply design a brand and start using it. In these states a brand must be registered with the state brand inspector's office. It must also meet certain criteria and can't duplicate another brand already in use. For example, the Arizona Department of Agriculture won't register a brand incorporating unusual designs, anything designed within an enclosure, or any single letter or number (all brands must be of two or more characters). Neck brands, jaw brands, and dual location brands are also unacceptable. Brands must be of a design that is easily identifiable and one that won't blotch.

Freeze branding requires a specially made freeze branding iron and cannot be done with old-fashioned hot branding irons. These irons are crafted of heavy copper or bronze with slightly rounded faces. For animals the size of miniature cattle and adult class B miniature horses, the brand should be about 3 inches (7.5 cm) tall, ⅜ inch (1 cm) thick, and 1 inch (2.5 cm) deep. Keep in mind that brands applied to young animals will "grow" with their wearers. Talk to the company that builds your iron to custom-size an iron for the species and age of the animals you intend to brand.

Microchips

Microchips are tiny devices designed to be implanted under an animal's skin. Based on Radio Frequency Identification (RFID) technology, they contain no power source. Microchips are composed of three parts: a silicone chip, a coil inductor, and a capacitor. The silicone chip contains a unique identification number and the circuitry needed to relay that information to a handheld scanner; these components are contained in a bio-compatible glass housing about the size of a large grain of rice. The chip is inserted using a syringe with a 12-gauge needle. The process appears to be nearly painless, though large animals are usually mildly sedated, just in case.

To read an implanted chip, the operator passes a microchip scanner over the implantation site. The scanner emits a low radio frequency that provides the power needed to transmit the microchip's code and positively identify the animal. Unfortunately, until recently, chips and scanners weren't standardized and not every scanner can read every chip. The new Euro chips, however, can be read by newer scanners manufactured by all three major chip makers — AVID, Destron, and HomeAgain — making them the best choice for compatibility.

Earlier chips had a disquieting habit of migrating away from the injection site, but newer chips incorporate Bio-Bond materials that are fully compatible with animal tissue, making migration problems a thing of the past.

Chips never go bad (they have a 25-year life expectancy) and cannot be altered or easily removed, making them arguably the best permanent means of

MANY MICROCHIPPED HORSES

Horses have been routinely identified with RFID technology since the early 1980s; microchip vendors estimate more than 600,000 horses have been successfully implanted with microchips in the United States alone.

positive identification. The downsides are that microchipping is comparatively expensive, and chips can't be read without a scanner.

If you are adept at giving injections, you can implant your own microchips (if not, a veterinarian can implant them for you). Each microchip comes in sterile packaging complete with a single-use, preloaded syringe and a package of stickers printed with the unique code programmed into the chip. These stickers can be affixed to registration papers and similar records, and the code recorded with appropriate registries and databases such as the ones managed by AVID, HomeAgain, the International Equine Recovery Network, and Stolen Horse International's NetPosse Identification Program.

Injection sites and methods vary from species to species. Equines are implanted on the left side of the neck into the nuchal ligament just below the mane, about halfway between the poll and withers, about ½ to ¾ inches (1.25 to 1.9 cm) below the mane (see illustration below). Pigs are implanted on the side of the neck behind the base of the ear. Sheep and goats are implanted subcutaneously in the manner of implanting dogs and cats, near the base of the ear or between the shoulder blades. The International Llama Registry recommends implanting llamas at the base of the left ear (other sites such as the base of the right ear, shoulder, rump, and base of the tail are also used); the Canadian Llama and Alpaca Registry prefers the right side of the base of the tail or the base of either ear. To avoid microchips surfacing in cuts of meat, approved sites in cattle include out-of-the-way locations such as the base of an ear and the neck.

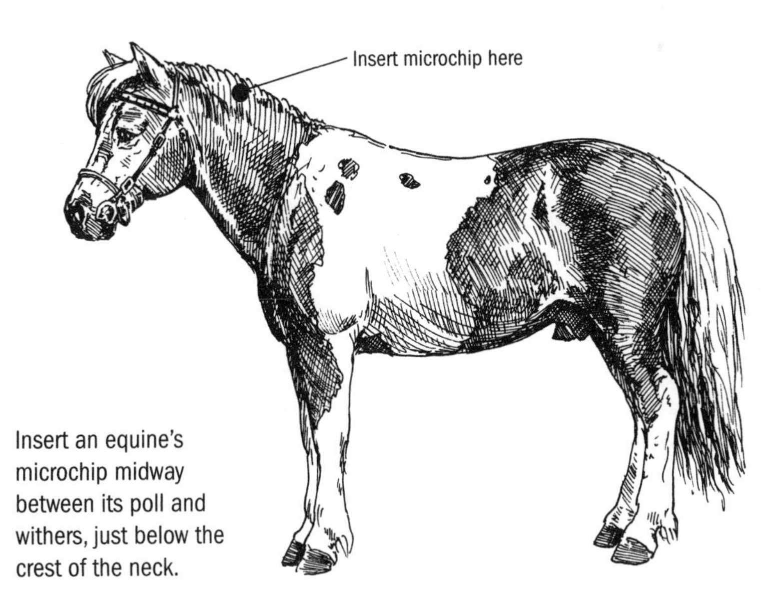

Insert an equine's microchip midway between its poll and withers, just below the crest of the neck.

BREEDER'S STORY

Lonnie and Brenda Short of Short ASSets Ranch, Texas

LONNIE AND BRENDA SHORT raise superior miniature donkeys on their 289-acre ranch near Eddy, Texas, 60 acres of which is home to 100 head of well-bred, beautifully conformed, and extremely well-cared-for little donkeys.

The Shorts' battery of breeding jacks includes homebred show-stoppers like Short ASSets Star's War, a 30½-inch dark brown jack who, as a yearling, was Grand Champion Jack at the 2000 National Miniature Donkey Association Show in Albany, Oregon. He was also Best of Breed, National Champion Jack, and National Coon Jumping Champion (jumping 42 inches!) at the 2004 National Miniature Donkey Association Show in Waco, Texas.

An especially interesting feature of their sales program is the Short ASSets Ranch Bray-a-Way Plan, whereby the Shorts hold a buyer's choice of donkeys for up to six months, interest-free and at no charge, allowing the purchaser to pay what he or she can afford each month. The donkey must be paid for in full before it can leave the ranch.

When we asked Brenda to tell us more about the Shorts' experience raising miniature donkeys, this is what she shared.

"We started to breed miniature donkeys in 1995. Whenever we were out driving around and Lonnie saw a donkey out in a pasture, he would say, 'Look at that jackass, isn't that the prettiest thing you ever saw?' One Christmas I didn't know what to buy the man who had everything, so I decided to buy him a jackass. We ended up with a pair of standards — a spotted jack named Gorgeous George and a gray jennet named Weezie (just like *The Jeffersons*). According to Lonnie's mother, we now had three jackasses on the ranch! We had so much fun with the donkeys, and everyone who came to visit wanted to see them.

"About eight months later, our newspaper ran a picture of a lady holding a two-week-old jennet. It was the cutest thing that I had ever seen. We had no idea that miniature donkeys existed. I called the lady, and she told me that these donkeys were better than Valium. I told her that I needed a lot of them! She told me the Legend of the Donkey's Cross; I still get goose bumps when I hear it or tell it. I told Lonnie that I wanted to go see the miniature donkeys. Both of us fell in love with these wonderful little animals. We bought the baby that she was holding and several others. No one warned us that they are like potato chips: 'You can't have just one.' We now have around 100 miniature donkeys!

"We have been very pleased with our decision to purchase those minis. In fact, we wish we had found out about miniature donkeys earlier.

"Breeding good conformation is our first priority. We want to see good, straight legs and nice, round butts! Our preference is for a heavy-boned, short-bodied donkey. Color doesn't matter to us, though spots are my personal favorite. Everyone likes something different, though, so we've always tried to have something for everyone. Color plays a role in determining price, but we try to have a donkey in all price ranges. We believe we have some of the best bloodlines in the country. We have more than a room *full* of ribbons and awards, which indicate that others agree with us.

"All of our miniature donkeys are unique. They love people, are a special gift from God, and carry a special peace about them. After all, God chose the donkey to carry Mary to Bethlehem and then again to carry Jesus to Jerusalem. He could have picked any other animal, but he chose the donkey.

"There is a strong market for our miniature donkeys. More and more people like the heavy-boned, short-bodied donkeys.

"Our jennets are gray-dun, sorrel, dark brown, black, and spotted. They have nice round butts, straight back legs, wide chests, good bites, and short, pretty heads. They range in height from 28 to 34 inches. Each jennet is paired with the jack that we feel will give us the best all-around foal.

"It is our belief that you shouldn't scrimp on your herd sire. Therefore, we have purchased and bred what we think are some of the finest jacks to be found. They are out of excellent bloodlines, have wonderful conformation, and are producing outstanding foals. Our herd sires have done very well in the show ring, and many of their offspring are carrying on their sire's winning tradition. We have six breeding jacks and offer outside breeding to select jennets.

"We have planted coastal hay on about 50 acres of our ranch, which we cut and bale ourselves. The donkeys' hooves are trimmed every three months. Our farrier and his wife come every month to work on a scheduled group of donkeys. The donkeys are dewormed every other month, and we rotate between different types of dewormers. Every spring we vaccinate the entire herd for eastern, western, and Venezuelan encephalomyelitis; tetanus; rhinopneumonitis; influenza; and strangles.

"I tell new breeders to advertise. A Web site is a must, in my opinion. Educate people that there exists a miniature donkey. One thing that we always hear is 'I didn't know there was such an animal as a miniature donkey.'"

How to Implant a Microchip

1. When implanting an animal you didn't raise yourself, scan the area to make certain that a microchip isn't already present.
2. Scan the uninserted microchip to make sure it works.
3. Close-shave the area where the chip will be implanted, and swab it with antiseptic solution.
4. Position the needle at a 45-degree angle to the skin surface, and insert it to the hub. Push the thumb button forward all the way, then slowly withdraw the needle, making certain the chip was implanted and isn't being withdrawn along with the needle.
5. Apply pressure to the site for a few seconds.
6. Scan the site to confirm the implant.

Temporary Identification

Colored plastic neck chains or neck straps can be used to identify classes of livestock according to age, ancestry, birth dates, productivity, and so on — and you can use them to lead tame animals. To reduce the risk of collared animals hanging themselves on fencing, brush, or another animal's horns, choose lightweight plastic items that easily break in an emergency.

For very temporary markings, use chalk or wax markers, paint brands (inexpensive numerical branding irons dipped in marking fluid), paint sticks, or aerosol spray paints designed for use on livestock. Chalk and paint sticks last about one hour; wax markers last about four; and aerosol spray paints and paint brands last up to six weeks. They're easy to read from a distance and are great for marking animals in a herd as you vaccinate or deworm them. Some producers use spray paint or paint brands to mark females and their offspring with matching numbers. The first family is marked with a "1," the second with a "2," and so on. This has the added advantage of giving the producer the approximate age of each youngster at a glance.

Plastic chain collars are strong enough to lead a sheep or goat but break away if the animal gets snagged on something.

10

Transportation

TRAVELING IS STRESSFUL for animals of all livestock species unless they are safely and sanely hauled on a regular basis, such as in the case of seasoned show horses, and even a majority of these develop stress-induced ulcers.

Stress-full Statistics

In an Iowa State University study that mimicked the life of a recreational show horse, test subjects were transported, just once, to a different environment, kept in a stall adjacent to familiar horses, lunged for 20 minutes twice daily for four days, and were then hauled home. A full 70 percent developed stomach ulcers, also known as equine gastric ulcer syndrome, or EGUS. This syndrome causes painful stomach ulceration that contributes to weight loss, recurrent colic, chronic diarrhea, and suboptimal performance, and it costs a great deal of time and money to cure.

Well-kept, old-style horse trailers make first-class mini haulers.

Hauling is tremendously stressful for other species as well. According to data collected from members of the National Institute for Animal

BUILD A TRAVELING FIRST-AID KIT

Put together a traveling first-aid kit to augment the kit you keep at home in the barn. Pack it in a lidded 5-gallon plastic pail (restaurants and fast-food restaurants will often give you one for free), and keep it in your truck or trailer at all times. If you use something, replace it as soon as you get home. Having a well-equipped first-aid kit and knowing how to use it can make the difference between life and death when you're on the road and far from the closest vet. At the bare minimum, a kit should include:

- Betadine solution to flush fresh wounds
- Sterile gauze sponges
- Telfa nonstick absorbent pads to cover wounds
- Several individually wrapped sanitary napkins to use as pressure pads to stop heavy bleeding
- Several rolls of Vetrap or a comparable self-adhesive bandage
- A roll of 2½-inch-wide sterile gauze bandage
- 1- and 2-inch-wide rolls of adhesive tape
- Antibiotic ointment
- Saline solution and topical eye ointment
- Rectal thermometer and lubricant
- Tweezers or a hemostat
- Probios or a comparable probiotic gel
- Banamine — an injectable pain reliever and anti-inflammatory drug you get from your vet. If you take it, pack it in your cooler and add a vial of epinephrine, just in case.
- Don't forget to add some bandages and nonprescription pain relievers for yourself

Agriculture (reported in "Economic Factors Associated with Livestock Transportation," by Speer et al.; see Resources), 80,000 market hogs die of hauling-related injuries and stress each year.

Hauling also elevates cortisol levels that in turn weaken the animal's immune system. Recently hauled livestock are far more likely to suffer from pneumonia, colic, and enterotoxemia than their stay-at-home kin.

Animals don't handle change very well. Factor in the noise, confusion, and crowding that are part and parcel of trucking and it's little wonder stress kills.

Whether you're hauling your first animals home to your farm, traveling to shows for the hundredth time, or trucking young stock to market, it's important to transport livestock with the utmost care.

Plan Ahead

To cut down on in-transit injuries, post-transport weight loss, and stress-induced illness, plan ahead to make hauling easier on your livestock. Here are some things to consider:

- Avoid hauling sick or injured animals and pregnant females in the last trimester (especially the final few weeks before they give birth); start out with sound, healthy animals, and do your best to keep them that way. And since most animals are social creatures, haul along a companion if you possibly can.
- Map the route in advance. According to Richardson, braking and cornering cause 75 percent of in-transit falls; crossing bumps and acceleration cause the rest. Stop-and-start driving causes hormones and blood components to fluctuate and can drive heart rates up to twice their norm. If the most direct route means dealing with rush-hour traffic or possibly hitting scores of red lights, choosing a longer but easier route is always a wiser choice.
- Allow enough time to drive carefully. Accelerate slowly and smoothly, and do your best to stop that way as well. Ease up on the gas well ahead of turns, and don't take corners too abruptly. Factor in load checks, too; stop twenty minutes after departure to check your load and at least once an hour after that.
- Factor rest stops into long journeys. Locate points along the route where you can safely off-load your living cargo at least once every 24 hours. Offer familiar feed, give them water, and check each animal for signs of injury and excessive stress. Pack along a well-appointed first-aid kit, and to cut back on digestion-related illness, include enough probiotic gel to dose each passenger at least once a day.
- Load livestock with care. Don't drag them along or lift them by their horns, head, ears, hair, or legs. Cover the nonslip floor of their conveyance with an adequate amount of dust- and mold-free bedding, or improvise by covering a slippery floor with several inches of damp sand and topping that with conventional bedding. If loading at night, provide plenty of interior lighting, as all species move more easily from darkness into light.

- Allow sufficient room for each animal. On short jaunts, individuals should be allotted enough room to stand without constantly slamming into the rest of the load; on longer hauls, animals hauled loose inside a conveyance need enough space to comfortably lie down.
- Uncastrated males and other aggressive animals often act out their annoyance at being transported by biting, kicking, butting, and horn-hooking timid and smaller compartment mates. If possible, pre-fit your conveyance with interior dividers so you can partition animals into compatible groupings based on sex, size, age, and/or aggressiveness. Barring that, halter and tie ornery passengers, making sure tie ropes are long enough to allow them to get back up if they fall but not so long that they can choke or get tangled in the slack.
- Whatever sort of conveyance you choose, cut down on vibration by using lots of bedding, and reduce noise levels by padding gates, loading chutes, and partitions with pieces of rubber matting or old blankets.

Weather Kills

Weather extremes head the list of long-term stressors. As temperatures and humidity inside their traveling compartment rise, animals become restless and start to pant. Individuals that are gasping for air are in deep distress, as are any who have fallen and refuse to get back up.

To compensate for sizzling, steamy weather, reduce loading capacity by 15 to 20 percent (overcrowding rapidly leads to excessive heat buildup); create additional ventilation by opening windows or replacing solid upper walls with sturdy, closely spaced pipe or heavy wire mesh; travel only at night or during cooler morning hours; and keep the number and length of stops to the barest minimum. Never park a loaded truck or trailer in direct, summer sunlight. If lots of summer travel is in your agenda, spring for a power inverter that connects to your truck battery and allows a full-size household box fan to operate on truck power.

Bring a cooler packed with jugs of ice cold water and bagged ice cubes. Offer water to parched livestock whenever and wherever you stop. Gently trickle cold water on the backs of distressed animals' heads and apply ice cubes wrapped in cloth to the animals' armpits or groins.

If an animal is suffering from heat exhaustion, you must cool him down — fast. The best way is to hose him down using lots of cold water and then place him in front of a fan to finish cooling. Just getting him wet isn't enough; you need to use *cold* water and then let it evaporate. Don't moisten only the surface of heavily coated species such as wool sheep, Pygora goats, or miniature llamas.

Doing so traps heat close to the animal's skin; you must saturate his fleece all the way through.

In a pinch, cool down a badly distressed individual as quickly as you can, then stuff him in the cab of your truck, turn the air conditioner on high, and drive around so the air conditioner cools more efficiently.

Cold kills, too. Animals are susceptible to frostbite and loss of body heat. It's important to keep in-transit livestock dry. Cover openings to block rain and wind. Add more bedding. Allow extra space so animals can move away from chilling wind. To keep their bodies warm, fit them with blankets specially cut for their species.

Choosing a Hauling Conveyance

Miniature livestock can be safely hauled in goat totes (see below) and trucks with caps, stock and horse trailers, and even in crates in vans and pickup trucks. It is dangerous to haul them in completely enclosed cargo trailers or any sort of conveyance with openings large enough to escape from (a determined animal can jump higher than you think and squeeze through tiny openings). When choosing a livestock hauling conveyance, keep these points in mind:

- All species die quickly from carbon monoxide poisoning, so make certain engine exhaust doesn't enter the conveyance.
- All livestock species are susceptible to respiratory problems, and they easily expire from overheating. Good ventilation is an essential element.
- Poor footing leads to scrambling, physical injuries, and rampant stress. Damp wood and some rubber flooring can be slick as glass. Invest in nonslip trailer mats, and use dust-free disposable bedding on top of the mats.

Goat Totes

A number of leading livestock-trailer and goat-handling-equipment companies make slide-in truck-bed carriers designed for hauling full-size goats. The best of these work admirably well for miniature sheep, goats, and pigs, as well as for calves, crias, and foals. Some are enclosed units with adjustable ventilation features such as securely screened, sliding windows; others are essentially cages with sliding back doors.

Enclosed units are heavier, sometimes requiring a fair amount of muscle power to hoist them on the truck. Look for aluminum slide-ins if weight is an issue. Enclosed units are more secure than cage-style goat totes and provide more protection when hauling valuable livestock.

Commercial or homemade goat totes are big enough and safe enough for most miniature species, even young stock such as calves.

Cage-type slide-ins work well, too, but should be used with covers to protect their passengers from inclement weather.

And double-latch those doors! It never hurts to add additional clips; an open door at 60 miles an hour is incredibly bad news.

Trailers

If you frequently haul a lot of livestock, you will probably want to buy a new or used trailer. If you plan to show your miniature livestock frequently and at large venues such as the American Miniature Horse Association Nationals or the National Western Livestock Show, by all means, opt for an impressive new trailer. However, if you haul your animals infrequently or show at less prestigious shows, a used trailer might do the trick. Used trailers range from barely touched last-year's models to golden oldies from decades past. Whether you're buying a new, gently used, or vintage trailer, certain elements remain the same. These are some things to consider:

- Trailers come in two basic floor plans: (1) horse trailers with interior partitions set up to accommodate individual equines, and (2) stock trailers with open floor plans, sometimes fitted with swinging gates to divide the load. Unless you buy a horse trailer designed for miniature equines, the partitioned areas inside a standard horse trailer are likely too roomy for your needs. Keep in mind that some horse trailer designs allow the easy removal of stall partitions, rendering them stock trailers for all practical purposes.

- Trailers come in bumper-pull and gooseneck models. Bumper-pull trailers attach to a ball on the towing vehicle's bumper. Gooseneck trailers attach to a more elaborate hitch mounted in the bed of the towing vehicle (always a truck). Each style has its proponents; neither is intrinsically better than the other.
- Aluminum trailers are popular due to their lighter weight, but steel is stronger, and strength equates with safety on the road. Hybrid trailers with steel frames and nonstructural features crafted of aluminum or fiberglass strike a middle ground.
- Some trailers come fitted with ramps; others don't. While step-ups work fairly well for seasoned full-size animals, reluctant animals, especially of diminutive proportions, could easily slide a leg under the trailer itself while being loaded. Before buying a trailer with a ramp, ask everyone likely to have to work the ramp to raise and lower it. Some ramps are beautifully balanced; others require plenty of muscle.
- No matter which design you choose, opt for a trailer with strategically placed escape doors big enough for an adult to easily climb through. Having been trampled (several times) in trailers that lacked easily accessible escape doors, I can't recommend this safety feature strongly enough!
- If you aren't mechanically minded, take along a savvy friend when viewing used trailers. The most important parts, safety-wise, are a trailer's frame, floor, hitch, and brakes. Don't buy a used trailer that is compromised in these areas.
- Weakened wooden floors can be replaced, but do it before using the trailer. Check aluminum floors in used trailers for metal fatigue by carefully examining welded areas and seams both inside and under the trailer. In either case, buy a trailer with good-quality mats or add new ones.
- Proper ventilation is of prime importance. Do windows and top vents slide open, or is the trailer fully enclosed? It's better to blanket animals in inclement weather than to haul them in a trailer with inadequate ventilation.
- Make certain the towing vehicle you plan to use can safely pull the trailer. Most two-horse, full-size trailers, even when converted to open-floor-plan miniature livestock trailers, require a class three hitch (a rating determined by overall trailer weight). Check your towing vehicle's specs before choosing a trailer too heavy for it to pull safely.
- New or old, carefully inspect the trailer for sharp edges and splinters that can injure livestock while being loaded or heading down the road. High-end trailers generally feature carefully de-burred metal and nicely sanded wooden parts, but cheaper trailers sometimes don't.

TRAILER MAINTENANCE 101

The safest trailer in the world won't stay that way unless you take care of it. Proper upkeep maintains its value, too, so keep it in the pink.

After every use

- Pull out the mats, hose and scrub off the floor, and allow it to dry thoroughly. Before replacing the mats, carefully inspect the floor for damage or rot. Check welds on metal floors, and poke suspect areas of wooden flooring with a penknife blade. Then, replace the mats.
- Check for signs of interior damage such as splinters or ripped padding.
- Oil the door hinges and latches with spray lubricant such as WD-40.
- Check the lug nuts using a lug wrench. You shouldn't be able to wiggle them at all.
- Grease the trailer ball with automotive grease. Cover the greased ball with an appropriately sized can or a tennis ball that has been sliced open.

Before the next use

- Check for hornets' nests both inside and under the trailer.
- Check the tire pressure; don't forget to check the spare tire.
- Check the lights, and make sure brake and turn signal lights are working.
- Check the brakes!
- Make certain spare equipment (don't forget spare halters and lead ropes), the first-aid kit, horse trailer jack ramp, and at least one fire extinguisher are where they're supposed to be.

Once a year

- Take the trailer to a good mechanic or trailer repair center for a professional checkup. Don't omit this important step!

When storing the trailer

- Wash and wax the exterior. Make certain the floor is meticulously clean and uncovered. Store mats elsewhere to allow air circulation to the floorboards.
- Oil and grease all moving parts.
- Ideally, trailers should be stored under cover when not in use. Park on blacktop, dirt, or stone instead of grass (grass contributes to unwanted dampness and moisture issues).

For your animals' safety and your own peace of mind, buy the best trailer your budget will allow. Consider a well-maintained, quality used trailer over a poorly crafted brand-new model. Good used trailers hold their value surprisingly well; cheap, shoddy trailers just don't.

Older models of full-size horse trailers are a good option. Today's full-size horses are taller and wider than horses ridden a few decades ago. Because these modern steeds require extra-tall, extra-wide trailers for hauling, old-style smaller used trailers aren't practical for them. However, if the older models were fitted with removable center partitions (and most were), structurally sound two-horse models are easily transformed into comfortable, safe, open-floor-plan stock trailers for miniature livestock — and at minimal cost.

Dog Crates

One item every miniature livestock owner or producer needs is at least one extra-large dog crate. We like heavy-duty wire dog crates best; they're ideal for housing bottle babies in your home, and they're great for hauling lambs and kids in the back of a van. Crates fit nicely in even the smallest van or SUV if its back seat is temporarily removed.

Choose a crate with a flip-open top and a swing-open, tightly secured end door. Pad it with old, easily washable bed blankets. It's that easy!

Fastidious folks with nice vans prefer airline-style plastic crates that don't let urine and worse leak onto the floor. However, we've never had a problem with our wire crates leaking (even on a 500-mile jaunt in a small van with a fully grown Miniature Cheviot ram in the crate), though you may want to put down a plastic tarp just in case. The trick is changing bedding before it gets nasty and having a place to put soiled blankets until they can be laundered (think rooftop luggage carrier and a nylon duffle bag).

Miniature lambs, kids, and piglets travel in style in airline-approved dog crates.

Crates are also useful for hauling small livestock in the bed of a truck. Airline crates are best for that application. If you use a wire crate, cut down on wind stress by wrapping a blanket around all but the end of the crate facing the tailgate of your truck. Substitute a plastic tarp if it's raining, but avoid plastic coverings when it's sizzling hot. Don't haul any type of animal in a crate in freezing weather; a single animal in such a close enclosure can't generate enough heat to stay warm.

Working with a Livestock Transporter

In many cases, transporting your own livestock isn't cost-effective. If you're relocating to a new home a thousand miles away and taking your animals along, or if you buy your first passel of breeding stock from a breeder in another state, you may elect to pay a livestock transporter to bring them to you.

Transporters usually charge a hefty fee for the first animal picked up per stop and considerably less if you ship more than one. Obtain estimates from at least three or four companies; prices vary dramatically from hauler to hauler.

Unless you contract for the whole load, routes are rarely direct, and your animals may be on board for days or even weeks. Since animals are easily stressed, it's important to choose a transporter who knows your species and goes the extra mile to keep his charges safe and well. When comparing rides, find out:

- What sort of trailer and hauling unit does the hauler use? What are his contingencies in case of breakdowns?
- Is he willing to partition your animals away from other livestock? If so, will they have any sort of physical contact with other livestock when off-loaded at rest stops?
- Will he pack a supply of your animals' accustomed feed (and feed it to them), or does he insist they eat what he provides?
- What is his policy for dealing with sick or injured passengers and with lactating dairy cows and goats?
- If something goes amiss, will he contact both shipper and receiver so they know what's going on? Will he phone if he's running late? Will he make his cell phone number available to involved parties?

When shopping for livestock transporters, ask for references and check them out. Better still, post to e-mail groups and ask other subscribers to contact you (off-list) with their recommendations or tales of woe.

Call your vet to find out what sort of tests and health papers are necessary for your animals' interstate shipment, and have the paperwork in-hand when the hauler arrives to pick them up.

And don't necessarily choose the cheapest or most well-known transporter; find one you're comfortable with based on experience or recommendations. Transporting livestock species is fraught with risk, so hire the best hauler you can find and afford.

11

Breeding

WHILE BREEDING PRINCIPLES VARY among species, some apply across the board. I'll discuss these commonalities in this chapter; specifics to individual species are covered in section 2. Some of the following material has been adapted from *Storey's Guide to Raising Meat Goats* by Maggie Sayer and *The Donkey Companion* by Sue Weaver (see Resources).

A Breeding Stock Primer

When you want to raise quality miniature livestock, the most important thing is to start with the best breeding stock you can afford and then take good care of it. These are some things to think about.

LEADING CAUSES OF INFERTILITY IN BREEDING MALES

- Disease or injury
- Physical defects (congenital or acquired)
- Too fat or too thin
- Improper feeding
- Poor libido
- Stress (including heat stress and the temporary aftereffects of high fever)
- Age (too young or too old)

Breeding Males

Never scrimp on your stud llama, stallion, jack, buck, ram, or boar. Depending on the species, females give birth to one to three or so offspring per year, but your male puts his stamp on a whole generation. Use a quality male, and have your vet give him a breeding exam six to eight weeks prior to breeding season to allow plenty of time to correct abnormalities.

The male should complement your females (small bucks with small females, for instance) and your breeding program, or investigate breeding by means of artificial insemination (AI). Choose bucks, rams, and boars from large, productive litters that are likely to sire multiples. Check a buck or bull's teat structure; he has as much influence on his daughters' teat structure as do their dams.

Up your breeding male's plane of nutrition heading into breeding season. Don't make him fat, but do make sure he has a lot of energy; he needs plenty of vim and vigor to do his job, especially if he's expected to serve many females. In the heat of breeding season, vigorous hardworking males often forget to eat. If he's lean going in to the breeding season, he won't have the stamina he needs to do his job well.

I can't say this often enough: If you keep breeding males, miniature or not, never, ever take them for granted. Rams and bucks in rut have maimed and killed humans; boars and bulls are notorious for the damage they can do.

THE TIE THAT BINDS

Every male llama, goat, sheep, pig, and bovine is born with his urethral channel and the glans of his penis attached to the inside of his prepuce (sheath) by the frenulum membrane. As his body begins producing testosterone (sometimes when he's only a few weeks old), he will begin "practice breeding" his dam. Some males, such as baby bucklings, semi-squat and then repeatedly thrust their hips. These sessions break down the adhesion of the membrane and allow the young male to extend his penis. When that happens, he's probably capable of impregnating females. Young boars break the membrane at approximately 4 to 6 weeks of age; bucklings and ram lambs at around 6 weeks to 6 months; bulls at about 8 to 12 months; and llamas at 1 to 2 years of age. Male equines do not go through this process.

Friendly males are especially dangerous because handlers loosen their guard around them. Whenever possible, stay out of loose housing, pens, corrals, and pastures where breeding males such as bulls and rams in rut are kept. If you must go in, carry something you can use to defend yourself should the need arise, and never allow a breeding male to cut you off from your only avenue of escape.

Breeding Females

The females you select for your breeding program should have good confirmation and be easy keepers, healthy, and productive. Then, they must be bred to a quality male.

Feed females correctly. No matter which species, unless they're old or exceptionally poor keepers, dry (open; nonpregnant, nonlactating) brood females don't need grain. Pregnant females don't need grain until the second half of their last trimester of pregnancy.

Have your vet give bred females an ultrasound to confirm pregnancy and to see how many offspring they're carrying. Ultrasound is effective and inexpensive, and it allows you to divide your females into groups and feed them according to their breeding status.

Vaccinate your females and give boosters a few weeks before they give birth so they're sure to pass antibodies to their young through their colostrum. A neonate's own immune system doesn't begin functioning for weeks (the exact amount of time is species specific), so it needs to receive antibodies from its dam to stay healthy. At the same time we boost our does and ewes, we inject them with a selenium and vitamin E supplement called Bo-Se. This is a common practice when raising livestock in selenium-deficient parts of the country; check with your vet or county extension agent to see if it's needed in your locale.

Be there when they give birth. Too much can go wrong to leave this to chance. We'll talk about that again in a moment.

And protect those babies from predation. Even a fox can carry off a newborn miniature lamb or kid; coyotes love them (and foals and calves as well). Use your best guardian animals in pastures where neonates are present, or keep youngsters penned near the house until they're older.

Finally, monitor babies' health by visually inspecting them at least twice a day. It doesn't take long for a youngster to die from coccidiosis, enterotoxemia, *E. coli*, bloat, or another fast-acting, life-threatening disease.

A WELL-STOCKED BIRTHING KIT

We pack our birthing supplies in a Rubbermaid toolbox-stool. It's roomy, it has a lift-out tray for small items, it's easy to lug to the barn, and it's so much nicer to sit on than an overturned 5-gallon bucket. It holds:

- A bottle of 7 percent iodine solution used to dip newborns' navels. Due to government regulations that went into effect on August 1, 2007, this has been difficult to obtain over the counter, but we prefer it and feel it's worth the hassle. Some producers substitute solutions with a lower percentage of iodine, and the University of Minnesota recommends a navel dipping solution concocted by combining one part chlorhexidine (Novasan) with three parts water. Commercial iodine-based teat dips are less than 1 percent iodine and *not* sufficient as a navel dipping solution.
- A shot glass (to hold navel dip)
- Sharp scissors (for trimming umbilical cords prior to dipping them in iodine); keep these scissors disinfected and stored in a sturdy resealable plastic bag
- Dental floss (if needed to tie off a bleeding umbilical cord)
- A digital veterinary thermometer
- A bulb syringe (the kind used to suck mucus out of human infants' nostrils)
- A rubber leg snare (cord snares work well, too)
- Shoulder-length obstetrical (OB) gloves
- Two large squeeze bottles of lubricant (we prefer SuperLube from Premier1 — see Resources)
- Betadine scrub (for cleaning up prior to assisting)
- A sharp pocket knife (you never know when you'll need one)
- A hemostat (ditto)
- A lamb-carrying sling (these work for miniature lambs, kids, crias, even tiny calves, and they really save your back from wear and tear)
- An adjustable rope halter
- Two flashlights (we like to have a backup in case the first flashlight fails)
- Two clean terry cloth towels

A Short Course in Livestock Midwifery

Many livestock keepers approach birthing season with a combination of terror and delight. There is nothing more satisfying than healthy newborn baby animals. Still, things can go wrong at birthing time — terribly wrong — so the more you know about birthing the species you raise, the better off you (and your animals) will be.

It's important to know your limits. If something is wrong and you have any doubt at all that you can handle it, *call your vet!* Minor things are easy to correct; a calf firmly wedged in its dam's birth canal is not so easy. We won't go into pulling calves with chains and tackle — that's way beyond the scope of this book. But you should know the basics in case your vet isn't available and you have to handle common dystocias yourself.

Keep in mind that most sheep, goat, and pig owners become quite proficient at this task, whereas correcting dystocia in larger miniature species such as llamas, equines, and cattle is a whole different story. Nonetheless, sometimes you have to try or your animal will die, and the following information is meant to help guide you through this process.

We'll talk more about individual species' birthing protocols in section 2 of this book. No matter what species you're monitoring, keep the delivery environment low-key. Avoid bright lights, noise, and dogs. Snap photos if you like but do it from a distance. Unless something goes awry, don't interrupt the female as she's giving birth. Even then, work quickly and quietly, then get back out of the way.

Be Prepared

Spend some time learning about the reproductive tracts of your females before your first llama, mare, jenny, doe, ewe, sow, or cow gives birth. You need to know where everything is located before you assist for the very first time.

While most births don't require intervention, be ready to help if the need arises. Keep your fingernails clipped short and filed throughout birthing season (you won't have time to do this when an emergency arises). Know how to determine the configuration of a "stuck" baby and how to correct it. If you think you might forget, photocopy and laminate instructions and keep them in your birthing kit.

Teach your fingers to see; this is especially important when assisting species prone to multiple births. To give yourself some practice, borrow a pile of plush toy animals. Put one in a paper bag not much larger than the animal. Without looking, stick your hand in the bag and figure out what you're feeling. Switch animals. Find a larger paper bag and put three stuffed animals inside; see if you can sort out triplets.

Most breeders who monitor their animals' births prefer that the event occur indoors. It's easier to keep track of a stalled animal, and if something goes wrong, there are lights and running water close at hand. However, if you allow your females to give birth outside in a pasture, paddock, or corral (especially if you won't be there), make certain it's under hygienic conditions, the area is predator-proof, and the enclosure is such that she can't lie down close to the perimeter and deliver her young under the fence.

A birthing stall is a better choice for all species except sheep. The stall should be completely stripped as needed, all surfaces sprayed with dilute bleach solution, and deeply re-bedded with dust-free straw (shavings and sawdust cause respiratory problems in neonates). It must be roomy enough for the female to walk around, roll, and stretch out in comfort. There must be no protrusions along the walls (nails, splinters, bucket hooks) that might hurt a newborn. It should provide privacy for the female but offer an unobtrusive spot from which an attendant can observe her giving birth. Water containers should be small enough or hung high enough that a wobbly cria or bouncy lamb can't tumble headfirst into a bucket and drown.

Sheep have different needs. Ewes prefer to choose their own birthing spot near but somewhat secluded from their flock; when isolated in a birthing stall, they fret. One option is to bring a compatible flock mate to the birthing stall (the birthing ewe's dam or a grown daughter is a good choice), or pen her where she can see other sheep.

And be sure to program appropriate numbers into your cell phone so you can phone several vets or mentors at the touch of a button.

Normal Birth

First-stage labor begins with uterine contractions, includes dilation of the cervix, and concludes as the baby enters the birth canal. Although each species reacts somewhat differently to the onset of first-stage labor, all females become restless and attempt to isolate themselves from the rest of their flock or herd. First-timers are generally more restless than experienced dams. Females tend to lie down, get up, and reposition themselves frequently, and most kick or glance back at their abdomens.

The second stage comprises passage of the baby (or babies) through the birth canal, and third stage is defined by the passage of fetal membranes (the placenta) and closure of the cervix.

Once second-stage labor begins, the first thing to appear at the female's vulva is a fluid-filled, water balloon–like sac called the chorion — one of two separate

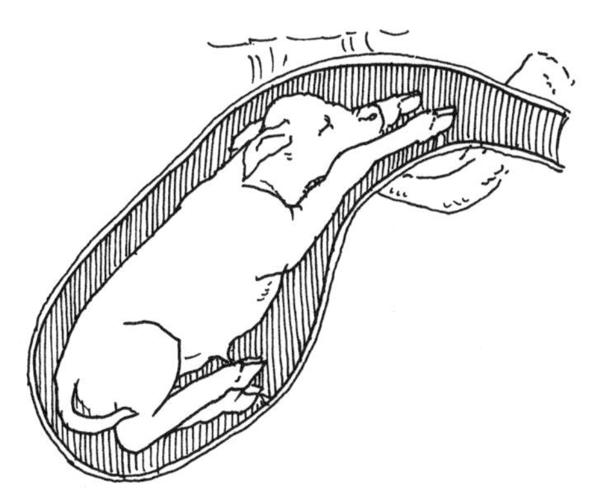

A calf in the normal headfirst birthing position

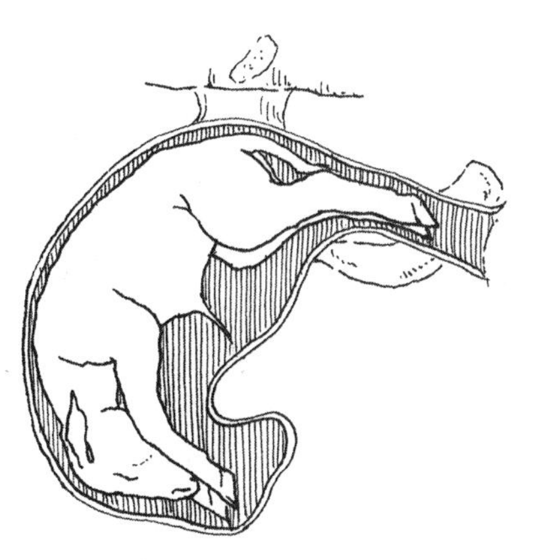

A lamb in the normal butt-first birthing position

sacs that enclose a developing fetus within its mother's womb (the other sac is the amnion). Either or both sacs can burst within the female or externally as the baby is delivered. This fluid provides lubrication for the balance of the delivery.

Normal delivery position is a front-feet-first, diving posture. A foot appears inside the chorion (or directly in the vulva if the chorion has already burst), followed by another foot slightly behind the first one, and then the baby's nose. Once the head is delivered, the rest of the baby quickly follows.

Also normal in some species (sheep, goats, pigs, and cattle) but not others (especially equines) is the caudle position in which the baby presents spine up but backward with his hind legs extended behind him. In a normal hind-feet-first delivery, two feet appear, followed by hocks. Because the umbilical cord is pressed against the rim of the pelvis during this delivery, it's wise to *gently* help the baby out once his hips appear.

Handling Dystocia

If it becomes apparent that you need to assist in a delivery, before you slide your hand inside a female, make certain your fingernails are short and that you've removed your watch and rings. Wash her vulva using warm water and mild soap or a product like Betadine scrub. Pull on an OB glove if you have one; if you don't, scrub your hand and forearm with whatever you used to clean her, and then liberally slather your hand and arm or the glove with lube. Now pinch your fingers together and gently work your hand into the vulva. (Remember, you're trying to form your hand into the smallest, most streamlined shape possible; don't make a fist, rather keep your fingers stretched straight out and grouped tightly together.) When working with goats, the doe will probably scream (and possibly keep screaming the entire time you're helping her — goats are notorious drama queens), so don't be surprised (or intimidated) when she does.

Determine which parts of the baby are present in the birth canal. Closing your eyes and visualizing what is at your fingertips can help. If the baby's toe points up and the big joint above the toe bends down, it's a foreleg. If the toe points down and the major joint above the toe bends in the same direction, it's a hind leg. If you can, follow each leg to the shoulder or groin, making sure the parts you're feeling belong to the same baby. When you're certain they do, if you can manipulate the baby into a normal birthing position, do so (see instructions for dealing with each abnormal position below), and then help pull it out (by this time the dam will usually be too exhausted to do it all by herself). If you can't reposition the baby in 10 minutes, stop trying and call your vet.

Because of anatomical differences, you can't pull the young of most species with tackle the way some producers pull a badly stuck calf. If you must pull, use lots of lube, be judicious, and pull only while the female is having a contraction. Don't pull straight back; pull out and down in a gentle curve toward the birthing mom's hocks.

If you've purchased a rubber lamb or kid snare, be aware that you can't use it to actually *pull* a lamb or kid. It's made to fasten around its legs so that you don't lose track of them while correcting dystocias. If you pull, the snare stretches — this is not a good thing. To actually pull a baby of any species, grasp the legs with your hands, preferably above the pasterns but below the knees or hocks, then pull.

The inside of the reproductive tract is extremely fragile, and if you or the baby tears part of it, the dam will almost certainly die. When repositioning a baby, cup your hand over sharp extremities, such as hooves, and work carefully and deliberately; your female's life depends on your gentle technique.

If you have to pull one baby and there are more present, pull them all. Your female will be exhausted; simply help her get them out so she can rest.

Any time your hands must go into a female to assist her, you *must* follow the birth with a course of antibiotics. Ask your vet for instructions.

These drawings (opposite) illustrate dystocia in goats, but they apply to all of the species covered in this book, except pigs. A sow's reproductive system is considerably different from the rest of the females we're discussing, so we'll talk about assisting pigs in section 2 of this book.

True breech presentation (butt first, legs tucked forward). Some does and ewes can give birth in this position and others can't. Larger species such as equines and cows are unable to give birth like this. If your animal can't, call your vet. It's best not to try to do this yourself; however, if you must attempt to reposition this baby, try to elevate the female's hindquarters before you begin. Push the baby forward, work your hand past its body (it's a tight squeeze), and grasp one hock. Raise up the hock and rotate it out away from the body. While holding the leg in that position, use the little and ring fingers of the same hand to work the foot back and into normal position. Repeat on the opposite side. The umbilical cord will be pinched, so pull the baby as quickly and safely as you can.

One leg back. Many does and ewes can deliver in this position, but larger species usually can't. Push the baby back just far enough to allow you to cup your hand around the offending hoof and gently pull it forward.

Head back. A small lamb or kid with its head bent back to its side can sometimes be pulled, but do your best to correct the position before pulling. Don't try to pull larger species when the head is turned to the side. To correct the position, attach a cord or a rubber snare to the front legs so you don't lose them, then push the baby back as far as you can and bring the head around into position. Alternate problem: Sometimes the front legs are presenting but the head is bent down. This is more difficult to correct. Attempt to correct this in the same manner, but call your vet and make sure he's on his way before you try it.

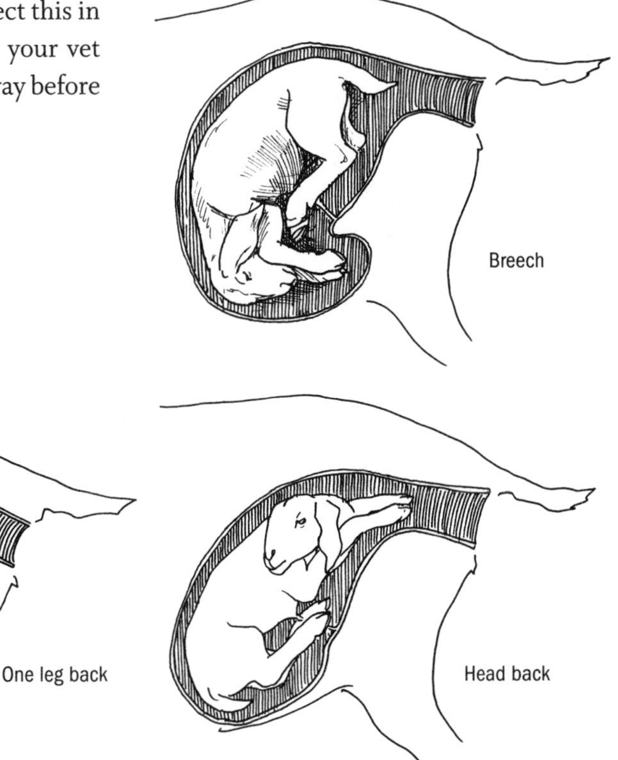

Crosswise. This is a bad one, so call the vet without delay. Until he arrives, push the baby back as far as you can (elevating the dam's hindquarters will help) and determine which end is closer to the opening. Then begin manipulating that end into position. These babies are usually easier to deliver hind feet first.

All four legs at once. Attach cords or a rubber snare to a set of legs, making certain you have two of the same kind (front or back), then push the baby back as far as you can. Reposition the baby for either a diving position or hind-feet-first delivery, depending on which set of legs you've captured.

Twins (or triplets) coming out together. Attach cords or a snare to two front legs of the same baby (follow the legs back to their source to make sure they're attached to the same individual). Push the other baby (or babies) back as far as possible, and bring the captured individual into the normal birthing position.

Twins coming out together with one reversed. Follow the same protocol as for twins coming out together, but since it's generally easier to do so, pull the reversed baby first. If both are reversed, pull the baby closest to the opening.

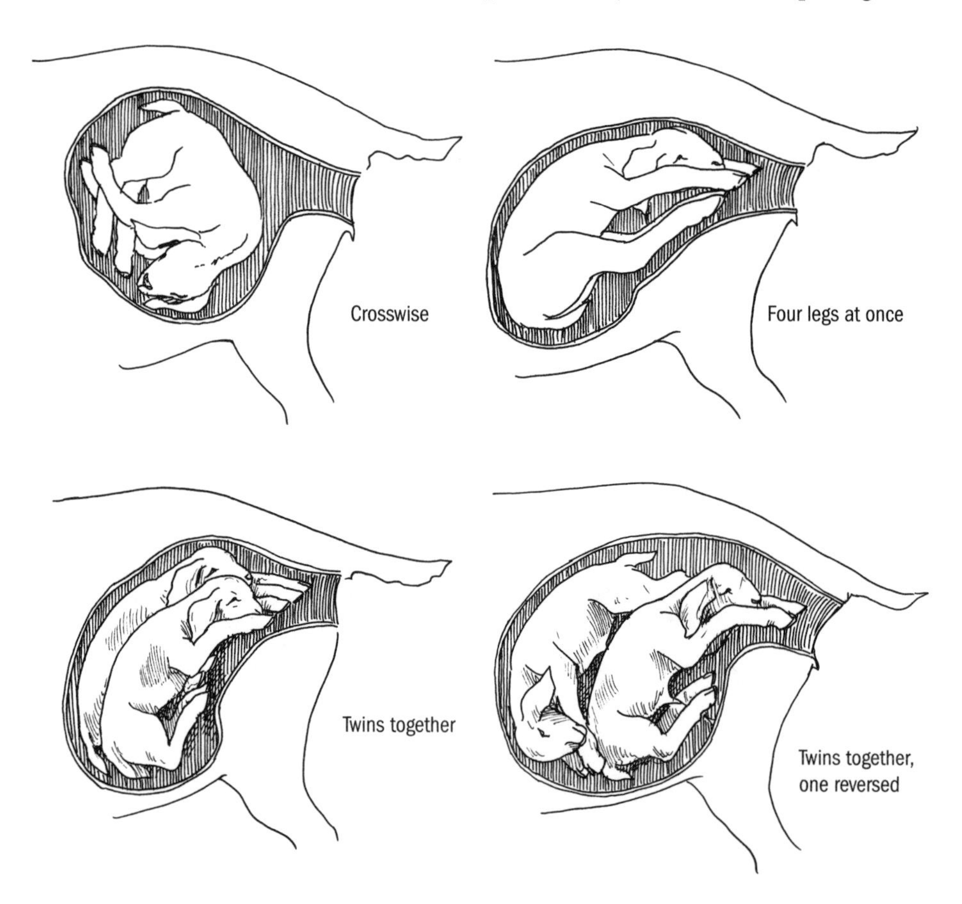

Here Comes Baby!

When the little one arrives, remove birthing fluids from its nose by applying pressure and sliding your fingers along the sides of its face. If the baby is struggling to breathe, use the bulb syringe from your birthing kit to suck fluid out of the nasal passages. If it's really struggling to breathe or not breathing at all, firmly grip its hind legs between the hocks and pasterns, place your hand behind its neck (near the withers) to support it, and swing it in a wide arc to jump-start its breathing. Be advised that the baby will be slippery, so hold on tight.

It's worth mentioning, for beginners' sakes, that an alarming number of newborn kids, lambs, and piglets appear to be dead. In most cases their hearts are pumping but they haven't started to breathe. Swinging usually helps them to breathe, and tickling the inside of their nostrils with a piece of straw or hay works well, too. Don't give up. If you keep stimulating these babies, chances are they'll start to breathe and be perfectly okay.

Once the baby is breathing, place it in front of its dam so she can begin cleaning it. The taste and scent of her baby creates a maternal bond. At some point she may leave the baby to deliver another one. This is normal. Simply place both babies in front of her after the second has arrived.

Once all babies have been born (if in doubt, go inside the dam and check), think "snip, dip, strip, and sip." Snip the umbilical cord to a manageable length if it's overly long (2 inches is a good length in most species). Fill a shot glass, film canister, or dairy cow teat dip cup with 7 percent iodine solution, hold the container to the baby's belly so the cord is completely submersed, then tip the cup back to be sure the cord is fully covered; don't omit this important step. Strip the female's teats to make certain they aren't plugged and that she indeed has milk, then make sure the baby sips its first meal of vitally important colostrum within an hour or so after birth.

Feed Colostrum

Colostrum is a thick, yellowish milk produced by female mammals for only 48 hours or so after giving birth. It's packed with important nutrients, but more important, colostrum contains antibodies that neonates need in order to survive.

The lining of a baby mammal's intestine can absorb maternal antibodies for roughly 12 to 24 hours after birth. Babies that don't ingest sufficient high-quality colostrum lack immunity to disease until their own immune systems kick in, and they rarely survive; colostrum is that important.

If a baby can't nurse from its dam but she's still available, milk her and tube- or bottle-feed her colostrum to the baby in species-specific portions, a little at a

WEIGHING MINIATURE BABIES

You'll need a bathroom scale and a flat board at least 2 feet x 2 feet in size.

1. Bring the scale and the board to the baby.
2. Place the board on the ground and the scale on the board. Weigh yourself and make a note of the weight.
3. Quietly approach the animal, reassuring it that you aren't going to hurt it. Catch it with a firm hold.
4. Hop back onto the scale and note your combined weight.
5. Put the animal down, return to the scale, and recheck your own weight in case things moved while you were weighing the animal.
6. Deduct your weight from the combined weight to get the newborn animal's weight. Done!

A miniature calf being weighed

time, until the baby has ingested 10 percent of its total body weight in fluid (see box above for estimating weight). If you can't use colostrum from the dam, fresh or frozen colostrum from another female of the same or a similar species will do (horse colostrum for a donkey or vice versa; cow, sheep, or goat colostrum for any small ruminant).

You can freeze colostrum for up to one year. Freeze it in single feedings in double-layered resealable plastic sandwich-size bags for easy access. Avoid storing it in self-defrosting freezers; constant thaw and re-freeze cycles affect its integrity. *Never* microwave colostrum. This kills the protective antibodies. Instead, immerse the container in a hot water bath until the colostrum registers normal body temperature for that species (see page 127). Measure the temperature with a thermometer to be certain.

If colostrum is unavailable, the baby may survive if promptly given a weight-appropriate injection of an immunoglobulin (IgG) replacer such as Goat Serum Concentrate (for sheep and goats) or feed Seramune Equine IgG (for horses, donkeys, and mules). These are *not* the same thing as the inexpensive, powdered supplements based on cow colostrum that you'll find on the shelf at your local feed store; if you have nothing else, try them, but most who do report very limited success.

Colostrum-deprived babies should be exposed to disease as little as possible. Keep them in the house (in the case of bottle babies) or house them away from other livestock (in the case of dam-raised youngsters).

Aftercare

Your female will ideally deliver her placenta shortly after the birthing. In some species, such as with equines, it's vitally important that it be delivered promptly, as a retained placenta quickly leads to toxemia and laminitis. Females of most mammalian species are hard-wired to eat the placenta. Try not to let that happen lest the female choke. Instead, wearing gloves, pick up the placenta and dispose of it by burning or burying it.

Neonates are delicate creatures, so don't allow them to become wet or chilled. Keep newborns inside during inclement weather. If one should get soaked, bring it in, dry it off, and warm it up as quickly as you can.

Babies delivered in winter generally require supplementary heat to get them warm and to keep them from getting cold. Avoid hanging a heat lamp in your barn, however. Improperly hung heat lamps cause fires. For a better and safer alternative, fit neonates with body coverings that hold in heat. Commercial kid, lamb, calf, cria, and foal blankets are readily available, but it's easy and cheaper to make your own. Snip holes in a man's large woolen sock to make a stretchy sweater for miniature lambs, kids, or piglets. Or buy an infant-size cardigan sweater at your favorite secondhand clothing store, snip off the sleeves, and fit it to a calf, cria, or foal so it buttons along the baby's spine. Also, deeply bed the maternity stall with straw so babies can hunker down into its warmth.

The first manure a newborn passes is a black, tarry substance called meconium (after that expect yellow-tan stools of puddinglike texture). If the baby seems to be straining or if you don't see traces of black goo on its butt or in the stall, give the youngster all or part of a Fleet enema for human infants, depending on its size.

Stay safe. Whenever possible, warm neonates with coverings instead of heat lamps.

The young of certain species require specific immunizations and supplementations (for instance, piglets are born needing iron supplementation). Ask your vet what newborns of your species, raised in your locale, need.

Bringing Up (Bottle) Baby

For many reasons—a ewe has more lambs than she can feed, you'd prefer to drink your Miniature Jersey's milk yourself, or you simply enjoy raising your stock from day one — raising bottle babies is often part of a breeder's life.

Raising bottle babies should not be a part of a llama raiser's life, however. Bottle-feeding baby llamas has been strongly implicated in the development of aberrant behavior syndrome (ABS) (see page 339 of the llama chapter). If you absolutely must do it, discuss the project with someone familiar with this problem in llamas, such as the Southeast Llama Rescue (see Resources).

Brace large neonates like equines and calves against your leg when teaching them to nurse from a bottle.

What to Feed

Bottle babies need colostrum for the first 24 to 48 hours of their lives. After that you'll have to choose between milk replacer and real milk.

Folks who raise a lot of bottle babies almost invariably prefer real milk. Though milk replacers are carefully formulated for each species, they still often lead to

SCOURS ALERT!

If a baby (bottle- and dam-raised alike) develops diarrhea, take its temperature. If it's more than a degree or two above or below normal, call your vet. If it has liquid or explosive diarrhea, call your vet. Otherwise, temporarily reduce the amount of powder used to mix a bottle baby's milk replacer and dose both bottle- and dam-fed kids with Kaopectate at the rate of 25 cc per 100 pounds of body weight every 2 hours for 24 hours, or until its stools firm up. If it still has diarrhea at the end of 24 hours, call your vet.

MILK-BASED FORMULAS FOR KIDS AND LAMBS

If you can't find fresh goat milk (canned is not the same), consider a formula based on store-bought, full-fat cow's milk. We've used these two formulas for kids and lambs and recommend them highly.

4 parts milk
1 part dairy half & half

Mix well.

— or —

1 gallon milk
1 twelve-ounce can of evaporated milk (*not* condensed milk)
1 cup buttermilk

Remove 2½ cups of milk from the gallon of whole milk and put the evaporated milk and the buttermilk into the gallon. Mix well.

scours and bloat. We've used pure, wholesome goat milk; real milk formulas; and milk replacers — and that's how we'd rank them in descending order.

If you plan to raise bottle babies every year, invest in a dairy goat (actually, buy two because goats are social creatures and need friends). The young of virtually every species discussed in this book can thrive on goat milk, though obviously cow's milk would be the first choice for calves. Or find a goat dairy to supply your needs; even in states where the sale of goat milk for human consumption is prohibited, it's often available for use as animal feed.

You could also feed a high-quality milk replacer formulated for your species. *Do not* feed multispecies milk replacers supposedly designed for newborn mammals of all kinds, and *do not* feed soy-based products. The former won't provide adequate nutrition and may lead to scouring (serious diarrhea); the latter is difficult for neonates to digest. Always mix the product you choose according to directions on the label and measure portions; it really does matter.

How Often to Feed

Some people say it's okay to feed neonates two times a day. Don't believe it! Baby animals require small amounts of nourishment that is fed often and at regular intervals as follows:

- **Days 1 through 7.** Feed every two to four hours around the clock (feed weak neonates smaller amounts more often).

- **Days 8 through 14.** Feed every four hours around the clock.
- **Days 15 through 21.** Feed four times per day, omitting the nighttime feeding.
- **Days 22 through weaning.** Feed three or four times per day. About two weeks before weaning, omit one feeding per day; one week before weaning, omit another feeding; at weaning time quit feeding altogether.

How Much to Feed

It's impossible to make a blanket statement about how much to feed such a wide range of species. Talk to your vet or someone experienced in raising youngsters of your species. Here is how we feed a typical Classic Cheviot lamb:

- **Days 1 and 2:** 1 to 2 ounces of colostrum per meal
- **Days 3 and 4:** 2 to 4 ounces of milk per meal
- **Days 5 through 14:** 4 to 6 ounces per meal
- **Days 15 through 22:** 6 to 10 ounces per meal
- **Days 23 through 35:** gradually work up to feeding 12 to 16 ounces per meal, depending on the lamb's size
- **Week 12:** cut meals back to 8 ounces each (by then the lamb is eating grass or hay and nibbling a small ration of commercial pelleted feed)
- **Between 12 and 14 weeks:** gradually wean the lamb

How to Bottle Feed

Stand babies of larger species like foals and calves in front of you so you're both facing the same direction. Brace its butt against your legs, and brace its forequarters with your elbows (you may have to sit on a bale of hay or straw to do this). Open its mouth with your left hand, and insert the nipple of the bottle. Place your palm under its jaw, and use your fingers to keep the nipple aligned with its mouth. With your right hand, elevate the bottle just enough to keep milk in the nipple, tilting the bottle more as the baby empties it. If the baby doesn't suck, gently squeeze the bottle so a *tiny* amount of milk goes into its mouth. Allow time for the baby to swallow.

To start a baby of a smaller species, like a kid or lamb, sit on the floor with your legs crossed. Sit the baby on its butt in your lap, facing away from you. Open its mouth with your left hand, and insert the nipple. With your palm under its jaw, use your fingers to keep the nipple aligned with its mouth. With your right hand, elevate the bottle just enough to keep milk in the nipple, tilting the bottle more as the little guy empties it.

Feeding Paraphernalia

Every producer has his or her favorite bottle-feeding gear. We feed our lambs and kids with Pritchard teats attached to reused plastic water bottles. You could also use them for starting miniature donkey and horse foals. The Pritchard teat is an oddly shaped red nipple attached to a yellow plastic cap ring that is sized to fit standard screw-top household containers (including recyclable 12-ounce and 1-liter soda bottles). Its plastic base incorporates a "valve" that makes fluids flow smoothly through the nipple.

Pritchard teat

Other folks prefer human baby bottles, especially brands with ergonomically designed nipples (NUK is a good one).

Soft rubber lamb nipples from the feed store work well for most species. However, lamb nipples push rather than screw on, so eager eaters such as foals and calves sometimes yank them off the bottle at inopportune times.

Whatever type you use, enlarge the hole in the nipple so the milk flows better as the youngster sucks; otherwise the baby might get discouraged and quit before it finishes its meal. Don't make the hole so big that milk pours down its throat and the baby aspirates fluid into its lungs.

Wash the nipple and the plastic soda or water bottle after each feeding in hot, soapy water and rinse them thoroughly. At the end of a 24-hour cycle, recycle the bottle. If you don't use a clean bottle every day, sanitize the one you use at least once a day using a solution of 1 part household bleach to 10 parts water and then *very* thoroughly wash and rinse it out. But don't bleach nipples; bleach degrades them very quickly. Simply scrub them well with good, old-fashioned soapy water.

When Baby Won't Nurse

There are few things as frustrating as a bottle baby that refuses to take a bottle. These are generally youngsters that nursed their dams and are pretty sure the stuff you are offering them is poison. However, the baby has to eat, so here are some tricks to try:

- Try different nipples until you find one it likes.
- Lightly coat the nipple with fruit yogurt or sugar.
- Cup your hand above its eyes to simulate the darkness of its dam's groin.
- When the baby won't nurse after you've offered it a bottle, put the bottle away and try again later. It may take three or four of these sessions until it's finally hungry enough to eat. In the meantime, tube-feed it (have your vet tube-feed foals) if necessary. The baby *must* stay hydrated to survive.

Whatever you do, be very careful not to force milk down a bottle baby's throat; if it aspirates fluid into its lungs it may die. Be patient and persevere. In time, most youngsters do nurse.

Tube Feeding

Most everyone approaches a first tube-feeding session with fear and trepidation. Don't worry; the process is easier than you think.

You'll need the proper equipment: a feeding tube and a 60 cc syringe with the plunger removed. Soft plastic tubes designed specifically for the job work best, but an open-ended, 6-mm-diameter piece of 20-inch-long plastic tubing (ask for it at the hardware store) will also do in a pinch. Always sanitize the feeding tube and syringe between uses, and warm the tube just prior to use by immersing it in a bowl of clean warm water. Here are instructions for tube feeding a goat (the procedure is largely the same for all species):

1. To tube-feed a kid, place the tube alongside his body with one end at the animal's last rib, as shown in the illustration. Mark the tube near the animal's mouth.
2. Sit in a comfortable position with the kid facing away from you, its shoulders restrained between your knees and its body dangling down between your legs. The goal is to insert the tube into the kid's esophagus, not the trachea (windpipe), and with its body in this position, the tube is almost certain to go where it's supposed to.

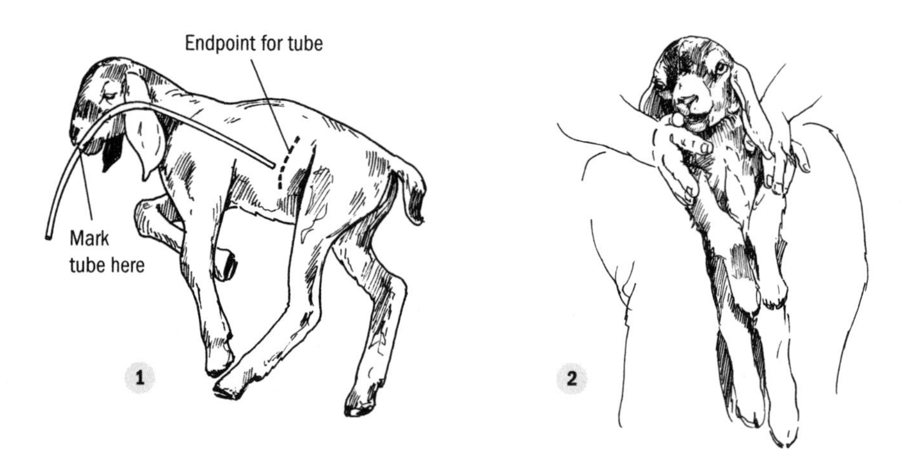

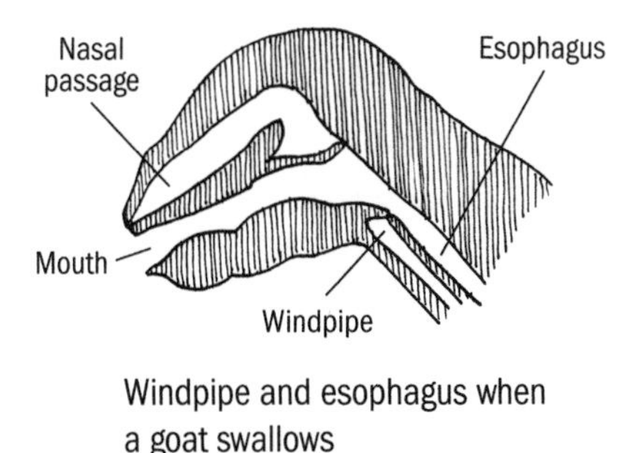

Windpipe and esophagus when a goat swallows

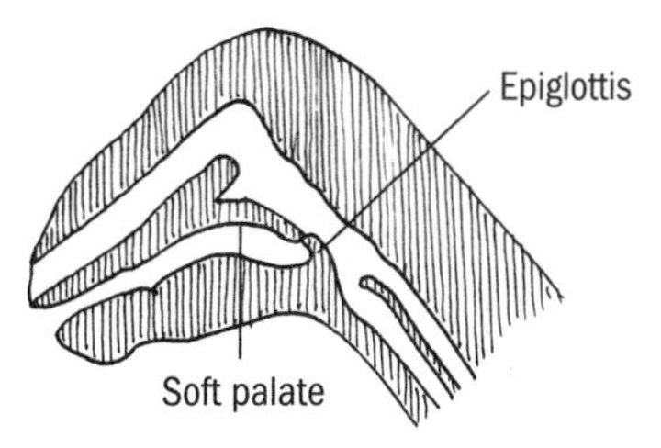

Windpipe and esophagus when a goat breathes

3. Take the tube out of the warm water, insert the rib end in the kid's mouth, and slowly feed the tube down the baby's throat. In this position, he will readily swallow the tube.
4. Check to make certain the tube is in the esophagus, not the trachea. Here's how you can tell it's in the esophagus: (1) As you insert the tube, you can see it advance along the left side of the kid's neck; (2) the kid shouldn't gag or cough (although it may briefly struggle); (3) the tube can be easily inserted all the way to the mark you made on it (a tube inserted into the trachea can't advance that far).

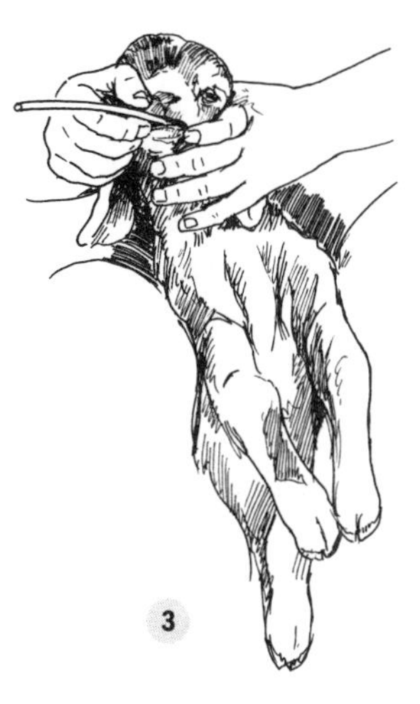

5. Attach the mouth of the feeding tube to the empty syringe. Fill the syringe with warm fluid and allow it to gravity-feed into the kid's stomach.

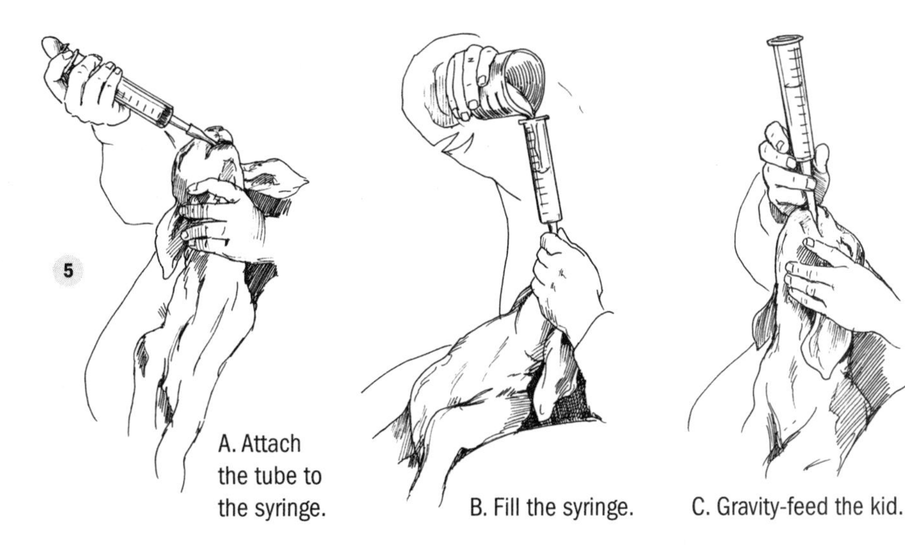

A. Attach the tube to the syringe. B. Fill the syringe. C. Gravity-feed the kid.

6. Next, detach the syringe and crimp off the end of the tube, then quickly and smoothly withdraw the tube (don't allow fluid to leak from the tube as you pull it out).

Pan Feeding

Some people prefer to pan- or bucket-feed piglets, calves, and foals because it's faster and easier than bottle feeding.

Most calves and foals learn to drink from a pan quite readily if you place your finger in its mouth, then, while it's sucking, raise up a small bowl containing milk or milk replacer to its muzzle. Slowly remove your finger from its mouth while it's drinking. If it stops drinking, repeat the procedure of sticking your finger in its mouth and letting it suck until it's drinking by itself. Always bring the milk up to the baby; never force its head into a bucket. Once the baby has mastered the skill of drinking, you can hang a shallow bucket in its stall and replenish the milk at every feeding. Cleanliness is important: buy two buckets so you can thoroughly wash and disinfect one while the other is in use.

Piglets are easier still. Simply hold the piglet and gently dip its snout in a shallow container of milk. It will fling its head and fuss, but in just a few tries most piglets catch on.

12

Got Milk?

WHETHER OR NOT YOU MILK a miniature cow, doe, or ewe for your own consumption, if you're interested in raising miniature livestock, this chapter should interest you. That's because pure, raw milk, especially goat's milk, is the ideal milk to feed young livestock raised on a bottle. Goat's milk is readily digestible by most species, even human babies.

There are countless milk replacers designed for baby livestock, both species-specific and generic, and some are infinitely better than others. But the bottom line is that even the best milk replacers don't agree with a surprising number of neonatal animals. When on milk replacer, these youngsters struggle with scours and bloat and often fail to thrive. That doesn't happen when they're given real, wholesome milk. Full-size and miniature cows and goats provide high-quality nourishment for the young of every species listed in this book. Why not add a milk-maker to your farm? A bonus: Your family can reap the benefits of real milk, too.

MILK: IT'S NOT JUST MOO JUICE

When most people think "milk," they visualize full-size dairy cows. However, miniature cattle, large- and small-size goats and sheep, yaks, water buffalo, reindeer, moose, horses, donkeys, llamas, and camels provide wholesome milk and dairy products, too.

According to United Nations figures published in 2001, 84.6 percent of the world's milk was produced by cattle; 11.8 percent by water buffalo; 2.1 percent by goats; 1.3 percent by sheep; and the remaining 0.2 percent by an assortment of other mammals for a total of 585.3 million liters of milk.

Milk for Your Family's Table

Some people keep dairy animals because they prefer to put unpasteurized, nonhomogenized, raw milk on the family dinner table. Those who drink it praise raw milk; opponents claim it's unsafe to consume, even in end products such as butter and cheese. You'll have to reach your own informed decision, but like millions of other dairy consumers around the globe, at our house we drink it raw.

Others keep dairy livestock for peace of mind. They know the health status and medical history of the animals that produce the milk they serve to their families. They know how the milk was handled and stored and how and when it was worked into butter, yogurt, and cheese.

But if you don't have a very large family and have no plans to sell your milk, a full-size dairy cow will give you more milk than you need. An average full-size Jersey, one of the smallest of the full-size breeds, produces five to six gallons a day. That's a *lot* of milk! What's a health-conscious, do-it-yourself dairy lover to do? Enter miniature dairy cattle such as Miniature Jerseys, Belfairs, and Dexters. Their output is ideal for a small family's needs. If it's sheep or goat milk you prefer, the full-size versions usually meet a family's needs.

Miniature cows, goats, and sheep all furnish luscious, homegrown milk.

Getting Started with Dairy Stock

No matter what type of dairy stock you choose, certain facts apply across the board:

- To begin producing milk and keep producing it, an animal must periodically give birth. Ordinarily, dairy livestock deliver offspring once a year.
- Some individuals "milk through," meaning they're capable of milking for a longer time than the norm (sometimes for two to three years and longer) before being rebred, but eventually they will have to be bred and you'll have to deal with the logistics of raising or selling the resulting offspring.
- Dairy animals must be milked every day at the same time(s), in the same place, preferably by the same milker; animals never take long weekends or sleep in.

That said, let me qualify the statement slightly. For maximum output, dairy livestock is milked every day, twice a day, at 12-hour intervals. If this doesn't suit your lifestyle, however, you may allow your dairy provider to raise her babies (instead of separating them and feeding them by bottle), and milk her just once a day. The usual protocol for milking once a day is to pen the offspring separately at night and milk the mother first thing in the morning. After milking, her young rejoin her and nurse until evening, when they're shuttled off to their separate quarters again.

Is this cruel? Not at all. Modern dairy animals are bred to give considerably more milk than their natural offspring require. Babies can be fed quality feed in their own hideaway, and in many cases they will grow faster and bigger than if they were raised solely on mother's milk.

TRY BEFORE YOU BUY

All breeds and types of cattle, goats, and sheep give milk. You don't have to buy a specialized dairy cow, goat, or sheep if you don't need a bountiful supply of milk. For instance, one year we milked a friendly, heavy-milking Miniature Cheviot (a sheep breed kept mainly for meat and fiber). She didn't give a whole lot of milk, but it was rich and ultra-yummy.

So why not pick an animal from your flock or herd, then tame and milk her? Choose an easygoing individual with a large, lump-free udder and good-size teats. She might provide all the milk you need; if not, you'll know if you enjoy dairying before springing for specialized dairy stock.

No cow, doe, or ewe provides maximum output year-round; her milk volume will decrease as her lactation progresses. She'll also require a period of downtime between lactations when she won't be milking at all. So to count on a continual supply of fresh milk, you'll need more than one dairy animal.

To milk them you will need proper equipment (see page 215) and a milking area separate from their living quarters. You can't cut corners to produce quality milk.

Cows

For all of their appealing qualities, such as beauty, personality, and the ability to give a lot of milk, cows also have their drawbacks. Even a miniature cow is relatively large and bulky, and when cows are feeling cantankerous, they can be a lot more beef than you care to wrestle with.

A small cow like this cute Dexter provides enough milk for a typical family of three or four.

Cows also produce a lot of loose, fly-attracting manure. Their large hooves quickly sink deep pocks in damp pastures and turn small enclosures into stinky, mired messes. Unless you practice impeccable sanitation, close neighbors may object to a cow.

Miniature cows are relatively expensive to purchase and feed, they require more room and better pasture than goats and sheep, and relatively few mature cows are accustomed to hand milking, so if you're a first-time milker, expect to conduct a lengthy search for a trained house cow (or choose something smaller like sheep or goats).

Goats

Milk and dairy products from healthy, well-fed goats are delicious. Contrary to popular opinion, properly handled goat milk neither smells nor tastes "goaty"; in fact, it tastes like full-cream, home-processed cow milk.

The nutritional differences between cow and goat milk are negligible. Goat milk is slightly higher in calcium, milk solids, and a few vitamins and minerals, but their protein and carbohydrate counts are practically the same.

Smaller fat globules make goat milk easier to digest. And goat milk is whiter because it lacks the carotene that turns the fat in cow's milk a pale, creamy yellow (goats convert carotene to vitamin A).

A miniature doe can provide enough milk for two people, but because goats are social animals and pine without company, it's better to keep more than one goat. Excess milk can be made into wonderful goat milk cheeses. Some, like chèvre and goat queso blanco, are so easy that anyone can make them.

Because of miniature goats' general joie de vivre and all-out affection for their caretakers; their compact, manageable size; low space requirements; and ease of milking, they are arguably the beginner's best choice for efficient, user-friendly, home dairy animals.

Goats produce far less manure than cows, they don't attract flies, and their hooves don't stir up a mess. Because they are browsers rather than true grazers, they thrive on pasture where cows would starve, ridding fields of brush, briars, brambles, and hard-to-rout noxious weeds such as star thistle, leafy spurge, and multiflora rose.

However, intelligent, ingenious goats require secure fencing to contain them. For their safety, never tether any animal in lieu of proper fencing!

A milking stand makes milking goats and sheep a whole lot easier.

DON'T KEEP JUST ONE

Virtually every animal is happiest when kept with others of its species. A single cow or horse is sad, but it learns to cope. Other species such as sheep, goats, and llamas *need* companionship, preferably others of their kind. Sheep, especially, being a flocking species, will fret and stress unless provided with a pal or two. Please be kind: Never keep *any* animal all by itself.

MILKING TRIVIA

- Cows have four teats; most sheep and goats have only two.
- Cows and goats are generally milked from the side, but sheep are milked from the rear. Goats are traditionally milked from behind in parts of Europe.
- Between milkings, milk accumulates within the udder in structures called alveoli before passing through a series of ducts into the gland cistern, the udder's largest collecting point.
- The gland cistern is connected to the teat cistern, a cavity within the teat where milk pools until milking time.
- A group of circular sphincter muscles surrounds the orifice at the tip of each teat. When an external force (a calf's mouth or a milker's hands) overcomes the strength of the sphincter muscles, they open and stored milk begins to flow.

Sheep

Producers in the United States rarely think of milking sheep, yet many of the world's greatest cheeses are crafted of sheep's milk. Sheep give less milk than cows or goats do, but what they give is marvelously rich, nourishing, and ideally suited for making cheese and ultra-creamy yogurt. High-protein, calcium-rich ewes' milk contains 6.7 percent fat and 18.3 percent solids, compared with 3.5 percent and 12.1 percent for cows' milk and 3.9 percent and 11.2 percent for goats' milk. It takes 10 pounds of cows' milk and only 6 pounds of sheep's milk to craft a 1-pound brick of cheese. And a bonus with sheep: You can harvest a wool crop, too.

Because their udders are situated farther back than modern dairy goats' udders, sheep are traditionally milked from the rear.

Milking Basics

Hand milking is a team effort between a milker and the animal he milks. When the milker preps his animal by washing her udder, the hypothalamus in her brain signals her posterior pituitary gland to release a hormone called oxytocin into her bloodstream, causing tiny muscles around milk-holding alveoli (hollow structures in her udder) to contract. In other words, she "lets down her milk."

If the animal becomes excited or frightened or experiences pain, her adrenal gland secretes adrenaline, which constricts blood vessels and capillaries in her udder and blocks the flow of oxytocin needed for effective milk letdown. Good hand milkers are efficient and patient. They approach milking in a low-key manner, and they practice good milking techniques.

Let's imagine you're milking a goat, but whatever the species, the same basic protocol applies. You will need:

- Freshly washed hands with short fingernails
- A recently sterilized, seamless, stainless steel milking pail
- Udder wash and paper towels (or unscented baby wipes work just fine)
- Teat dip or an aerosol product such as Fight Bac
- A teat dip cup or a pair of disposable 3-ounce paper cups (not needed if you use Fight Bac)
- A strip cup with a dark, perforated insert
- A milking stand set up in your milking area with grain waiting in the feed cup

It is a good idea to practice before the fact. Collect a strong latex glove, and follow these steps adapted from my favorite foaling book, *The Complete Book of Foaling: An Illustrated Guide for the Foaling Attendant*, by Karen E. N. Hayes, DVM, MS (see Resources): Poke a tiny hole into one of the glove's fingertips (this is your makeshift teat), and then pour in 6 ounces of water. Hold the latex "udder" with your free hand. Now, follow the protocol listed below. That's it! Continue milking till the glove runs dry.

How to Milk

1. Lead the doe to the milking stand, ask her to hop up, and secure her head in the stanchion.
2. Cleanse her udder using your favorite prepping product. Dry each half using a paper towel (omit this step if you're using baby wipes), then massage her udder for 30 seconds to facilitate milk letdown.

3. Squirt the first few streams of milk from each teat into the strip cup, and examine it for strings, lumps, or a watery consistency that might indicate mastitis (an inflammation of the udder usually caused by infection [see page 136]).
4. Place the milking pail slightly in front of the goat's udder, sit down next to her, and grasp a teat in each hand (when milking a cow, most people milk the nearest teats first).
5. Trap milk in each teat by loosely wrapping your thumb and forefinger around its base (see illustration 5a). If the teat is larger than a handful, grasp it closer to the orifice end, taking care not to catch the animal's bag. Gently nudge the doe's udder with the upper edge of the same hand, and then close off the teat with thumb and forefinger (never allow milk to backflow into the udder). Now, still holding the teat closed, squeeze with your middle finger (see illustration 5b), then your ring finger, and then (if you haven't run out of teat) your pinkie, in one smooth motion, to force milk trapped in the teat down into your pail (illustration 5c; never *pull* on an animal's teats). Relax your grip to allow the cistern to refill and do it again.
6. Alternate teats, squeezing one teat while the other refills.
7. As her teats deflate and become increasingly more flaccid, gently bump or massage the goat's udder to encourage additional milk letdown. Don't finish by stripping the teats between your thumb and first two fingers; this hurts and annoys the goat.

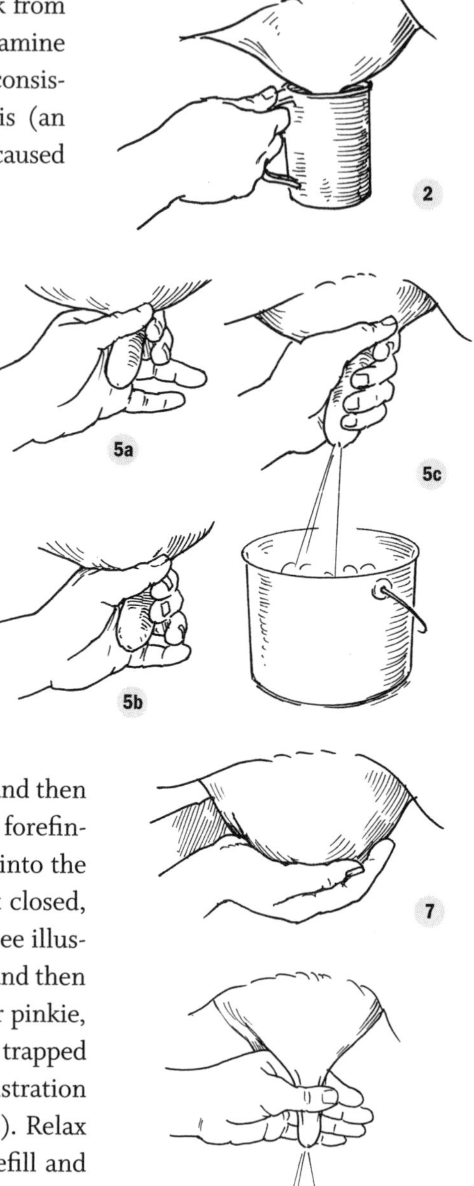

NEVER pull on an animal's teat.

8. Pour enough teat dip into the teat cup (or paper cups) to dip each teat in fresh solution and allow the teats to air dry. Alternately, spray the end of each teat with aerosol product until a bead of fluid forms on each tip. Then release the goat and let her jump down.

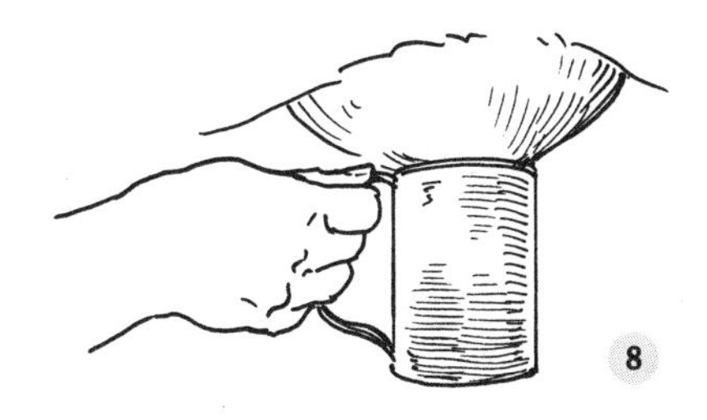

Processing Tasty Milk

If you have more than one cow, doe, or ewe to milk, pour the milk from each into a covered stainless steel milk tote, proceeding through the group until you're done. Then go straight to wherever you process your milk and get started.

Pour the milk through a stainless steel strainer lined with a milk filter (you can use coffee filters, but they cost more and don't work nearly as well) into sparkling clean, covered glass containers (lidded canning jars work well), then cool the milk as quickly as you can. Some folks place containers in the freezer for 10 minutes and then transfer them to the refrigerator. Others immerse them in a sink or bathtub of ice water. Secondhand plastic food containers that have been filled with water and frozen make dandy large ice cubes for this purpose. Handled in this manner, raw milk should stay good in the fridge at 40°F (4°C) or cooler for five or six days.

Cleanliness Counts

If you've ever tasted nasty, "goaty"-flavored goat milk or punky-tasting raw cow's milk, you're probably thinking, "Why would I go to a lot of bother for *that*?"

Rest assured, properly handled raw milk is so fresh and rich tasting that once you've tried it you'll never want store-bought milk again. The key words are: properly handled.

Everything that comes in contact with milk must be squeaky clean. The animal's belly should be brushed and her udder wiped before milking. The stainless steel bucket into which you milk and the items used to process fresh milk must be sanitized in 10 parts water to 1 part bleach solution after every use.

Milk should be processed as quickly as possible. Milk, go directly to the kitchen, and set to work. Strain the milk and cool it down fast.

Watch What You Feed

The forage and feed a cow, doe, or ewe eats and the scents she inhales can flavor the milk she gives you. Keep a buck in rut with your doe and she'll give nasty milk; housing her in a stinky barn will produce the same effect. According to the University of California Cooperative Extension's bulletin "Milk Quality and Flavor," 80 percent of the off-flavors in goat milk are feed related. This is true for cow's milk, too.

The best way to avoid off-flavors is to eliminate suspect plants from your milk producer's diet, and don't give feedstuffs that are known to sometimes flavor milk (such as alfalfa, soybeans, rye, rape, turnips, cabbage and kale, and clover) within five hours of milking. Plants to avoid entirely include bitterweed, buckthorn, buttercups, chamomile, cocklebur, cress, daisies, fennel, flax, horseradish, marigolds, mustards, onions, pepperwort, ragweed, sneezeweed, wild carrot, wild garlic, wild lettuce, and yarrow.

And keep in mind that when milk cows, goats, and sheep ingest certain plants that are poisonous to us but not to them, residues end up in your milk supply. Abraham Lincoln's mother died after drinking cow's milk tainted with white snakeroot. It can still happen, so be careful.

Raw versus Pasteurized?

All milk and the products made from it have the potential to transmit disease-causing organisms to humans, among them pathogens that cause diseases such as tuberculosis, brucellosis, diphtheria, scarlet fever, listeriosis, *E. coli*, and salmonella.

Yet millions of people around the world consume raw milk every day without dire consequences. Visit some Web sites or read books on both sides of the raw-versus-pasteurized milk controversy before you decide.

Pasteurizing Milk at Home

If you choose in favor of pasteurized milk, buy a pasteurizer; if you follow the manufacturer's instructions, you'll have properly processed milk every time. Companies like Hoegger Goat Supply, Caprine Supply, and Hamby Dairy Supply sell good ones (see Resources under Miniature Goats).

Barring that, there are several ways to pasteurize milk using everyday kitchen equipment. According to Michigan State University Extension's publication "Milk Pasteurization," the safest method is this:

- Heat the milk to 165°F (74°C) in a double boiler, and hold at this temperature for 15 seconds while stirring constantly.
- Then cool it very quickly. Set the top of the double boiler in cold water, and stir until it reaches 145°F (63°C).

- Add ice to the cooling water to cool the milk further, stirring occasionally, until the temperature of the milk falls below 40°F (4°C).
- Store the cooled milk in clean, covered containers at a temperature below 40°F (4°C) until used.

Save Some for Later

When your dairy animal produces more milk than your family or bottle babies can use, freeze some for later. Raw milk freezes better than pasteurized milk, and sheep and goat milk freeze better than the milk from cows. But all milk can be safely frozen with minimal loss in quality — if the milk is handled right.

While we don't usually endorse storing milk in plastic (it is harder to clean, and some plastics flavor milk stored in the refrigerator), glass containers and freezers don't necessarily mix. We freeze our surplus milk in secondhand 20-ounce drinking water bottles for several good reasons:

- They are readily available (we drink bottled water anyway).
- They can be rinsed and recycled after a single use.
- They store well in the freezer.
- The bottles are the perfect size for our two-person family or for feeding young lambs and kids.

When using plastic water or soda bottles, briefly immerse them in bleach solution, then thoroughly rinse and allow them to dry, upended, in a clean dish strainer for 30 minutes to an hour before filling.

If you freeze milk as soon as it's cooled and you thaw it slowly in the refrigerator, little or no separation of solids and whey will occur. Don't freeze milk that's been sitting in the refrigerator for more than a few hours, and never thaw milk in hot tap water! If you do and the milk separates, it's still safe to drink but it will look somewhat disgusting.

THIS FARMER'S TAKE

We freeze excess fresh milk in plastic water bottles. When my husband makes the 500-mile trip north twice a year to visit his siblings, he packs along a supply in coolers to share with them. Packed in coolers, it remains frozen solid for as long as two full days.

13

The Business End

IF YOU'RE GOING TO SELL your miniature livestock, you'll need to market your business. If you don't have a lot of money in your budget for large-scale advertising purposes (and few of us do), there are lots of ways to publicize your business without breaking the bank. The first thing you should do is to establish an Internet presence.

The Big Billboard in the Sky

According to usage statistics posted at Internet World Stats (see Resources), roughly 74 percent of North Americans use the Internet, and some of them want to buy what you have to sell. If you're in business, you need a Web site. The Internet is a customer's quickest way to locate the goods he needs. Make it easy for him to find you.

With a Web site you can show your livestock and present your message to an unlimited number of people from all around the globe, without leaving your home or paying for a single expensive ad. You can present much more informa-

CHOOSING A BUSINESS NAME

Before creating a Web site or mapping out a business plan, choose a business name. Make certain people can remember, spell, and pronounce it. Don't choose a name another producer within your breed is already using (contact appropriate registries and ask). Google "trademarked names search" to find out if your chosen preferred name is chosen; if it isn't, consider having it trademarked in your name.

tion than you could in an ad or a typical brochure, and with most Web site hosts these days, you can change it whenever you want, from home, 24 hours a day. You'll automatically target an audience interested in what you have to sell (if they weren't interested, they wouldn't visit your site).

Maintaining an Internet presence is a remarkably cost-effective way to market and promote your livestock, especially if you do the work yourself. A comprehensive, do-it-yourself Web site takes an hour or two a week and $200 to $400 per year to build and maintain, although you'll have to master certain skills to do it well. If you're a technophobe, you can always hire a developer to create and maintain your site. Whichever route you choose, make sure the site is done well.

Build It Yourself?

If you'd like to hire an outside party to build your Web site, first surf the Internet and pick out other sites you like. Contact whoever built them. The company or person's name is generally found at the bottom of a Web site's home page. If it isn't, e-mail the business and ask.

Ask each of these Web designers about their rates and policies. A base fee gives you how many pages and how many pictures per page? What else is included in the base price and what, if any, features are available at extra cost? How many updates are included? How much are additional updates? Will the developer work with you or will he do everything his way (some designers favor flashy tools, but technology doesn't sell livestock). How long will it take to get your Web site up online? Will the developer list your site with major search engines or is that your job? Does he offer a satisfaction guarantee? Ask for URLs (Web site addresses) to other sites he's built. E-mail their owners. Are they happy with his work, and if they aren't, why not?

If you're not computer illiterate and have the time, however, why not build your own site? You know what you want, and you're much more likely to get it if you do the job yourself. It also costs less if you build your own site; updates are easy and free; and the satisfaction of implementing your own design is priceless.

There are several avenues you can take to build your own Web site. Schools and libraries offer free or inexpensive weekend or evening computer classes, including classes on site building and Internet commerce; county extension services hold Internet farm-marketing seminars; the Small Business Administration (see Resources) helps farm businesses get online; and there are dozens of up-to-date, easy-to-understand books designed to guide beginners through the intricacies of crafting a Web site.

HOW DO THEY FIND YOU?

According to the United States Department of Agriculture Marketing Service publication "How to Direct-Market Products on the Internet," when Internet customers were asked how they located sellers' Web sites, 57 percent of the respondents replied that they used search engines. Twenty-eight percent used e-mail messages, 35 percent other Web sites, and 28 percent word of mouth.

And there is WYSIWYG (What You See Is What You Get) Web page–building software. With this software, you don't need to use the complicated programming language of HTML. This sort of software is a must-have for novice site builders, and you needn't break the bank to buy the kind you need. Popular programs such as Dreamweaver and Microsoft Expression build wonderful high-tech Web sites, but for many small businesses that's a lot more bang than you need. For our purposes, free or inexpensive programs such as Trellian WebPage and PageBreeze work just fine. I build and maintain our Web sites on a vintage Mac using Claris Home Page 3.0 software that was written in 1997 and purchased outdated but brand new at eBay for $12.

Web sites sell something or distribute information; the best of them do both. Decide what yours will be all about. Plan each page on paper, based on the following principle: It's easier to build from a blueprint than to backtrack and fix things later on.

Choosing a Domain Name

Your domain name is the "who" segment of your Web site's Internet address. Register it right now—catchy names are purchased and taken out of circulation every day, so don't lose yours because you waited too long. Use your business name if you can. If your business name — let's say it's Ouachita Lowlines — is already taken, try customizing it by adding dashes (Ouachita-Lowlines), underscores (Ouachita_Lowlines), numbers (OuachitaLowlines22), or additional descriptive words (OuachitaLowlineAngus). If it still doesn't work, brainstorm a clever, easy-to-remember name relating to the type of stock you breed. As we go to press, these names are up for grabs: TastyLowlineBeef.com, BigProfitsSmallCattle.com, and BestBeefOnEarth.com. Whatever you decide, be sure to keep it simple. This will make it easier for those visitors who type in URLs to access internal Web pages.

If OuachitaLowlines.com is taken, you might be able to register it as Ouachita Lowlines.net, OuachitaLowlines.biz, or OuachitaLowlines.us. To see if your favorite choices are already taken, visit Instant Domain Name Search (see Resources).

Keep in mind that when typing domain names into a browser bar, you needn't stop to capitalize its components; to the Internet, www.ouachitalowlines.com is the same as www.OuachitaLowlines.com. However, when adding your URL to e-mail signature lines, directories, or printed promotional items such as business cards and brochures, capitalizing helps folks recognize your name.

Before registering your domain with an independent registration service, remember this: Some hosting servers include domain name registration in their monthly service charge. Check first, and save yourself some bucks.

Finding the Right Hosting Service

The cardinal rule of Web site construction is to never use free hosting services to host your business Web site! People despise the advertising banners and pop-ups that are part of sites on freebie services. They're annoying, and they "freeze" some computers.

Freebie services equate with cumbersome URLs. Which will your customer remember: www.OuachitaLowlines.com or www.freeserve/~freesites/Ouachita Lowlines.com? Also, many editors encourage authors to omit resources with cumbersome URLs (check my resources — you'll see).

And they're not always free! "Freebie" services have been known to abruptly switch to paying status. If it happens, you'll have to shell out some bucks or see your site go down the drain.

Choose a hosting service you can live with. If the server you choose is constantly down, no one can view your Web site. Call or e-mail the customer service person at hosting services you might like to deal with. Ask these questions before you decide:

- How much bandwidth do you get for your money? Bandwidth is the number of bytes (usually expressed in gigabytes or GB) your customers' visits consume each month. When you run through your monthly allotment, your site goes down until the next billing period begins.
- How fast does the service connect to the Internet? Connection speed varies; some services are faster than others.
- What is their support system? Will you have access to clearly worded online documentation when you have questions? Can you contact support

staff by e-mail or phone? At what times? How fast are e-mails answered? Are tech support personnel willing to walk newbie designers through problems, using everyday language in lieu of technical jargon?

An easy-to-use hosting service is priceless, and switching from one to another is a pain in the pants. Do the homework up front so you don't have to do it again.

Designing the Site

Whether you build your own Web site or a developer does it for you, you'll want it to be eye-catching, effective, and user-friendly. Most Web sites aren't.

According to the USDA Agricultural Marketing Service publication *How to Direct-Market Farm Products on the Internet* by Jennifer-Claire V. Klotz, 75 percent of visitors to business Web sites will revisit the site if it has high-quality content, while 66 percent will revisit the site if it is easy to use. Fifty-eight percent won't revisit slow-loading Web pages, and 54 percent avoid dated sites. Only 12 percent revisit a site to view cutting-edge technology. By the same token, Netsmart found that 97 percent of consumers go to commercial Web sites to be informed, not entertained.

As farm-marketing guru Ellie Winslow, author of *Marketing Farm Products: And How to Thrive beyond the Sidewalk* and *Growing Your Rural Business from the Inside Out,* says, "Over the last decade Web sites in general have gotten more glitzy. If the Web site is for business, don't distract your visitor and don't wear him or her out with all the moving icons, streaming banners and other technically advanced stuff that doesn't actually promote your marketing goals."

For selling purposes, Web sites should present products and facts in a plain but attractive, easy-to-use manner. Keep in mind these important considerations:

Don't assume your visitor is Internet savvy. Many people aren't. If they don't understand mouse-over images and pull-down menus, they'll miss a lot of content, so place important information where people can find it easily.

TYPICAL TIME ONLINE

According to Nielsen Net Ratings compiled in 2005, each month the average Internet user visits 59 domains; views 1,050 pages, allocating 45 seconds for each page; and spends about 25 hours online. Each surfing session lasts 51 minutes.

Take measures to prevent your Web site from crashing. If your Web site crashes a visitor's computer, he probably won't be back. This is good reason to avoid free hosting services and fancy designs that include such things as animations and music.

Avoid using frames. Sites that use frames (material that is on every page of the Web site, usually going down the left side) are cluttered, confusing, and rarely print out well. They load poorly on some computers, often overfilling the screen and sometimes blocking access to your site's best features. If you think you must use frames, don't place contact information near the bottom; that's the part that generally doesn't fit. Or better yet, design two sites linked to your home page, so visitors can choose the version they prefer.

Choose fonts with care. Make certain they're large enough to be easily read on both Macs and PCs and in all of the standard browsers. Avoid nonstandard fonts; if in doubt, use Arial or Times New Roman.

Avoid patterned backgrounds. They tire visitors' eyes and copy gets lost in the morass. Strive for clear contrast between font and background colors. Light-colored copy against a dark background also tires many viewers' eyes, and while it looks good online, it doesn't print well from older versions of popular browsers.

Your home page should state what your business is about and what it has to offer your customers. The page should be welcoming and uncluttered and load in a snap. You can present only one first impression, so make it count.

Provide a site map and link to it from all of your other pages. Visitors stay longer when they know what's there to see.

Compose headlines and titles for pages. Search engines frequently link to internal Web pages; if there is no obvious description of the content on the page, your visitor feels lost.

Place your name, contact information (including physical address), and logo on every page, and link the logo back to your site map or home page. This will help visitors who don't enter your site via the home page, and it will help boost local business. Florida buyers might not be interested in buying livestock from Oregon, but if you're only a county or two away from a buyer, he's much more likely to call and arrange a visit.

Write good copy. Buy a marketing book, visit a Web site, or take a class on effective sales ploys before you put up your site. I recommend reading Ellie Winslow's excellent farm-marketing books (see Resources).

Triple-check all copy for typos and misspellings. Don't make nitpickers grit their teeth. Typos and misspellings are also interpreted as a lack of professionalism or care; you want to do everything you can to make the best first impression.

Add educational content — the more the better. Give people a reason to revisit your site. They'll be back the next time they need information, and they'll tell their friends. Prime advertising? You bet!

Register your site on major search engines. Don't wait for search engine Web crawlers to find you.

Keep your content sharp and fresh. Tweak your sales lists, upload new images, or add new late-breaking news or educational items. Change something on your home page every day to keep visitors coming back in a timely manner. For example, post a daily photo (it's easy in this age of digital cameras) or a witty or inspirational quote (there are books and Web sites devoted to this sort of thing). You'll develop a following, and they'll tell their friends.

Spend time each week checking links, especially those that lead to outside sources. One hour a week spent on Web site maintenance pays huge dividends in customer approval.

Respond as soon as you can if someone e-mails by way of your Web site. When marketing through e-mail, punctuality counts. According to a study reported in "How to Direct Market Farm Products on the Internet," 40 percent of the top e-commerce Web sites did one of the following: took longer than five days to respond to e-mail, never replied, or simply weren't accessible by e-mail. People want results right now.

Aggressively promote your site in every way you can. People are much more likely to visit your site if they've heard about it than if they stumble upon it. Add the Web site address to your e-mail signature line so it appears on every e-mail you send. Use it on your business cards, brochures, and road signs. Letter it on your truck and have it emblazoned on the T-shirts you wear to the store. Add links to the Web sites of other livestock producers, and have them in turn link to your site. As a farm or ranch entrepreneur, your Web site is your most important marketing tool. Make it a great one, and consider your time (and money) well spent.

Loading Times

According to recent studies, the average Internet user spends 9 minutes a day — that's a full 55 *hours* per year — waiting for Web pages to load. For each person who patiently waits 6 minutes for a gizmo-laden Web page to load, another will allow 30 seconds, then leave.

BLOG!

Another good (and inexpensive) way to market on the Internet is by blogging.

The word *blog*, for those who don't know, is short for weblog. A blog is an online journal found on the author's site or at a blogging site such as Blogger, LiveJournal, or WordPress.

Blogs consist of a single page (usually an attractive template chosen from the blog site's menu) onto which the blogger uploads new content on an ongoing basis. New entries appear at the top of the page, recent posts appear below the new content, and old posts are archived and listed along one side of the page. Most bloggers add images and video to their written content.

Blogs are interactive, meaning visitors can post comments to which the blogger replies in turn. This is the beauty of blogging as a promotional tool: It's an attractive venue whereby livestock producers can talk about their animals and breeding programs, upload recent pictures and videos, and answer questions where others who might be wondering the same things can read them. All entries are archived so they can be seen at any time by new users, and the blog usually costs nothing but the blogger's time.

Some free blogging sites run ads on their blogger's pages. Visit sites and review their policies until you find one you'd like to work with. Larger entities like Blogger walk new subscribers through the process of subscribing and putting up a blog. Most offer a slew of tutorials, too.

Use your blog as a promotional tool, not simply as an online diary. Tell readers how to do something, and upload images to show how. Discuss your miniature stallion's famous ancestors, and add pictures so readers know what they looked like. Find references to your breed in antique books, type them out, and upload them. Review a new item of tack purchased for showing your miniature donkeys. Show readers how to make queso blanco from your Belfair cow's rich milk.

However, entries needn't — and *shouldn't* — be too long. Make them short and punchy; cut them into segments (part one, part two, and so on). This type of content keeps readers coming back for more.

The most important thing to do is to post often. To help readers find your farm blog, print its URL on your business card, provide links from your Web site, and add it to your e-mail signature line.

Many Web developers assume every Internet user has Direct Satellite Link service. They don't! In rural areas (in many cases, precisely the regions where you want to market your wares) dial-up is still the norm, often through antiquated phone lines.

And since many livestock owners hang on to dated computer equipment, preferring to use their dollars to purchase livestock and equipment instead of faster hardware, it's important to design fast-loading pages that grab visitors' attention lest they leave to peruse another seller's site.

Savvy Web designers recommend that a Web page's elements add up to single page downloads no greater than 180 K. This precludes huge pictures and fancy technical applications, so plan your layouts within those constraints.

Or if you love bells and whistles, experts suggest building a dual Web site linked through a common home page. This allows visitors to click a button to enter a no-frills version or one resplendent with music, glitz, and animations. Add hit counters; you'll be surprised which one most visitors choose.

Images

Resize large photos and other graphics so they're attractive yet quickly downloadable. Software such as Adobe PhotoDeluxe or PhotoShop, Microsoft Photo Editor (it's part of Microsoft Office), or any of dozens of free and inexpensive photo-editing software downloadable online will help you prepare your images in style. Resolutions in the 72 pixels-per-inch range are ideal for Web site applications. That size looks great on the computer screen yet is small enough to load quickly.

Plan your pages carefully to quash photo glut. Start with a spare, uncluttered introductory page for each product (in this case perhaps an individual animal) and link to additional pages of photos and information.

The cardinal rule for using photos on your Web site is to choose good ones! Anyone can learn to take good pictures if they're patient and persistent. Barring that, hire a professional to shoot clear, well-composed pictures of your livestock. Or if you see photos you like on other Web sites, call or e-mail the owners to ask permission to use them. Don't assume everything online is up for grabs.

Every Good Business Needs a Card

Short of a great Web site, dollar for dollar, the most effective advertising tool is a top-notch business card. Cards are inexpensive and exquisitely portable, and if designed right, people hold on to them for a long time. And there are scores of ways to promote your business with good cards. Buy lots — the more you order (usually in multiples of 500), the less your cost per card. One thousand business

Choose well-designed business cards printed on quality stock to make a good first impression.

cards is a working minimum if you plan to promote aggressively using cards. Never leave home without business cards (and keep a stash in your car or truck in case you run out)!

Purchase inexpensive, take-one business card holders, and ask businesses to display your cards near their cash registers. Target veterinarians and farm stores in your area, but place them in other frequently visited local businesses, too.

Tack cards to every bulletin board you pass, be it at the supermarket, coin-operated laundry, work, library, feed store, or sale barn. Use pushpins so you can stack them; this encourages interested parties to take one. Return often and replenish the supply.

Tuck a business card into every piece of mail you send out — personal correspondence, invoices, even bank payments and the electric bill.

Use them as calling cards by handing them to people you meet. Ask them to take several cards and give the extras to their family and friends. And take the time to *look* at cards you're given in return. Put them away carefully in an attractive holder; don't cram them in your pocket or purse.

Buy a conference-style name tag holder, insert your card, and wear it on your lapel. Scan and add it to your e-mail as an attachment. Use your card as a camera-ready ad for publications; it'll save you money on setup fees!

Quality Counts

The only thing worse than not using business cards is distributing bad ones. Soiled, poorly designed, el cheapo cards affixed to bulletin boards or stacked on the counter at the feed store create a poor impression of your business. And while it's tempting to settle for free cards you can order online, don't. The advertising on the back of free cards definitely distracts from your image. "Cheap"

WHICH BUSINESS CARD STYLE IS BEST?

Business cards can be ordered in a wide range of styles. The ones most commonly seen are:

- Colorful cards on glossy stock. These eye-catchers work especially well on bulletin boards where they have to compete with lots of other business cards. They usually feature photos or fancy artwork. Very nice!
- Tasteful, no-frills cards with raised print on matte or linen stock. They have one or two colors of ink, just the facts and perhaps a logo; nothing more. They're for folks who like an understated, traditional look. (These are the cards I use.)
- Folded cards, usually colorful and usually on glossy stock. These are good for when you need just a little more room than what's available on the traditional cards.
- Magnetic cards of either style. Buy them already printed or affix your regular card to peel-back adhesive business card magnets. I do the latter. You'd better position it correctly the first time because you won't get a second chance. Almost no one throws away magnetic business cards; they hold the children's art to the refrigerator door just so (and they're easy to find when you need them).

and "shoddy" aren't words you want associated with your operation, so go the extra mile and pay for quality cards.

If you spring for good card stock (not perforated punch-outs) and you understand design, you can probably print your own business cards. Many word processing and most desktop publishing programs offer create-a-card capability or use downloadable templates available from suppliers such as Paper Direct. If you can do a professional-looking job, do it; if not, you're better off ordering business cards printed by the pros.

Before designing your card or hiring a professional to do it for you, collect business cards that speak to you. They needn't be farm-related; you're looking for elements that engage your eye and can be incorporated into your own card's design. What do you like about them? The typeface? Ink colors? Paper stock? What do they say about the business they purvey? These questions will help you form an image of the type of card you would like to build or have built.

Designing and Building Your Own

When designing your own card, keep in mind that you have 7 square inches to tell your story (14 square inches if you use the back). Here's what you need to know to do it right:

- Include all contact information on the face of your business card. This includes your name, farm name, mailing address, e-mail address, phone and FAX numbers, and your Web site's URL. Your name (or farm name) should be the largest text on your card.
- Keep it simple. Avoid fancy, hard-to-read fonts, and use at most two font-type families per card.
- Use your logo but otherwise avoid nontext elements unless they're good ones. A bad photo is worse than no photo, and run-of-the-mill clip art lends nothing to the image you're trying to project. If using a photo, its focus should be spot-on.
- Devise a way to make your card stand out. Textured card stock and interesting cutouts (use a scrapbooking paper punch) help make a card unique.
- Choose cardstock and font colors that complement one another. Color adds interest, but a little color goes a long way.
- Create visual interest at minimal cost by carefully rubber-stamping a small design element on basic black-and-white business cards, using colored or metallic ink.
- Stick to standard-size 3.5 × 2-inch cards. Because other sizes don't fit ready-made cardholders, they're likely to be discarded.

THIS FARMER'S TAKE

One day, soon after moving to the Ozarks, my husband and I stopped at a local café for lunch. On our way in I spied a bulletin board. Pinned to the upper left-hand corner was a short stack of business cards with a horse logo on them. I took one, and before we finished our grilled cheese sandwiches, I'd called and made an appointment to visit Ozark Miniature Dream in Camp, Arkansas. A month later we owned six miniature horses — one purchased from Ozark Miniature Dream — and had five new friends, all because of a well-placed business card.

- People will save your card if you print something useful on the back, such as a gestation or measuring table, important phone numbers, a map to your farm, your farm's Code of Ethics, contact information (phone numbers, URLs) of the registries or breed promotion groups you do business with, URLs of favorite species- or breed-specific Web sites, your best animal's major show wins, a list of the males you stand at stud, a calendar, a time zone chart — the list goes on and on!
- Keep your business card up to date; don't scratch through or white out text and write in corrections — get new cards. Likewise, don't hand out tired, worn cards you've had in your wallet for months. Use a business card case and keep them crisp.
- If you sell several species of miniature livestock, consider printing more than one card. Instead of cramming a glut of information onto a single business card, spread it out — specialize!
- Most important: Run a spell-checker program over your finished card, and/or ask someone to proofread it, before you print. Nothing says "amateur" more succinctly than advertising replete with misspelled words.

Say It with Flyers and Brochures

A flyer is usually printed on one side of standard 8½ × 11-inch paper and designed to be tacked to bulletin boards, taped to the inside of windows, and distributed as handouts. It's an ideal venue to announce an upcoming production sale, spotlight the stallion you stand to outside mares, or simply indicate you have livestock to sell.

A brochure (also called a pamphlet) incorporates more detailed information. It can be as simple as an 8½ × 11-inch or 8½ × 14-inch sheet of paper folded in half or thirds or as detailed as a multipage, spine-stapled booklet. An effective brochure makes the reader want to open it and read what's written inside. It must be visually appealing and well written and satisfy a need or send readers a specific message.

While you can certainly have these items professionally designed and printed, in this age of desktop publishing software, high-end printers, and quick-print shops, if you're a fairly articulate writer you can create effective flyers and brochures right at home. Showing you how to do it is beyond the scope of this book, but keep these tips in mind:

- Determine what you want your flyer or brochure to achieve and focus on that goal. If you raise Pygmy goats, Babydoll Southdown sheep, and Miniature Longhorn cattle, don't try to market them all through a single brochure.

SAY IT OFTEN AND MAKE IT COUNT

Even if you don't have a huge advertising budget, consider running small ads in species- and breed-specific periodicals. These magazines usually offer inexpensive classified and "breeder's card" advertising, and these can be outstanding places to market your wares. But you must advertise often. Marketing experts agree that 12 small ads garner more results than one large ad placed once a year; you must constantly reinforce your presence lest buyers forget who you are. One-shot ads, no matter how big and how flashy, won't do an effective job.

- When writing headings, avoid artsy fonts that are difficult to read. Basic sans serif fonts work best (serif fonts have small strokes at the end of each letter; sans serif fonts don't); some good ones are Helvetica, Arial, and Geneva.
- Text should be written in a standard serif font such as Times, Times New Roman, or New Century Schoolbook.
- Don't use script fonts or ALL CAPITALS anywhere. They're distracting and difficult to read.
- Keep sentences short and to the point. Strive for simplicity and clarity. If you're not sure if your writing is clear, ask someone unfamiliar with what you're selling to read it before it goes to print.
- Don't crowd too much onto a single page. Double-spaced text is easier to read than single-spaced text. Adequate margins on all sides of each page add elegance and also increase readability.
- Always proofread your work. If you've written "Cute Nigerian Dwarf weathers for sale," a spell-checker can't know that you meant to type "wethers." Always run a spell-checker over everything you've written, but after you do, proofread to catch small errors the program might have missed.

Creative Ideas That Won't Break the Bank

The best ways to promote your miniature livestock are often the cheapest. For instance, you don't need a big, fancy commercial sign to get people's attention. Erect a large, legible road sign by your driveway. If you live off the beaten path, provide directional signs with your farm name on them. Or if you live on a well-traveled byway, display your best stock in a pasture adjacent to the highway. If your females have youngsters by their sides, put them by the road; nothing stops traffic better than adorable baby animals.

Consider the following ideas to promote your business.

Become Involved in Your Community

Take friendly livestock and display booths to gatherings of all kinds. Show your animals. Train representative animals and drive or lead them in parades with farm signs attached to their carts or panniers. Haul a friendly animal to hospitals and nursing homes to visit shut-ins. When you have fun, gain exposure, and make people happy, everyone wins!

Give talks and demonstrations. Prepare an educational program, and take it to area schools. Civic groups need speakers for meetings and events; let it be known that you're available and interested. Volunteer to speak at seminars. Hold an open house. Sponsor or lead a 4-H miniature livestock project.

WORDS FROM THE WISE

This sage advice is from *Storey's Guide to Raising Meat Goats,* but the words apply to getting started in *any* new phase of livestock keeping. We asked Mona Enderli of Enderli Farms in Baytown, Texas, "What advice would you give someone thinking about starting out in meat goats?" This is what she said:

1. Have a business plan, and know your markets.
2. Make sure you have the time and energy to commit to the endeavor. This business takes dedication.
3. Raising goats isn't easy. If you're not prepared to work long hours in all kinds of weather, you may want to consider a different business.
4. It isn't cheap! You can't scrimp on pasture, feed, housing, or medicines. If you're not prepared to provide your goats with the best care possible, again, you may want to consider another business. What you may save by scrimping will be lost when your goats die from lack of proper care.
5. This business is 24 hours a day, 7 days a week, 365 days a year, and it's not for the fainthearted. Would you be able to deliver a kid that is breech, too large, or possibly dead in the uterus? If seeing blood makes you feel faint, you don't like being dirty, the thought of having goat poop stuck to your shoes all the time is disgusting to you, or you expect to be able to take a normal vacation at the drop of a hat, then you'd better rethink your plans.

Sponsor an event. If you raise Guinea Hogs, sponsor a Kiss a Pig booth at the county fair and donate proceeds to your favorite charity. Host a trail ride or driving rally culminating with a potluck supper; a spinning, shearing, or cheese-making field day; a training clinic; or a business seminar. Hold them at your farm.

Read stories to Head Start or kindergarten classes, from books featuring your type of livestock, of course. Take along a living example, and tip off your local newspaper in advance.

Participate in e-mail lists, and be the first to field questions. People buy from people they know and respect, and you can advertise on many lists for free.

Contact your state adopt-a-highway program and pick up trash along the byway. They'll erect a sign with your name on it. Take a goat, calf, or other manageable animal with you and have someone photograph your cleanup crew (with someone holding your animal friend) by the sign. Send the photo and a story about highway adoption to your local paper. Get ready to see it in print.

Offer free, individual consultations to folks interested in raising your type of livestock. Invite them to your farm. Give them good, solid advice without a sales pitch, but show them your stock as part of the visit.

Investigate co-op advertising with other folks producing your type of livestock, your breed, or your bloodlines. Scope out the Smallest Horse Group (see Resources, under Miniature Horses); enjoy major exposure at minimal cost.

Serve homegrown products if you raise animals for food (beef, pork, goat milk, cheese crafted from your Miniature Jerseys' rich milk) whenever and wherever you can, especially at public functions and to people who may not have tasted it before. Distribute recipe sheets with your contact information printed on them.

Spread Your Name Around

Invest in truck and trailer lettering. Incorporate your phone number, e-mail, and Web site address into the design. Turn nonbusiness vehicles into rolling billboards by affixing magnetic farm signs. Don't forget the back of your truck — put signs on the tailgate or back window.

Think bumper stickers! Everyone who follows you down the road will be exposed to your URL or your message, and you can give them out to customers, family, and friends. A slew of companies offer design-online capability with no minimum order, in vinyl and magnetic styles.

Buy T-shirts, jackets, and hats printed or embroidered with your logo and farm name and wear them everywhere you go. Order extras for your customers, family, and friends.

Opt for custom-printed checks, invoices, and other business forms imprinted with your logo and business information.

Create Visual Appeal

Shoot good video footage of your animals, especially your breeding males. Edit it well, add music if you like, and burn it to CD or DVD. Distribute copies for free or at cost via your Web site, print advertising, or through your booth at shows, fairs, and livestock expos.

Upload lots of videos for instructional and entertainment purposes to YouTube and MetaCafe. Show off your stallion or teach viewers how to yoke Dexter oxen. End each video with a short talk about your farm and include contact information.

Publish a calendar featuring photos of your livestock. Send it to customers for free or sell it at cost from your Web site or your booth at shows, fairs, and expos. If you can't shoot publication-quality photos yourself, invite shutterbugs to a photo-ops day at your farm. Spruce up your show stock, and provide handlers to pose them. Bring out the cute, fuzzy babies. Hold a photo contest later on, awarding prizes for one-time use of the winning pictures.

If you shoot great photos, let them work for you. Order quality enlargements of your best photos or burn them to a disk, then send them to your breed associations and trade publications, stipulating they can use your photos however they wish, but subjects must be identified and you must be given recognition. Writers and editors often contact organizations seeking publication-quality photos; your photo could end up in a book like this one!

Use the Power of the Pen

Publish an electronic newsletter, also known as an e-zine. Include educational content so others want to read it. Add a business-related signature to your outgoing e-mail; incorporate your farm name, your location, a tag line describing your services, and your farm's Web address.

Blog. Add new content every week.

Write an educational column for an area newspaper. Call it "Goat Talk" or "Ewe Read It Here" or "Cattleman's Corner." Write it for free with one caveat: your contributor's byline must include your farm's contact information.

Write magazine articles for farm and breed periodicals with the same stipulation; many are eager to work with new writers.

Write reports and e-books about training, breeding, or showing the breeds you raise; about their history or bloodlines; or about utilizing the products they produce. Publish them in PDF format, and sell them for a modest amount or offer them as free downloads from your Web site. Include a page about your business, of course.

SECTION 2

The Species

14

Miniature Cattle

THE MINIATURE CATTLE MARKET IS HOT. Mini cattle are the most popular alternative livestock in North America. If you're looking for a money-making project, raising miniature cattle fills the bill.

Miniature cattle are scarce (some breeds more so than others), making breeding stock production a lucrative business. And there is a big demand for many beef breeds, so excess bull calves are an asset as well.

Miniature beef breeds, such as Lowlines and Miniature Herefords, have many attractions over their full-size counterparts. They need less feed — many miniatures eat roughly half to one-third less feed; they mature earlier, meaning that they can be slaughtered at an earlier age (some as early as 12 months); they can be stocked at a higher rate per acre, thus producing more beef on the same amount of pasture; their smaller size means their muscles are shorter, resulting in what some consider to be tenderer meat; and they produce a more manageable amount of food for an average-size family.

There is also a growing interest in dairy stock among hobby farmers, be they dairy goats or a family cow. An average 1,500-pound Holstein produces between 8 and 10 gallons of milk (plus a *lot* of manure) every day, at the same time taking up a great deal of barn and pasture space and eating a lot of concentrated feed. Miniatures, such as the Miniature Jersey or the Dexter, are much more appealing choices for the hobby farmer. The cute, petite, 600- to 650-pound (272–295 kg) Miniature Jersey stands just 38 to 42 inches (97–107 cm) at the hip and gives 2 to 4 gallons of rich milk that is high in butterfat. The Dexter cow is the same size as the Miniature Jersey. She gives 1 or 2 gallons of milk a day and has calves that make fine freezer beef. *These* are hobby farm cattle.

CATTLE PHYSIOLOGY (ADULTS)

Temperature: 100.4 to 103.1°F (37.8–39.4°C)
Heart rate: 40 to 70 beats per minute
Respiration: 10 to 30 breaths per minute for beef cattle; 18 to 28 breaths per minute for dairy cattle
Ruminal movements: 1 to 3 per minute
Natural life span: Up to 20 years, typically (Big Bertha, an Irish Dremon cow born on St. Patrick's Day in 1944, died just three months short of her forty-ninth birthday. She boasts two entries in the *Guinness Book of Records*: she is the oldest cow ever recorded, and as mother to 39 calves, she holds the record for lifetime calf production.)
Heat cycle: 18 to 24 days
Heat duration: 6 to 18 hours
Ovulate: About 32 hours after the onset of heat
Length of gestation: 283 days on average
Number of young: One (occasionally twins)

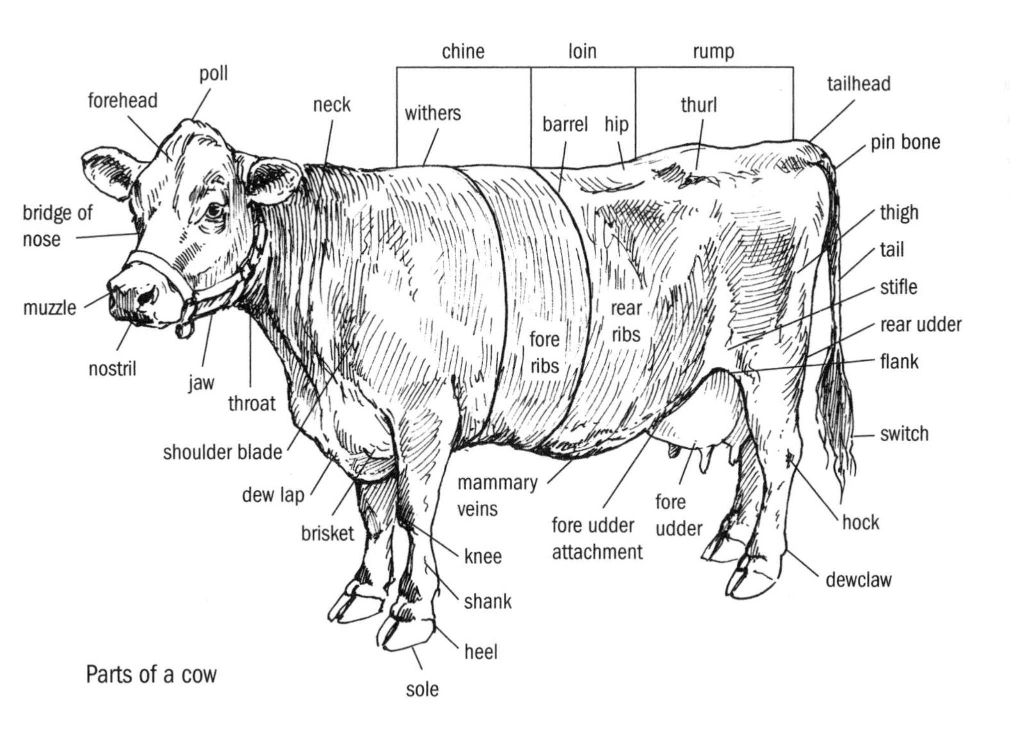

Parts of a cow

Miniature cattle, whether beef or dairy, are also in great demand as traffic stoppers and pets. Breeders of nonbeef breeds consider the pet market to be their best venue for steer sales. Miniatures, particularly tiny ones like Miniature Zebu cattle, are far more docile than their full-size counterparts, making them great backyard buddies and ideal cattle for petting zoos, agri-tourism attractions, and pet-assisted therapy programs.

HOW TO MEASURE A COW

Unlike llamas, goats, and sheep, which are measured in the manner of most equines — to the highest part of the withers — most miniature cattle are measured from the pin bone on one side of their hips straight down to the ground. Miniature Zebus are measured from a spot immediately behind their humps to the ground.

Most cow breeds are measured from their pin bones to the ground.

Measure Zebus, however, from a point directly behind their humps.

Compare a typical, full-size Holstein with a diminutive Dexter. Miniatures are much, much smaller!

Getting Started with Miniature Cattle

Everything written in chapters 3 and 4 applies to getting started in miniature cattle. Pay special attention to the section on dwarfism when purchasing breeds based on Dexter genetics; have these animals DNA tested before buying to save yourself heartache later on.

Buying Miniature Cattle

Miniature cattle are scarce and often expensive. Choose stock wisely: know your breed thoroughly, insist on vet checks prior to purchase, and always buy from reputable sellers. Here are some ways to get started.

Adults or Calves

The quickest and most reliable way to get started is to buy fully grown, purebred breeding stock. It's also by far the most expensive. In some breeds, mature stock simply isn't available, so you'll have to pursue other options.

Calves are generally more readily available than adults, though heifers may be in very short supply. Calves may cost less but usually don't come with breeding guarantees. If they don't produce as adults, you're out your money.

Embryos

You may be able to purchase the embryos of some breeds. Before considering this, read the university publication "Embryo Transfer in Cattle" (see Resources) to see how it's done.

There are two ways to purchase embryos: frozen or already implanted in a certified pregnant recipient cow. Even factoring in the cost of the recipient cow (who

BRING ON THE BULL (OR NOT?)

Do you need a bull to raise purebred calves? If you have reliable access to fresh cooled or frozen semen, probably not; but if your breed is extremely rare, you probably will. All bulls, even the gentle ones, require strong housing and fencing and should never be taken for granted; bull keeping isn't for the faint of heart. If you do keep a bull of any breed, heed these suggestions:

- Feed your bull before you interact with it. Don't bother it while it's eating or interested in a cow. Likewise, familiarize yourself with its normal behavior patterns, and when it's acting strangely, leave it alone.
- When you're in a pen or pasture with a bull, always know where it is in relationship to your best escape route.
- Don't allow children or vulnerable adults to interact with the bull, even if it's a gentle one.

is referred to as a *recip* in reproductive circles), the latter is the least costly and the safest. The success rate varies according to the quality of the embryos and the skill of the vet or technician doing the work, but it averages roughly 60 percent.

If you decide to try frozen embryos, touch base with a reproductive specialist dealing primarily with cattle and ask lots of questions. And choose your recips carefully. They should be reproductively sound and proven easy calvers with good mothering skills and milking capacity. Recips must be sized to miniature calves, with teats small enough for a tiny calf to suckle and close enough to the ground for the calf to reach.

In addition to the overall risks assumed when buying embryos, you can't choose the sex of the calves. It's an expensive and risky business, but in really rare breeds it may be the only avenue to pursue.

Upgrading

Some miniature cattle registries allow upgrading to purebred status. An upgraded animal is to your advantage, because in today's market, high-percentage registered cattle have more value than animals of unknown breeding.

The upgrading program offered by the American Miniature Jersey Association is typical. It registers percentage animals sired by Foundation (full-blood) Miniature Jersey bulls as one-half and three-quarters Miniature Jersey. Animals of seven-eighths or greater Jersey breeding earn the status Native Pure and are considered purebred but never full-blood Miniature Jerseys.

Breeders can use a Foundation Miniature Jersey bull or Foundation Miniature Jersey semen on standard-size Jerseys and then select for size, or they can use small cows of another small breed or a combination of breeds.

Check with your registry before planning to go this route. You'll want to consider whether it's worth pursuing because most never consider upgraded animals to be the same as bona fide full-bloods.

Fencing and Housing

Though adult miniature cattle aren't particularly hard to fence using any type of wire, plank, or pipe fencing, remember that miniature calves are *tiny*. They slip through gaps in fencing with alacrity and make a tasty package that predators as small as coyotes can carry away. Make fences secure enough to keep calves in and predators out — and a herd guardian (or two) is always good insurance.

Most miniature cattle breeds are quite hardy, and few require elaborate housing of any kind. Keep in mind that long-haired Highlands suffer in hot, humid climates (provide a pond if you raise them in the South) and sleek-haired Zebus require protection during northern winters.

Feeding

Again, your county extension agent is the best source of information about proper feeding for the breed and type of animal you produce. Most miniature cattle fare well on grass-based diets supplemented with a suitable mineral mix, but they usually require concentrates while growing, during late gestation, or while lactating. Dairy breeds in milk always require grain.

Think like a Cow

Cattle, even those that are small by conventional standards, are powerful and agile animals capable of accidentally injuring caretakers. Understand your animals' way of thinking to make handling them a safer and more rewarding experience.

How Cattle Perceive Their World

Cattle experience panoramic vision of 330 degrees; their only blind spot is directly behind them. Binocular vision, when the field of vision for both eyes overlaps, is 25 to 50 degrees depending on the size and placement of an animal's eyes. Cattle have weak eye muscles that limit their ability to focus quickly on objects, so give them time to examine new situations. Because of poor depth perception, cattle may balk at shadows or patterns on the ground and are best moved through diffused light.

Cattle distinguish long wavelength colors such as red, orange, and yellow much better than they see short wavelength colors such as green, blue, and gray.

Cows recognize their newborn calves by smelling them. Cattle also use scent to discern when cows are coming in heat.

WHAT IN THE WHORL?

In 1995, animal behaviorist Temple Grandin assessed whorl patterns (spots on an animal's face where hair growth twists) on 1,636 head of beef breeds (both unhumped *Bos taurus* and humped *Bos indicus*), dairy breeds, and their crosses. She noted lateral position (left, right, center) and height (high = above eye level, middle = at eye level, and low = below eye level) and gave each a point score of one through four based on the animal's reaction in an auction barn situation. One equated with calm and four with highly agitated. Animals rating three or four on that scale were rated on whether they displayed fight-or-flight behaviors. Normal whorls were defined as a single, spiral whorl at eye level. These are her findings:

- Facial whorls were absent in 10 percent of the cattle.
- 86 percent had a normal facial whorl.
- 47 percent had middle-whorl placements.
- Cattle with low whorls were more likely to have abnormal and off-center whorls.
- Animals with higher reaction scores had higher facial whorls.
- Females had more abnormal whorls than males; beef cattle had more abnormal whorls than Holsteins.
- Reaction point scores were higher for females and animals with high whorls than for males and animals with low or middle hair whorls.

According to Dr. Grandin, a fetus's brain forms at the same time as skin and hair — hence the connection. For centuries, horsemen have observed and written about the relationship between the position of hair whorls on a horse's forehead and its temperament. Animals with whorls positioned high above the eyes and with multiple whorls are said to be more likely to have a flighty temperament. The results of Dr. Grandin's study support this theory.

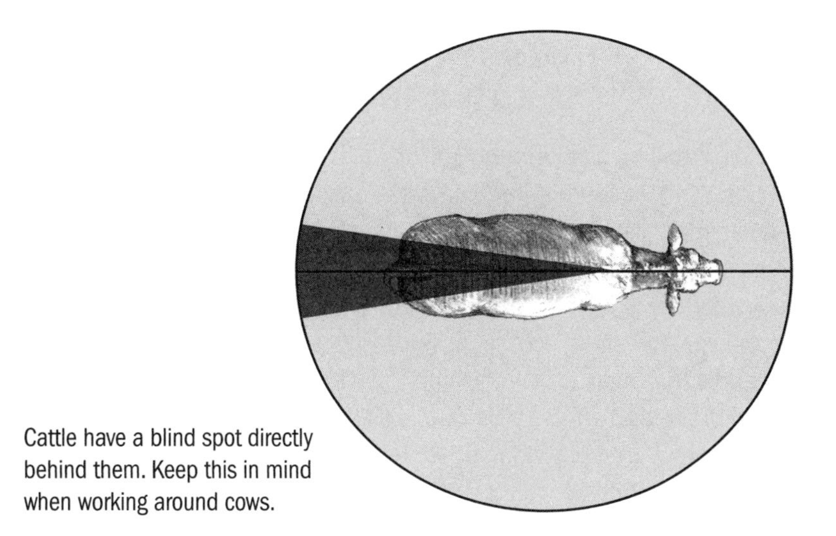
Cattle have a blind spot directly behind them. Keep this in mind when working around cows.

Cattle have sensitive ears and hear in a frequency similar to our own. They are easily stressed by yelling and high-pitched sound. Dairy cattle are more sensitive to sound than are beef cattle; white noise such as soothing music from the dairy parlor radio calms them.

Cattle have reasonably sensitive skin. Scratching an animal's neck or behind its ears is a special treat that calms cattle habituated to human touch.

Social Hierarchies

Cattle, regardless of size, establish hierarchies or pecking orders in any group, whether the group is made up of 2 or more than 100 cattle. An animal's rank within the hierarchy depends on many factors ranging from sex, age, and breed to height and weight.

In a study conducted with dairy cattle in 1955, British researchers M. W. Schein and M. H. Fohrman found that older, heavier cows ranked higher in their herd's hierarchy, and tall steers outranked the short ones.

Other researchers mapped an inverted U-shaped relationship between dominance and age in dairy herds. Cows rose in rank until about nine years of age, then their dominance declined with age-related weight loss.

Horned animals nearly always rank higher than non-horned animals.

Most large herds break down into a series of smaller hierarchies that begin forming as soon as calves are weaned. These consist of one or more groups of adult females, at least one juvenile group, and if more than one male is present, male hierarchies as well.

UNDERSTANDING FLIGHT ZONE PRINCIPLES

By understanding an animal's flight zone — and this applies to all prey species, not just cattle — we can more easily control its movements.

An animal's flight zone is the area around it that cannot be entered without causing alarm. The size of the flight zone depends on the tameness of the animal. Tame animals have no flight zone; that is, they will allow a person to approach and touch them. Wild, feral, and previously unhandled (or mishandled) animals have very large flight zones.

When a person enters an animal's flight zone, the animal moves away. If the person moves outside of the flight zone, the animal stops. The direction the animal moves depends on the position of the person in relationship to the animal's point of balance. If the person is in front of the point of balance, the animal will move back; if the person is in back of the point of balance, the animal will move forward.

An animal's flight zone can change very quickly. Consistent, quiet handling decreases the amount of fear an animal experiences when in close contact with humans, causing its flight zone to shrink.

Don't move directly toward an animal's head while it is grazing; this generates fear and increases the size of its flight zone. Instead, approach its shoulder (its point of balance) and don't look directly into its eyes.

High-ranking cows eat more feed but in fewer meals than low-ranking herd mates, perhaps because they know they needn't hurry. Unless feed is very limited, dominant cows rarely chase subordinates away from feed; however, feed bunks should be one-sided, as face-to-face feeding spurs aggression.

Aggressive interactions between cows appear to be ritualized and occur in this manner: approach, threat, and only then physical contact. Once two animals establish their relationship to each other, the lower-ranking animal retreats from the dominant one at the slightest threat, so there is no need for further combat.

An aggressive bull turns its body perpendicular to a perceived challenger to display its full height and weight. It then lowers its head and sometimes shakes it rapidly from side to side. Its eyeballs protrude and its hair stands up along its back. When a bull becomes seriously angry, it faces its target head-on with its head lowered, shoulders hunched, and neck curved. It bawls and paws the ground with

Moving parallel to stock while outside of their flight zones generates less fear than walking straight at them. Moving out of a parallel position and straight toward animals increases their flight zone.w

DIAGRAM OF FLIGHT ZONE

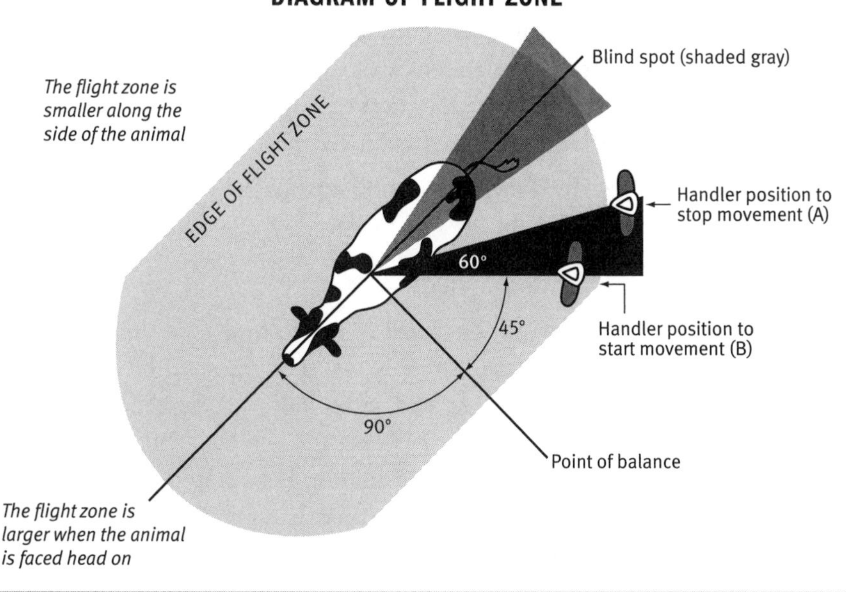

its forefeet, sending dirt flying behind it or over its back, and rubs or horns the ground. If an opponent, such as another bull or person, withdraws, the bull will usually turn away. If not, the bull will circle and initiate head-to-head or head-to-body pushing. Dairy bulls are generally more aggressive than beef bulls.

Herd Behavior

A typical day for a cow includes standing, walking, lying, feeding, drinking, self-grooming, grooming herd mates, and ruminating.

Under average grazing conditions, dairy cows graze about eight hours a day, and beef cattle graze for close to nine hours. Cattle are *crepuscular*, meaning they are most active at dawn and dusk. However, when flies are out or when hot, humid conditions prevail, cattle may graze mainly at night.

Grazing behavior among herd members is similar: each animal moves slowly across the pasture with its muzzle close to the ground, using its teeth and tongue to tear off grass, which is swallowed without chewing. Cattle ruminate when resting; time spent ruminating is about three-quarters of that spent grazing, so a cow that grazes for eight hours a day spends an additional six hours ruminating.

The amount of time spent resting depends on environmental conditions, breed type, and the amount of time spent grazing and ruminating. Cattle rest with their heads held up or drawn back along either flank. Resting cattle dislike noise and disturbance and will temporarily abandon a favorite resting place to avoid them. Zebu cattle remain outside in bright sunlight, whether resting or grazing, while British breeds tend to seek shade.

In cold weather cattle crowd together in their appropriate social group, thus sharing warmth and decreasing the surface area exposed to chilling winds.

In open areas, cattle gather into large groups, and distances between animals are smaller than in groups grazing in areas with shelter such as trees and shrubs.

Breeding Behavior

A bull can detect that a cow is about to come into heat (the time of sexual receptivity and fertility in females) roughly two days prior to the event. As a cow enters heat, the bull becomes excited and follows her closely, flehmening (lip curling), and licking and sniffing her external genitalia. Before breeding he snorts, paws the ground, and rests his chin on her rump before mounting and breeding. Breeding occurs in seconds, rather than minutes. Dairy bulls are generally more sexually active than beef breed bulls.

While entering the heat portion of her cycle, a cow becomes more antagonistic toward her peers, she urinates more than the norm, and as she becomes sexually receptive, she'll mount and be mounted by other cows.

BOVINE BREEDING TRIVIA

- A cow and her calf can develop a strong bond after just five minutes of contact. If you plan to remove the calf, do it right away to prevent undue stress to the cow. Such a calf should be fed colostrum derived from its dam or frozen colostrum from another cow as soon as possible, at least within a two-hour time frame.
- Studies indicate that dystocia is responsible for 33 percent of all calf losses and 15.4 percent of beef cattle breeding losses. If you raise cattle, you must know how to resolve difficult births.

READ IT, YOU NEED IT

Just as I recommend anyone involved in foaling out mares and jennies keep a copy of *The Complete Book of Foaling* by Dr. Karen E. N. Hayes in their foaling kit, I believe the *Essential Guide to Calving: Giving Your Beef or Dairy Herd a Healthy Start* by Heather Smith Thomas belongs on every beginning bovine breeder's book shelf. Read it; you need it; I mean it. You'll be very glad you did.

Maternal Behavior

A cow recognizes her calf's scent within moments after birth. Experienced cows stand within a minute or two of giving birth and immediately begin licking the calf to stimulate breathing, circulation, and elimination. Heifers and cows experiencing difficult births take longer to stand.

Most calves stand within an hour of birth and suckle within two to four hours. Between birth and seven months, calves suckle four to five times per day, for about 45 minutes total. Cattle are a laying out species like deer and goats; they hide their calves near their birthing sites, returning to suckle the calves at intervals. Cows may be very protective of their calves at this time and highly aggressive toward intruders (including humans that they know very well).

Within its first week of life, a calf starts following its dam. During the day, however, bunches of calves lie together in small groups, usually under the scrutiny of one or more "guard" cows, while their mothers graze.

A cow and her calf are strongly connected from birth until 100 to 120 days of age. At this time the cow's caregiving behaviors begin to decline and her milk output decreases. The calf becomes more independent and begins seeking additional nourishment.

Birthing Cattle

The principles covered in chapter 11 apply to calving out cows. However, since cattle are far more prone to dystocia than most other species, you need to know how to deal with those cases as well.

You can't learn to handle serious calving dystocia from a book. Line up a good vet or an experienced cattleman (preferably one with small hands) to help you learn the ropes. Pulling calves with tackle is a scary, risky business when calving out full-size cattle, and the problems and risks are magnified manyfold when dealing with very small cows.

THE INTERNATIONAL MINIATURE CATTLE BREEDERS SOCIETY AND REGISTRY

In the mid-1960s, young Richard Gradwohl was a new professor at Highland Community College in Seattle, Washington, where he taught entrepreneurship and marketing. He, his wife, Arlene, and their children decided to move to the country. They bought a small farm with a view of Mount Rainier, which they named Happy Mountain Farm.

The couple began by raising full-size Herefords, then added Angus and some Angus-Hereford crosses. Eventually they sold some of their land, and Gradwohl began thinking of raising smaller cattle tailored for today's smaller farms.

In the spring of 1971, the Gradwohls purchased a small herd of short-legged Dexter cattle. They got rid of everything in their original herd except their smallest Herefords. Gradwohl began developing new breeds of smaller bovines. As of mid-2009, the now-retired professor has created, registered, and aggressively marketed 18 breeds! His organization, the International Miniature Cattle Breeders Society and Registry (see Resources), registers an additional 8 breeds of miniature cattle, for a grand total of 26 breeds. What's more, the Gradwohls are arguably the world's greatest supporters and promoters of miniature cattle.

Trademarked breeds developed at Happy Mountain Farm include the American Beltie (Mini-Cookie), Auburnshire, Barbee, Belted Irish Jersey, Belted Lessor Jersey, Belted Kingshire, Belted Milking Dexter, Miniature Black Baldie, Burienshire, Covingtonshire, Four Breed Grad-Wohl, Five Breed Grad-Wohl, Happy Mountain, Justinshire, Kentshire, Kingshire, Panda, and Red Kentshire; breeds in development include the Mini Holstein, Red Panda, and White Dexter. The International Miniature Cattle Breeders Society and Registry also maintains herd books for Australian Kyrhet, Belmont (Irish Jersey), Dexter, Durham/Shorthorn, Hereford, Highland, Lessor Jersey, and Zebu miniature cattle.

Mini Panda

The Breed You Need

Breeders are working hard to produce new breeds of miniature cattle every year. If you have a favorite full-size breed, you could do the same. Following are the established miniature cattle breeds, most of which are supported by at least one breed registry or association. Which will you choose?

Dexters

Dexter

Registered by: American Dexter Cattle Association, Purebred Dexter Cattle Association of North America, Canadian Dexter Cattle Association, International Miniature Cattle Breeders Society and Registry
Use: Beef, dairy, draft, pets
Origin: Southwest Ireland
Size: Cows 38 to 42 inches (97–107 cm); bulls proportionally larger than cows
Color: Black, red, or dun with a small amount of white permissible on or near the udder or scrotum, and in the switch of the tail
Type: Short, broad head; short neck, deep and thick; full shoulders; wide hips; well-rounded rump; flat and wide across the loins; straight, level topline, with tail set level with spine; straight underline; good width between legs; short (especially from knee to fetlock), strong legs. The length of the back from shoulder to tail set should be the same or greater than the height of the animal at the shoulder.
Horns: Yes, though polled genetics occur. Horns are short and moderately thick, with an inward and upward curve; white with black tips on black cattle, white with red tips on reds, white with red or black tips on duns.
Dwarfing gene present: Yes

Dexter cattle descend from small, black Celtic cattle that Neolithic humans brought north from the Mediterranean basin during the Stone Age. It's uncertain when these cattle became today's Dexters and Kerrys; however, when Professor David Low wrote *On the Domesticated Animals of the British Islands* in 1853 (see Resources), he said of the Dexter, "It was formed by the late Mr. Dexter, agent to Maude Lord Howarden. This gentleman is said to have produced his curious breed from the best of the mountain cattle of the district. He communicated to it a remarkable roundness of form and shortness of legs. The steps, however, by which this improvement was effected, have not been sufficiently recorded and some doubt may exist whether the original was the pure Kerry, or some other breed proper to the central parts of Ireland now unknown, or whether some foreign blood, as the Dutch, was not mixed with the native race. . . . however the Dexter breed has been formed, it still retains its name, and the roundness and depth of carcass which distinguished it."

In 1882, Martin Sutton of Kidmore Grange in Oxfordshire imported a group of 10 Dexters to England. Their popularity skyrocketed, and 10 years later a Dexter/Kerry cattle society formed. Dexters were subsequently exported to the rest of Europe and to North America. Around 1905, wealthy American businessmen such as James J. Hill, Howard Gould, and August A. Busch kept traffic-stopping herds of tiny Dexters at their posh estates. The breed's popularity eventually waned, and they were all but extinct in the 1970s. Since then there's been a huge resurgence of interest in the breed, and though it's still listed as a Recovering breed on the American Livestock Breeds Conservancy watchlist (see box on pages 16 and 17), the diminutive Dexter's future looks bright indeed.

Dexters are gentle, fertile, and great mothers. They're also good milkers, dress out at 50 to 60 percent live weight, and produce extremely tasty cuts of succulent beef. Dexter steers make cute yokes of oxen that are amazingly strong for their size. Many breeders use Dexters to develop new breeds of miniature cattle, breeding full-size cows to diminutive Dexter bulls.

Dexters come in two types: short-legged and long-legged varieties (the latter are sometimes referred to as Kerry types). A recessive gene for chondrodysplasia (dwarfism) is common in this breed (see chapter 4) and should be fully understood before you buy breeding stock. An effective test for the gene is readily available, and by following sound breeding practices, you can avoid producing lethal bulldog calves.

In the 1800s Professor Low remarked that "When any individual of a Kerry drove appears remarkably round and short-legged, it is common for the country folk to call it a Dexter." At that time, and for many years later, Kerrys and Dexters were the same breed. The first volume of the *American Kerry and Dexter Cattle Herd Book*, published in 1921 (see Resources), registered Kerrys and Dexters in separate sections (to the tune of 87 Kerrys and 313 head of Dexter cattle). Kerrys are now a distinct and critically endangered breed.

WHAT'S A KERRY?

Kerrys aren't as tiny as most of the miniature cattle described in this book and hence aren't profiled here, but at 50 to 55 inches (127–140 cm) tall and 750 to 1,000 pounds (340–454 kg), they're smaller than conventional breeds. Like Dexters, they're black or red with a modicum of white allowed on tail switch and along their rear undercarriage. Unlike their dual-purpose cousins, however, Kerrys are mainly dairy cows.

Fine-boned, agile, hardy, and alert, the Kerry is a long-lived breed that calves well into its teens and produces a plentiful amount of milk high in butterfat. Unusually small fat globules in their milk render it (like goat milk) highly digestible for babies and invalids.

Professor Low liked the Irish Kerry so much that he wrote in his book *On the Domesticated Animals of the British Islands*, "In milking properties, the Kerry Cow, taking size into account, is equal or superior to any in the British Isles. It is the large quantity of milk yielded by an animal so small, which renders the Kerry Cows so generally valued by the cottagers and smaller tenants of Ireland. She is frequently termed the Poor Man's Cow, and she merits this appellation by her capacity of subsisting on such fare as he has the means to supply. . . . [T]he cultivation of the pure dairy breed of the Kerry mountains ought not to be neglected by individuals or public associations. The breed is yet the best that is reared over a large extent of the country."

Nonetheless, like many other small, native British breeds, the Kerry eventually fell out of favor and is now one of the rarest of heritage dairy breeds. There are fewer than 100 head of purebred Kerrys in all of North America.

Kerry

Guinea Cattle (Florida Cracker and Pineywoods)

Registered by: Florida Cracker Cattle Association, Pineywoods Cattle Registry & Breeders Association
Use: Dairy, beef, pets
Origin: Florida (Florida Cracker) and the pine forests of Louisiana, Mississippi, Alabama, and Georgia (Pineywoods)
Size: Standard animals are 600 to 1,000 pounds (272–454 kg); guinea cattle are one-third to one-half that size
Color: All
Type: Most are rangy, resembling Texas Longhorns but with less extreme horns
Horns: Yes, though polled lines exist
Dwarfing gene present: Yes (in lines that produce guinea cattle)

Florida Cracker and Pineywoods cattle are listed as critically endangered by the American Livestock Breeds Conservancy (see box on pages 16 and 17). The breeds share a common history: their ancestors are both of Spanish descent. When Spain sent conquistadors and colonists to claim the New World, about 300 head of cattle accompanied the voyagers. Cattle escaped or were abandoned, and in time a vast feral population roamed New Spain. They became the Longhorns of Texas, the Corriente of Mexico, and the Pineywoods and Florida Crackers of the southeastern states. Eventually, breeders "improved" these breeds by crossing them with heat-hardy American Brahman cattle, destroying most purebred lines. A handful of dedicated families in the Southeast maintained herds of pure, native stock, and in the 1990s the Pineywoods Cattle Registry & Breeders Association and the Florida Cracker Cattle Association were formed to preserve these breeds.

Florida Crackers and Pineywoods come in spotted and solid colors, including blacks, reds, yellows, and blues, and in speckles, linebacks, and roans. Some lines lean toward beefiness; others toward dairy type. Most Pineywoods and Florida Crackers are rangy; some are much smaller, with shorter heads and legs. Most Pineywoods and Florida Cracker cattle are horned, with horn types varying from short, curved, Holstein-like horns to large, upswept, somewhat Longhorn-type varieties.

Florida Cracker

Florida Cracker and Pineywoods cattle are self-sufficient animals that thrive on rough pasture and brush with minimal human intervention. They are remarkably easy keepers. Both breeds are long-lived and prolific. Guinea cattle are scarce but available.

Lowlines

Registered by: American Lowline Registry, Canadian Lowline Cattle Association, International Miniature Cattle Breeders Society and Registry
Use: Beef, pets
Origin: Aberdeen-Angus hail from Scotland, Lowlines from Australia
Size: Mature Lowline bulls range from 40 to 48 inches (102–122 cm) tall and weigh from 900 to 1,500 pounds (408–680 kg). Mature cows measure from 38 to 46 inches (97–117 cm) and weigh between 700 and 1,100 pounds (317–499 kg).
Color: Black (or rarely, red); white markings on the underline on the area behind the navel are not uncommon
Type: Short head with a broad forehead and prominent poll; the face may be slightly dished. Smooth neck of moderate length. Dewlap of modest size, with a full brisket. Straight, long topline; the barrel should be deep and uniform from end to end; a well-muscled rump. The Lowline should stand squarely on sound feet and legs of modest length, allowing for freedom of movement and length of stride.
Horns: No (all Angus are genetically polled)
Dwarfing gene present: No

Lowlines are small Angus. They were bred from Aberdeen-Angus (known to most Americans as simply Angus cattle). Aberdeen-Angus descend from naturally polled cattle (known as *doddies* and *hummels*) indigenous to counties Angus and Aberdeen in Scotland. Early-nineteenth-century Scottish breeder Hugh Watson is often considered the founder of the Angus breed. One of his notable animals

Lowline

was a cow named Old Granny, born in 1824 and said to have lived to 35 years of age, during which time she produced 29 calves. Many modern-day Angus trace their lineage to Old Granny.

During the mid- to late 1800s, Scottish breeders exported cattle to the rest of the world, including North America, New Zealand, and Australia. Between the early 1920s and the 1950s, the Trangie Agricultural Research Centre in Trangie, New South Wales, Australia, imported Angus from Canada, the United States, England, and Scotland, amassing one of the finest herds of Aberdeen-Angus in the world.

Beginning in 1974, the Australian Meat Research Corporation funded trials to establish whether large or small Angus were more efficient converters of grass into meat. These trials produced the Lowline breed.

Researchers selected one herd for high yearling growth rates and another for low yearling growth rates, with a randomly selected control group. They dubbed the herds High Line, Low Line, and Control Line.

The original Low Line herd comprised 85 cows. These were bred to yearling bulls also selected for low growth from birth to yearling age. From 1974 to 1993, the Low Line herd remained closed, with all the replacement bulls and heifers selected from within the line. After 15 years of selective breeding, the size of the Low Line herd stabilized. Low Lines were about 30 percent smaller than standard Angus cattle: the bulls matured at a height of roughly 43 inches

HOW THE LOWLINE COMPARES WITH FULL-SIZE BEEF BREEDS

Breed	Average number of breeding cows per 100 acres	Average pounds of carcass per head weight at 15 months old	Pounds of carcass weight per acre	Retail carcass yield: percentage saleable weight (%)	Pounds of retail product per acre
Lowline	54	418	203.1	76	154.3
Full-size Angus	33	543	161.7	68	110
Simmental	23	556	115.0	69	79.4
Wagyu	38	442	151.1	55	83.1
Shorthorn	28	532	132.4	65	86.1
Hereford	30	552	147.2	64	94.2
Murray Grey	32	547	157.7	67	105.7

Source: The New South Wales (Australia) Department of Agriculture

LOWLINE OBSERVATIONS

The New South Wales Department of Agriculture in Australia have recorded observations on the Lowline beef. Here's what they discovered:

- The average Lowline cow is only 40 inches (102 cm) tall and weighs 800 pounds (363 kg); the average mature bull is 42 to 44 inches (107–112 cm) tall and weighs 1,200 pounds (544 kg). Lowlines are 50 to 60 percent the size of most full-size beef cattle and need approximately one-third the amount of feed. Because of their size and easygoing temperaments, they are more readily worked than larger breeds, and they require no specialized handling equipment.
- Lowlines have a life span of 12 to 25 years; they are excellent foragers with terrific mothering ability; and they are hardy, easy keepers.
- Tests have proved that Lowlines are entirely free of dwarfing genes.
- Lowline meat has 5 percent more marbling than other breeds, half the backfat of full-size Angus, and 30 percent more rib eye per hundredweight than any other breed. Lowlines dress out at up to an amazing 76 percent live weight.

(109 cm) and the cows at about 39 inches (99 cm) or less (standard Angus cattle are 59 inches [150 cm] tall).

In the early 1990s the trials ended. The center had originally intended to send the herd to slaughter but were convinced by a group of Australian cattlemen to auction the cattle instead. At the first Trangie sale on August 8, 1992, seven purchasers formed the Australian Lowline Cattle Association, adopting the name Lowline for the breed. The final dispersal sale was conducted on October 30, 1993.

In 1996, the first Lowline Angus were imported into the United States; the same year, six Lowline heifers were brought to Canada and were placed in an embryo transfer program in Alberta. A few of the original importers referred to Lowline Angus Cattle as "Loala" cattle, but in 2001, American Lowline breeders voted to register their animals as Lowlines.

Miniature Bucking Bulls

Registered by: Miniature Bucking Bull Breeders of America
Use: Rodeo
Origin: United States
Size: 48 inches (122 cm) tall and smaller, measured at the withers (behind the hump in Zebus)
Color: Any
Type: Any
Horns: Some are horned, others polled
Dwarfing gene present: Depends on breed

The Miniature Bucking Bull Breeders of America registers bucking bull stock sized for junior rodeos. These little bulls sell for top dollar, and they're in great demand.

Miniature Bucking Bulls differ from their full-size counterparts in size, shape, and attitude. They are better suited to younger, smaller riders, whether beginners, intermediate, or advanced. The mini bulls help the junior riders gain valuable experience before challenging the large bulls at adult rodeos.

Cattle are registered in one of several classifications:

- Miniature: Both parents must be true miniatures of recognized breeds (such as Zebu, Miniature Hereford, Lowline Angus, Dexter, Miniature Jersey). Their papers are marked TM for True Miniature.
- Stunted: Animals that were bred from full-size cattle but did not mature to breed standard. Papers are marked SA for Stunted Animal.
- Grade Cattle: Grade cattle are full-size cattle used in downsizing programs. Papers are marked GC for Grade Cattle.
- Half blood: Cattle that have one Miniature (TM) and one SA or GC parent. Papers are marked HB for Half Blood.

Miniature Bucking Bull

THE BATTLING HÉRENS

Another interesting, naturally small European breed is the Hérens (called the Eringer in nearby Germany), named for the Val d'Hérens region of Switzerland. These intrepid alpine cattle come in black, brown, or dark red (often with a lighter stripe along the spine). They are often used in organized cow fights called *combats de reines* (battles of the queens), during which cows and heifers matched by weight engage in pushing and shoving contests. These events are major tourist attractions in the Valais.

There are only 13,500 Hérens left in Europe, and there is a small North American population registered by the American Hérens Cattle Association.

Recent DNA testing established that Hérens are an ancient breed unrelated to other Swiss cattle. The Hérens breed standard was written in 1884. These cattle have short, broad, dished heads and short-legged, muscular bodies; both sexes are strongly horned. Cows are 46 to 50 inches (117–127 cm) tall and weigh 1,000 to 1,300 pounds (454–590 kg); bulls are 49 to 52 inches (124–132 cm) and weigh 1,400 to 1,500 pounds (635–680 kg). Though primarily raised for beef, cows give a surprising amount of tasty, high-butterfat milk.

Hérens cow fights are shoving matches between two cows.

Miniature Herefords

Registered by: American Hereford Association, International Miniature Cattle Breeders Society and Registry

Use: Beef, pets

Origin: Full-size Herefords in Herefordshire, England; Miniature Herefords in the United States

Size: Most measure 38 to 43 inches (97–109 cm) at the hip
Color: Red (dark to yellowish red) with a white head, underbelly, and switch, with or without a partial dorsal stripe
Type: Strong, muscular, short-legged, proportionate
Horns: Both horned and polled genetics exist
Dwarfing gene present: Not in registered American Hereford Association cattle

Miniature Herefords are another logical breed for potential cattle raisers seeking small, productive, "name brand" cattle. Miniature Herefords are first and foremost Herefords; like their full-size kin they're registered by the American Hereford Association (see Resources). They share the same bloodlines that trace back more than 100 years to the first Herefords brought to North America. Many Americans consider the Hereford to be the quintessential beef breed.

Benjamin Tomkins of Herefordshire, England, is credited with creating the Hereford breed in 1742 by using a bull and two cows inherited from his father's estate. His breeding goals included producing animals that were easy keepers, able to grow and gain on grass and grain, hardy, early to mature, and prolific — all traits that still exist and are valued today.

In the 1700s and early 1800s, Herefords were used for draft purposes and were much larger than modern animals. Many mature Herefords of that era weighed 3,000 pounds (1,360 kg) or more. The size gradually decreased until the midtwentieth century, when the craze for short, blocky cattle resulted in the dwarfism issues discussed in chapter 4. Today's Miniature Herefords, however, aren't dwarfs. The Miniature Hereford breed is entirely free of genes associated with dwarfism.

Miniature Herefords originated at Point of Rocks Ranch in Fort Davis, Texas. In 1970, the Largent family began selectively breeding small, sound, dwarfism-free Herefords, beginning with a select bull and five cows. In 1981, the first true

Miniature Hereford

Miniature Hereford, a bull named LS Real MT 3 was born. In 1989, the Largents sold their first Miniature Herefords.

Miniature Herefords are measured in frame scores 0 through 0000 (Classic Herefords score 1 to 3 and modern Herefords 4 to 10). Charts for figuring cow and bull frame scores are similar but slightly different. They both use the animal's age in months and height in inches to determine frame score.

SHETLAND CATTLE

Historians believe Shetland cattle descend from cattle that Vikings brought to the Shetland Islands between AD 700 and 1100. Like other livestock species that evolved on the isolated, windswept Shetland Islands, they are small yet very productive animals that thrive on plain fare.

In the nineteenth century as many as 20,000 head of Shetland cattle grazed the islands, but their population fell to about 50 by the midtwentieth century. In 1981, the British-based Rare Breeds Survival Trust (see Resources) located remaining purebreds and reestablished a Shetland herd book. The organization still lists the Shetland as a watchlist category four (at risk) breed.

Shetlands are fine-boned, black-and-white dairy cows with short legs, deep bodies, and horns. Cows stand 44 to 50 inches (112–127 cm) at the withers and weigh 650 to 1,100 pounds (295–499 kg). They are registered by the Shetland Cattle Breeders Association (see Resources).

Though records indicate Shetlands were previously exported to North America, no North American registry currently exists.

Shetland cow

Miniature Highlands

Registered by: International Miniature Highland Cattle Association, American Highland Cattle Association, Canadian Highland Cattle Society, International Miniature Cattle Breeders Society and Registry
Use: Beef, pets
Origin: The Highlands of Scotland
Size: Under 42 inches (107 cm) when fully mature
Color: Black, brindle, red, dun, silver, yellow, and white
Type: Cobby, long-haired, with shaggy bangs and sweeping horns
Horns: Yes
Dwarfing gene present: For cattle registered with the American Highland Cattle Association or the Canadian Highland Cattle Society, no

At the beginning of the twentieth century, 300-pound (136 kg) Highland cows were common in Scotland; today, most registered Highlands run 800 to 1,800 pounds (363–816 kg). A number of modern cattle registered by the American Highland Cattle Association stand 42 inches tall or less, and most Miniature Highland breeding programs are based on these small Highlands. Other developers are creating a miniaturized Highland by crossing Dexter bulls with small, full-size Highland cows, then selecting further generations for Highland hair, horns, and body type.

The Highland is an ancient breed developed through natural selection in rugged, remote areas of Scotland as early as AD 600. Originally there were two types: the slightly smaller, black kyloe originating on islands off the west coast of northern Scotland and a larger, red version from the Scottish highlands. The types have merged to become the Highland as we know it today. Today's Highland more closely resembles the larger, red Highland version. The breed's herd book society was organized in 1884. Its first herd book listed 561 bulls; the second included 866 cows and 63 bulls; and the third was a record of awards to Highland Cattle at shows of the Highland and Island Agricultural Society from 1822 to 1884. Today, Highlands flourish throughout North America, Europe, Australia, and South America.

Miniature Highland

The Highland is primarily a grass-fed beef breed and at this it excels. Highland beef is lean, well marbled, and flavorful, with little outside fat. (Highlands are insulated by long hair rather than a thick layer of fat.) They are disease resistant, adaptable, and incredibly easy keepers. Their docile nature and appealing Ewok looks make Miniature Highlands popular pets.

Miniature Jerseys

Registered by: American Miniature Jersey Association & Registry, International Miniature Cattle Breeders Society and Registry

Use: Dairy, pets

Origin: The Isle of Jersey, one of the British Channel Islands in the English Channel, 14 miles off the shore of Normandy, France

Size: Miniature Jerseys must be 42 inches (107 cm) and under at three years of age; Mid-Size Jerseys range from 42 to 46 inches (107–117 cm) at three years of age

Color: Pale grayish fawn to very dark fawn to almost black, in solid or broken (spotted) colors, with darker shading across the hips, shoulders, and head

Type: Beautifully refined, dished head with huge, doelike eyes; long, straight topline with long, level rump; deep in the body, and full and deep in the barrel

Horns: Yes

Dwarfing gene present: For purebred Jersey cattle, no

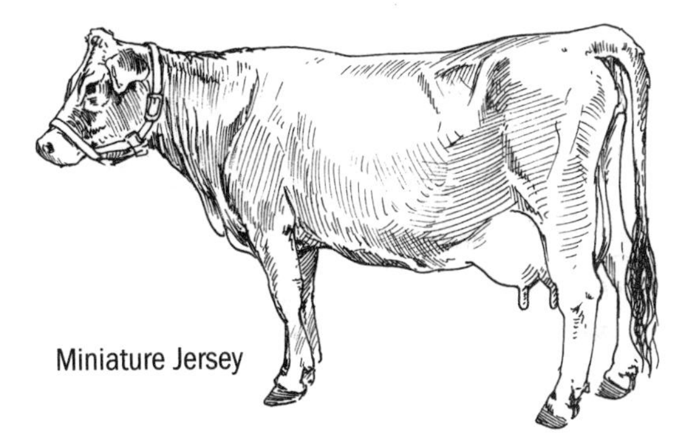
Miniature Jersey

MUSINGS ON THE JERSEY COW

"[When] Jerseys first entered the field of dairying in the United States, their appearance on the scene brought a new conception of cows. The Jersey was the pioneer in awakening the desire for better dairying stock. . . . [W]herever the milk of the Jersey cow was seen, whether in a pitcher, crock, or pan, the cream that rose to the surface proclaimed her superiority."

— *The Jersey: An Outline of Her History during Two Centuries, 1734–1935*, R. M. Gow, 1936 (see Resources)

The Jersey is one of the oldest dairy breeds; it was a purebred on the Isle of Jersey for nearly six centuries. It's thought to be descended from cattle brought to Jersey by the Normans and was first recorded as a separate breed around 1700. Imports of foreign cattle into Jersey were forbidden by law beginning in 1789 and continuing until 2009, in an effort to maintain the purity of the breed.

Jerseys were taken to the English mainland as early as 1741 and probably much earlier. At that time they were known as Alderneys (after the nearby Channel Island of Alderney) rather than Jerseys. The British Jersey Cattle Society was founded in 1878. One of the oldest herds in England is kept by Her Majesty the Queen at Windsor.

Jerseys first came to North America with George Poingdestre, his wife Susanna, and their children when they emigrated from England to Middle Plantation (now Williamsburg, Virginia) in 1657. The cattle came from Poingdestre's family home, Swan Farm, on the Isle of Jersey.

In 1817, Richard Morris, a member of the Philadelphia Society for the Promotion of Agriculture, wrote that "I have on my farm in the Delaware a cow of the Alderney breed. . . . She has been fed in the usual way on potatoes, and yielded eight pounds of butter [per week]." (The quote can be found in *The Guernsey Breed* by Charles L. Hill [see Resources].)

The American Jersey Cattle Association was founded in 1868, and by 1910 more than a thousand head were imported annually into the United States. Jer-

BELFAIRS AND BELMONTS

The Belfair is a dual-purpose, predominately dairy breed composed of 50 percent Dexter and 50 percent Miniature Jersey genetics. It was named and promoted by Robert Mock of Babydoll Southdown sheep fame and is now registered by the American Miniature Jersey Association & Registry (see Resources).

The International Miniature Cattle Breeders Society and Registry maintains a herd book on Belmont cattle, which are also bred from Dexters and Mini Jerseys and contain up to 75 percent of either breed. Belmonts must be sired by a registry-approved Dexter or Lessor Jersey (Miniature Jersey) bull. They must be 42 inches (107 cm) tall or under at maturity (three years of age) to be classified as Miniature Belmonts or between 42 inches and 48 inches (122 cm) at maturity to be classified as Mid-Size Belmonts.

MARKETING MINIATURE CATTLE

What is my best advice for anyone interested in raising miniature cattle for profit? Go directly to Professor Richard Gradwohl's International Miniature Cattle Breeders Society and Registry Web site (see Resources) and read the wealth of information posted there. If anyone knows how to market miniature cattle, it's Richard Gradwohl. Don't bypass the opportunity to learn from a master.

seys became America's favorite milk producers. Over time, however, Americans bowed to the "bigger is better" principle and Jerseys began to gain dramatically in size. Nevertheless, a few farmers held on to their small, old-fashioned Jerseys; these are the ancestors of today's Miniature Jersey cattle.

British Lieutenant General Sir John Le Couteur (1761–1836), himself a native of Jersey, described Jerseys thus, "The Jersey has always possessed the head of a fawn, a soft eye, an elegant crumpled horn, small ears, a clean neck and throat, fine bones, a fine tail; above all a well formed capacious udder with large, swelling milk veins." Miniature Jerseys still look like that today.

Miniature Jersey cows are quite rare. They produce two to four gallons of sweet milk that contains 5 to 6 percent butterfat. They are gentle, people-oriented cows and easy keepers.

Miniature Longhorns

Registered by: Texas Longhorn Breeders Association of America

Use: Beef, pets

Origin: Texas, descended from stock brought to the New World as a walking meat supply by Spanish conquistadores and as breeding stock by Spanish padres and colonists

Miniature Longhorn

Size: Animals registered with the mainline Texas Longhorn Breeders Association of America can be any size; most Miniature Longhorns are purebred Longhorns registered with this group. Miniature Longhorns registered with International

Miniature Cattle Breeders Society and Registry comply with their rules: cattle must be 42 inches (107 cm) tall or under at maturity (three years of age) to be classified as Miniature Longhorns or between 42 inches and 48 inches (122 cm) at maturity to be classified as Mid-Size Longhorns.

Color: Any

Type: Rangy with beautiful long horns

Horns: Yes

Dwarfing gene present: For purebred Longhorns, no

Miniature Longhorns have all the eye appeal of their full-size cousins but are smaller, thus easier to contain and work with than full-size Longhorns. Miniature Longhorns make fine pets, and some become outstanding bucking bulls. And anyone who wants to stop traffic needs a herd of tiny Longhorns in a roadside pasture.

Miniature Zebus

Registered by: American Miniature Zebu Association, International Miniature Zebu Association, International Miniature Cattle Breeders Society and Registry

Use: Pets

Origin: Asia

Size: Not to exceed 42 inches (107 cm) at three years of age, measured behind the hump; most cows weigh 300 to 500 pounds (136–227 kg), bulls 400 to 600 pounds (181–272 kg)

Color: Steel gray to nearly white, cream, red, and black in solid colors, spotted, and brindle

Type: Refined, angular — never square or boxy; athletic and fluid in their movement. Short, straight, slick coat. Coffin-shaped skull; convex profile, never dished; almond-shaped eyes. Medium-size ears held outward or slightly upward, not drooping. Moderate dewlap; moderate to well-developed hump. All body parts in proportion.

Miniature Zebu

Horns: Both horned and polled bloodlines exist

Dwarfing gene present: No

THE LITTLEST ZEBUS

The smallest Zebus on record may be the Bonsai Brahmans discussed in the National Research Council publication *Microlivestock: Little-Known Small Animals with a Promising Economic Future* (see Resources).

Beginning in 1970, Juan Manuel Berruecos Villalobos, former director of the Veterinary Medicine School at the National Autonomous University of Mexico in Mexico City, and his colleagues miniaturized cows by selecting the smallest specimens from a herd of normal-size Brahman cattle and breeding them with one another. After five generations, adult females averaged 325 to 400 pounds (147–181 kg) and adult males 440 to 485 pounds (200–220 kg). A few of the smallest cows were only 24 inches (61 cm) tall and weighed less than 300 pounds (136 kg). Some cows produced up to a gallon of milk a day, compared with a gallon and a half from their full-size Brahman counterparts.

Other small Zebu breeds include the Nepalese Hill cattle of Nepal (35 to 47 inches [89–119 cm]; 240 to 330 pounds [109–150 kg]), the Kedah-Kelantan of Malaysia (35 to 38 inches [89–97 cm]; 400 to 480 pounds [181–218 kg]), the Sinhala of Sri Lanka (35 to 38 inches [89–97 cm]; 440 to 550 pounds [200–249 kg]), and the Hill Zebu of northern India (42 to 46 inches [107–117 cm]; 500 to 900 pounds [227–408 kg]).

The smallest miniature cattle of all are Miniature Zebus. Supporters of the breed believe that the ancestors of today's American Miniature Zebus hailed from southern India and Sri Lanka, where they are known in Hindu as *Nadudana* (small cattle). Many Miniature Zebus are less than 32 inches (81 cm) tall.

The Miniature Zebu's distinctive looks and its friendly, gentle disposition make it a favorite cattle breed for pets, a popular addition to petting zoos, and a show stock animal (the American Miniature Zebu Association hosts a full complement of sanctioned shows each year).

These little cattle are, like their larger Zebu kin, extremely hardy and both heat and disease resistant. Compared with nonhumped cattle, Zebus of all sizes have looser and tougher skin; finer, shorter, and glossier hair; fully functional sweat glands; and well-developed panniculus muscles (used in twitching) that are functional over the entire body. Miniature Zebus are, however, slower to mature than most other breeds, and they require adequate winter shelter in cold climates.

BREEDER'S STORY

Dottie Love and Tom Sale of Fancher Love Ranch, Texas

DOTTIE LOVE and her husband, Tom Sale, are college art and computer graphics teachers who also raise outstanding Miniature Zebu cattle at their Fancher Love Ranch just south of Dallas, Texas. The couple aims to produce 10 tame, halter-trained, show-quality, DNA parental verified calves per year. Most are shown prior to leaving for new homes.

Dottie and Tom show their cattle at major livestock shows. In 2009, their herd sire, FLR Mini Takka, won Grand Champion and Champion Challenge Bull at the Houston Livestock Show; he's since become an American Miniature Zebu Association Permanent Grand Champion.

When asked to talk about Miniature Zebus and the Fancher Love herd, here is what Dottie had to say.

"My husband and I bought 5 acres in 2000 and could only have mini animals. We bought several species, but when I first saw a Miniature Zebu, I was magnetized. My fascination with them is due to their combination of exotic beauty and unique personalities. Our daily interactions are a study in bovine behavior: herd dynamics, motivations, fear/aggression/contentment signals — in short, I became a zebu whisperer.

"The *Bos indicus* cattle haven't been studied much. They are 'just cows' with a few physical differences from *Bos taurus* cattle, but their behavior has not been modified by selective breeding. They seem to be less domesticated and therefore appear unpredictable. But in my experiences with them, they're very predictable; just predictably different from *Bos taurus* cattle. The difference can be likened to the difference between horses and donkeys.

"When they're stressed, their fight-or-flight instinct is at a lower threshold. If a zebu gets panicked, all progress made that day evaporates. I like to boast that I can do anything with any of my cows as long as I work slowly, quietly, and alone. To move my cattle carefully and calmly, I use the cattle-handling techniques espoused by Dr. Temple Grandin and authorities on the Web. I begin working with calves as early as the mother lets me, so they're tame when I sell them. I've only had to throw a lawn chair at one aggressive mother! My two mature bulls are extremely tame, but I have to be very careful around their heads and never touch their polls.

"I started with a wild female and her heifer. Then I bought a bottle-baby bull calf sight unseen off the Internet, planning to breed him. Luckily I received advice — not from the breeder — to steer the bull calf, as a bottled bull calf will almost always become overly aggressive to humans at maturity. After that I bought a bull and a few cows from a breeder who became my Zebu mentor and a good friend. We formed the American Miniature Zebu Association to promote the breed with shows, seminars, and educational material. We now have an extensive Web site, an Internet chat group, and a registry and have sponsored many national shows and been accepted into several state fairs.

"I can't imagine a Zebu-less life! We have raised two bottle babies. One slept in a playpen at night, and I had to teach the other how to eat grass. I never keep more than 25 animals at a time, so I can spend time with each animal.

"There is definitely a strong market for Miniature Zebus. Whenever we have American Miniature Zebu Association shows, our aisles are packed with interested visitors: families with kids eager to get involved with shows, retired folks who'd like to have cows that are easy to handle, people who'd like to have a twentieth-century family-size milk cow, owners of small acreages who want a lawnmower that never breaks down. The Mini Zebu can give you a gallon of milk a day. The milk is high in fat and of excellent quality — I'd like to have a gourmet Zebu ice cream business!

"There is always room for new breeders. Hardly anyone knows that small cattle even exist. The prices that our cattle bring are tied not to the sale barn or beef markets but to functional conformation, docile temperament, and their exotic breed character. Zebus are not typically raised for beef, as the breed is quite lean. Junior rodeos, petting zoos, and pony rides that feature Mini Zebus are a big attraction.

"To a new breeder, I would say smaller does not automatically mean better. In fact, breeding only for a small size can cause birthing problems and delayed maturity. Predators can be a problem, especially if cattle have been dehorned.

"Don't buy the lowest-priced animal. Buy the absolute best herd sire you can afford — he's half your herd — and buy the animal with the best temperament. Some people believe that disposition is inherited, but I do not. Herd dynamics, along with the handling methods of the breeder, will influence the attitude of the animal. Buy from a breeder who knows his or her herd and will give you support after your check clears. A bottle-fed steer is the best pet and the most economical in price. Never keep a bottle-fed bull calf as a bull. Their emotional bond with you means that you're part of their herd and you become a threat to them at breeding time.

"Find a vet before you need one. Some vets are uncomfortable dealing with small or unusual breeds; don't be offended if they refuse your business — they want to protect you from their lack of experience."

15

Miniature Equines

IF DONE RIGHT, you can certainly make money raising miniature horses, donkeys, and mules. That said, miniature equines are appealing for reasons beyond their financial worth. They can do many jobs, from driving to showing, packing (with miniature donkeys and mules), and therapy for people who are confined indoors. They're easy to house, feed, and care for; and minis are priced for everyone's budget, ranging from $150 into the five-figure range.

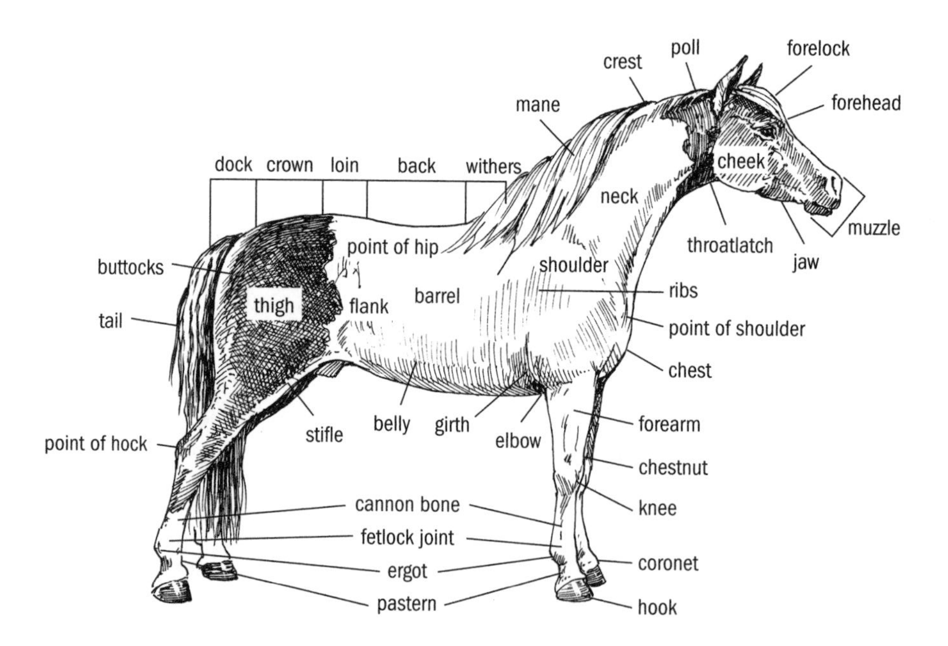

Parts of a miniature horse (same terms apply to donkeys and mules)

Think like an Equine

Miniature horses, ponies, donkeys, and mules think and behave exactly like their larger cousins.

Equine Senses

The manner in which miniature horses, ponies, donkeys, and mules see, hear, taste, smell, and experience sensation are all pretty much the same.

Vision

An equine's vision is its primary means of detecting danger, so sight is very important to horses, donkeys, and mules. Both eyes can be used together to focus on an object (binocular vision) or independently of one another (monocular vision). Equines have blind spots directly in front of and behind them. They can see very well to the sides and reasonably well toward the back, especially when the head is lowered. By raising or lowering his head or turning it to one side, a horse, donkey, or mule can more clearly focus on specific objects, whether near or far. Equines can quickly adjust their focus from distant to near objects, and they easily detect the smallest movement, even from far away — a trait especially important to wild and feral equines.

Equines have superior night vision, though it's not as acute as that of cats. Their eyes need time to adjust when moving from light into darkness, hence an equine's usual hesitancy when being loading into a dark horse trailer.

And equines discern some but not all colors (though researchers disagree about which colors they actually see).

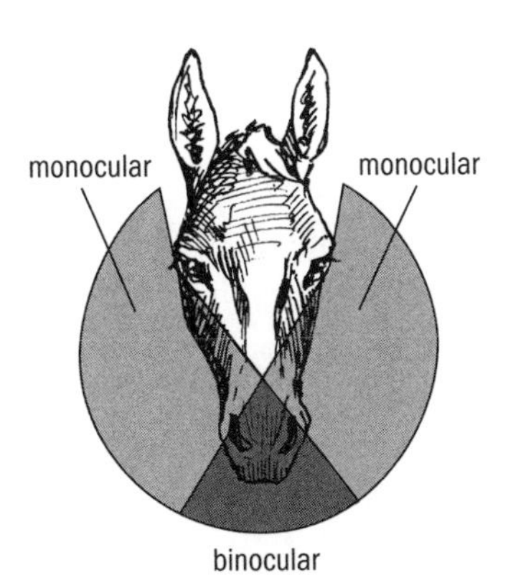

Equines use each eye independently to see to the sides and binocular vision to see to the front.

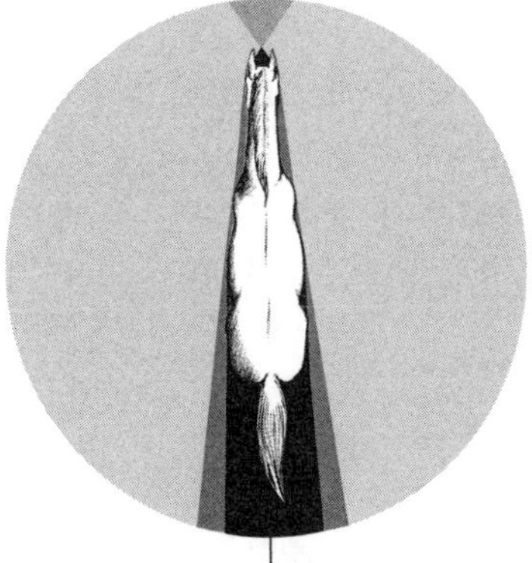

Equines see well to the sides, and reasonably well in the front, but they have a blind spot directly behind them.

THE FLEHMEN RESPONSE

When a horse, pony, donkey, or mule flips its lip up and back in a "horse laugh," it isn't amused; it's *flehmening*.

The flehmen response (from the German word *flehmen*, meaning "to curl the upper lip"), also called the flehmen position or flehmen reaction, is a grimace most ungulates, cats, and a few other mammals make to examine scent. Miniature llamas, equines, sheep, goats, pigs, and cattle all flehmen.

When an animal flehmens, it flips its upper lip upward and back to expose and draw odorants into its Jacobsen's organ, a pheromone-detecting organ located in the roof of its mouth. Intact males do this to determine if a female is in heat, but both sexes flehmen after sniffing unusual scents, especially the manure or urine of unfamiliar members of their own species.

An animal flehmens to draw scent into a pheromone-detecting organ in the roof of its mouth.

Sense of Smell

An equine's sense of smell is incredibly acute. Equines refuse foods that smell different from their usual fare, making it difficult to persuade them to eat feed laced with medication or dewormer. They frequently sniff objects and each other (as well as each other's urine and dung) to identify their surroundings. Mares and jennies recognize their neonatal foals by smell. Equines also learn to associate certain odors with friendly or scary situations. Some, for instance, misbehave when they smell the medicinal aroma of an approaching vet.

Hearing

Equines' sense of hearing is very similar to that of humans, but they can probably hear a higher pitch than we can.

Equines' ears are controlled by 10 separate muscles, allowing them to turn their ears in almost any direction. Their ears capture sound independently of each other, but sounds are processed and analyzed at the same time. When an equine hears or sees something unusual, it pricks both ears in that direction to capture the minutest sounds and funnel them to the ear canal and the brain.

Because they have relatively larger ears that capture a lot of sound, miniature donkeys and mules likely hear much better than shorter-eared miniature horses and ponies do.

Taste

Equines prefer sweet and salty to bitter or sour flavors. They rarely consume poisonous plants, owing to their bitter taste.

Touch

Equines are extremely sensitive animals with exceptional tactile perception. Their most sensitive areas are around their eyes, ears, and muzzle; their withers, ribs, flanks, and legs are also sensitive. Overall, horses and ponies are more sensitive than donkeys, with mules falling someplace in between.

Virtually all types of equines enjoy being scratched when groomed, especially if the spot is itchy.

Social Order

A well-defined pecking order or herd hierarchy exists within every herd of equines, large or small, and every newcomer has to earn a place in the group. Where an individual stands in the hierarchy depends on its age, sex, personality, aggressiveness toward other herd members, and, in the case of mixed herds, its species. Unweaned foals assume their dam's place in the order and often rank immediately below her after weaning.

Threats

Equines offer a warning before biting or kicking, except when they're startled or extremely angry. They threaten by offering to bite or kick their adversaries. Mild threats often occur at feeding time, when defending a favorite resting spot, or when mares are teaching their foals to behave.

Bite threats involve fierce glares, laid-back ears, abrupt upward jerking of the chin, or extending the head and neck toward the opponent. Stronger language involves pinned ears and a mouth slightly opened as if to lunge and bite.

When giving a mild kick threat, an equine lays its ears back and swivels its rump toward its opponent. Escalating threats include lifting or stomping a hind leg, lashing the tail back and forth, and, in donkeys and many mules, hopping on the hindquarters without actually kicking.

Flight or Freeze

Horses and ponies flee when faced with a scary situation, whereas donkeys and mules are wired to freeze in their tracks. Because of this trait, donkeys and mules are unfairly pronounced stupid or stubborn when the truth is that they're confused or afraid.

Also, unlike horses and ponies, donkeys and mules rarely put themselves in harm's way. Handlers misread this response and think they're being stubborn or cantankerous when, in fact, they are simply refusing to do something they think might endanger their life, like working beyond their capability or entering a dark, scary place they've never been before, such as a poorly lit or unfamiliar trailer.

Safety in Numbers

Equines feel safer in groups and fret or panic if isolated from their herd (or even a favorite companion). This tendency hails from the time when their ancestors ran free, and herd outliers and the slow or careless became some predator's lunch. Horses are more apt to panic than the far more stoic donkey, but separation anxiety takes a toll on both species.

Friendship

Equines create close bonds with one another and frequently with other creatures around them, such as barn dogs or humans who dish out their supper. Mutual grooming strengthens these bonds. Donkeys form especially strong bonds with their companions and often suffer deep emotional pain, leading to depression and refusal to eat, when a close companion is sold or dies.

Miniature Horses

Horses and ponies belong to the subspecies *Equus caballus* of the broader family Equidae. Ponies are smaller versions of horses, but miniature horse owners claim their Lilliputian equines are horses, not ponies. This isn't entirely correct.

DYSTOCIA STATISTICS

According to the American Association of Equine Practitioners, dystocia occurs in 15 to 20 percent of miniature horse births compared with 2 to 4 percent of standard-size horse foalings. This may be due to fetopelvic disparity (a large fetus attempting to pass through a small mare's pelvis) or fetal malpresentation (abnormal positioning of the foal).

Origin of the American Miniature Horse

A great deal of folklore surrounds the origin of today's miniature horse. Much of this lore is quite colorful but unverifiable.

We do know, however, that in 1880, young Lady Estella Hope and her sister, Lady Dorothea Hope, began breeding tiny Shetland ponies at their family's estate, Hopetoun House, in Scotland. By 1900 the ladies had sold ponies to both Queen Victoria and the Empress of Russia. Just prior to World War I, they and their ponies moved to Robertsbridge, in Sussex, from where they continued to breed and show ponies until the death of Lady Estella in 1952 at age 92. A photograph of Lady Hope driving four Shetlands less than 36 inches (91 cm) tall, featured in the February 1901 issue of the *Black and White Budget*, is one of the first photos ever taken of a British four-in-hand (a carriage drawn by four ponies).

We also know that by the mid-1800s shipments of "pit ponies" were regularly imported from Britain and the Netherlands to work in Appalachian coal mines. Most were shaggy Shetland ponies bred for working in the coal mines of Wales. Some were very small; in 1888, import records indicate one such pony named YumYum stood just 31 inches (79 cm) tall.

During the early 1900s, importer Norman Fields of Bedford, Virginia, began keeping and breeding the smallest of the animals he imported from Britain, calling them "midget ponies." He bred these for more than 50 years, and by 1964 his herd numbered 50 head.

Walter Smith McCoy of Roddenfield, West Virginia, also founded his herd with imported pit ponies around the same time. At one point he collected all of the ponies he could find that were less than 33 inches (84 cm) tall and used these to found his herd of midget ponies. In his self-published book, *The Story of the World's Smallest Ponies*, he wrote, "I discovered that the smaller the ponies were, the more they sold for — usually five or six times as much as the large ponies would bring." He went on to say, "I found 10 or 12 ponies that small [under 33 inches]. From these few ponies, by continually breeding the smallest down to the smallest, I now have 100 ponies under 32 inches tall, 20 of which are under 30 inches." The tiniest of all was his pride and joy, a mare named Sugar Dumpling. She was 20 inches (51 cm) tall and weighed only 30 pounds (14 kg). On September 16, 1967, McCoy held the first miniature horse sale, at which he sold an array of the "world's smallest midget ponies," many of which were used to found additional herds in North America.

So how did midget ponies, also sometimes called midget Shetlands or miniature ponies, come to be called miniature horses? To understand, we'll need to take a brief peek at the history of the American Shetland Pony.

HOW BIG IS A MINIATURE EQUINE?

MINIATURE HORSES

According to the American Miniature Horse Association, a horse is considered miniature if it stands 34 inches (86 cm) or less, measured at the last hairs of the mane. In American Miniature Horse Registry terminology, an animal standing 34 inches or less, measured at the last hairs of the mane (class A), or one between 34 inches and 38 inches (97 cm) tall (class B) is considered miniature.

MINIATURE DONKEYS

The American Donkey and Mule Society/Miniature Donkey Registry registers donkeys standing up to 36 inches (91 cm) tall, measured at the withers; the International Miniature Donkey Registry registers miniature donkeys up to 38 inches (97 cm) tall.

Oversized or Class B

The American Donkey and Mule Society/Miniature Donkey Registry registers donkeys 36.01 to 38 inches tall in their Class B studbook.

MINIATURE MULES

Miniature mules are the offspring of a miniature donkey jack bred to a miniature horse mare. The American Miniature Mule Society registers them in two sizes: Class A (under 38 inches [97 cm]) and Class B (38 inches to 48 inches [97–122 cm] tall).

The British Shetland Pony hails from the Shetland Islands off northern Scotland. Small equines have inhabited those windswept islands since the Bronze Age, and while their origin is uncertain, it's believed they are related to the ancient Scandinavian pony because the Shetland Islands were physically connected to Scandinavia until the end of the last Ice Age, around 10,000 to 8,000 BC.

Shetland Islanders used stout, short-legged Shetland ponies to pull carts carrying peat and seaweed. The first written record of these ponies was made in 1603 in the *Court Books of Shetland*. The Shetland pony became the first pony to have its own breed society when the Shetland Pony Stud Book Society was formed in 1890. Today's minimum height for a British Shetland is 28 inches, and its maximum is 42 inches (71–107 cm), or 10.2 hands; these sizes are also historically accurate.

Shetland ponies were first imported to the United States by Eli Elliot in 1885. Many importations followed, primarily to provide pit ponies for mines in the South. Better ponies provided foundation stock for the development of the American Shetland Pony.

In 1888, the American Shetland Pony Club was formed to, according to the club's current Web site, "preserve the bloodlines of the Shetland Pony while improving and refining the breed."

Ponies registered in the organization's first studbook were small, stocky beasts like their British ancestors; 27 percent were 38 inches (97 cm) or less (the current maximum height for Class B American Miniature Horse Registry miniature horses) and 36 ponies were between 28 and 34 inches (71–86 cm) tall.

Shetlands were frequently crossed with ponies of other breeds, including Hackney and Welsh ponies, to produce a taller, leggier, more streamlined pony that was wildly popular from the 1920s through the early 1960s. During this period, outstanding animals sold for astronomical prices. Primarily because of overbreeding, the market became saturated and the bottom fell out of the Shetland pony industry practically overnight. Ponies that sold for high four-figure prices only years before now exchanged hands for a few hundred dollars. In some places it was hard to give away the average Shetland pony.

As interest in Shetland ponies plummeted, an increasing number of pony people, looking for a way to recoup their losses, became interested in producing the type of extra-small pony that continued selling well even in the failing Shetland market. In 1971, a group of breeders approached the American Shetland Pony Club about establishing a separate studbook for midget Shetlands. The registry acquiesced, naming a committee to formulate the rules and regulations for the new registry. The committee, in a forward-thinking flourish of salesmanship, decided to distance the new breed from its failing Shetland forbears by dropping the words *Shetland* and *pony* and calling them "miniature horses" instead.

And that's how they're known today.

MINI EQUINE TRIVIA

- The first class for "Midget Ponies" was held at the Shetland Congress national Shetland pony show in 1948.
- Shetlands and American Miniatures come in a full palette of equine colors, including a wide range of dilute coat hues such as crèmes and duns, and any eye color is accepted. Shetlands never come in Appaloosa colors, but American Miniatures can and do.

PHYSICAL DIFFERENCES AMONG MINIATURE DONKEYS, HORSES, MULES, AND HINNIES

Characteristic	Miniature Donkey	Miniature Horse	Miniature Mule or Hinny
Back	Lacks sixth lumbar vertebra in spinal column	Most have a sixth vertebra	Varies
Chestnuts	Front chestnuts only; flat, smooth, leathery	Front and rear; thick and horny	Varies
Chromosomes	62	64	63
Ears	Long and thick-textured	Smaller, usually more shapely; thinner-textured	Intermediate in size and thickness
Estrous (heat) cycle/estrous portion	23- to 30-day interval/ average 6–9 days	21- to 25-day interval/ average 3–7 days	Most mollies cycle albeit erratically
Eyes	Larger; heavier, D-shaped, bony eye orbits (heavy brow ridges) set farther out on side of head	Smaller in proportion to head; round eye orbits, not as bony	Intermediate with somewhat D-shaped orbits and brow ridges
Fertile	Yes	Yes	No
Gestation	360–375 days (or more)	335–345 days (or more)	Nonfertile (although mollies can be used as embryo recipients
Head	Broader forehead; much deeper jaw	Comparatively narrower forehead; smaller, shallower jaw	Intermediate in size and shape
Hindquarters	Longer, more steeply angled hip bones; hindquarters higher, steeper, narrower; much more smoothly muscled than horse	More muscle mass than donkey and generally much more muscle definition	Intermediate; generally smoother-muscled than a horse but with more muscle mass than a donkey

The Horse or Pony Breed You Need

The American Miniature Horse isn't the only small horse breed available in North America. Additional small breeds of *Equus callabus* exist, among them the American Shetland Pony, the British Shetland Pony, the Critically Endangered Caspian horse of Iran, and the Falabella horse of Argentina.

American Miniature Horse

Registered by: American Miniature Horse Association, American Miniature Horse Registry, among others across the world

PHYSICAL DIFFERENCES AMONG MINIATURE DONKEYS, HORSES, MULES, AND HINNIES

Characteristic	Miniature Donkey	Miniature Horse	Miniature Mule or Hinny
Hooves	Smaller in proportion to body size, boxy, more upright (hoof angles average 65 degrees), tougher and more elastic; larger, well-developed frog with a thicker sole	Larger in proportion to body size; oval, less upright (hoof angles average 55 degrees); smaller frog with thinner sole	Usually boxy, steeper than horse
Male reproductive organs	Much larger than those of a stallion; rudimentary nipples on sheath; larger scrotal vessels and thicker scrotal skin; more prone to hemorrhaging during castration	Smaller than those of a jack; no nipples	Intermediate; some have nipples, some don't
Mane	Mini donkeys grow short, stand-up manes (if any at all)	Long, finer-textured manes	Most donkey hybrids have sparse, fall-over manes that are trimmed to stand up like a donkey's mane
Nasal passages	Smaller than a horse's; relatively nonflaring	Medium to large; flaring	Varies; generally smaller than a horse's
Ovulation	3–6 days after onset of estrus	12–24 hours before the end of estrus	Most mollies cycle, albeit erratically
Tail	Cowlike with long, coarse "swish" or "switch" on the lower one-third of the tail bone	Long and luxurious	Donkey hybrids have a more horselike than donkeylike tail, usually with shorter hair near dock; coarser than most horses' tails
Withers	Doesn't have horselike withers (makes back appear longer)	Usually somewhat pronounced	Varies

Origin: Developed primarily using Shetland pony genetics
Size: Depending on registry, maximum heights are 34 to 38 inches (86–97 cm)
Color: Any color is acceptable
Type: Two types: a refined, horselike miniature and a cobby, British Shetland-type mini

The American Miniature Horse Registry (AMHR), a subsidiary of the American Shetland Pony Club, was incorporated in 1971 to register miniature horses 34 inches tall and shorter (they later created two divisions: Class A for minis 34 inches or less and Class B for larger individuals up to 38 inches tall). Breeders dissatisfied with the registry's early policies soon founded a plethora of offshoot groups, including the American Miniature Horse Association (AMHA) in 1978. The American Miniature Horse Registry and the American Miniature Horse Association remain the two major players, although several smaller North American registries exist.

American Miniature Horse

Early breeders who used registered Shetlands in their breeding programs, and who wished to distance their breeding programs from those of the failing Shetland industry, often listed these animals' breeding as "unknown." Others, however, embraced their minis' Shetland heritage. For instance, Gold Melody Boy, grandsire of Boones Little Buckaroo (1983 AMHA National Grand Champion) was sired by and out of registered Shetland ponies. Buckaroo's maternal grand-dam was sired by a Shetland, too.

And miniature breeders whose animals can be traced to a particular group of Shetlands, Arenosa-prefixed Shetland ponies, are usually quick to publicize that fact. Audrey and Clinton Barrett founded the Arenosa Pony Farm of Victoria, Texas, in 1941. At a time when many people were crossbreeding their Shetlands to Hackneys, Audrey Barrett preferred to intensely linebreed her ponies to the Barretts' best-known sire, a 39-inch (99 cm), black-and-white pinto named Kewpie Doll's Oracle. The average height of ponies in the Arenosa herd was 40 inches (102 cm), though many were small enough to be registered as miniatures. Three Arenosa animals were entered in Volume One of the American Miniature Horse Registry Stud Book, all with their Shetland pedigrees intact.

Another noteworthy Miniature Horse foundation stallion of Shetland lineage was Rowdy, a 34-inch (86 cm) stallion foaled in 1973. Rowdy sired Lazy N Boogerman, the AMHA National Grand Champion Stallion in 1992 that subsequently sold for $110,000 at the NFC Dispersal Sale in 1993. Rowdy's sire was the registered Shetland Kewpie Doll's Sun, and his sire was Kewpie Doll's Oracle!

While today's show-winning American Miniature Horses are sleek, horse-proportioned animals, American Miniatures run the gamut from ultra-refined to miniature draft horse type. There's an American Miniature Horse to suit every taste. Furthermore, minis of all types are hardy, friendly, generally very laid-back, easy-care animals. Five miniatures eat less than one full-size horse eats, and they take up much less room. They're perfect for retirees who want to keep horses but lack the physical wherewithal to deal with bigger steeds or for small children who aren't a good fit on larger horses. They shine in a wide array of additional uses, including showing, pleasure driving, animal-assisted therapy, and keeping the back pasture nicely mowed.

If you hope to sell American Miniature Horses at high-end prices, choose double-registered AMHA/AMHR stock. Although animals can be "hardship" registered (by paying an extra fee because both of their parents aren't recorded in that registry's studbook), the costs of hardship breeding stock are steep and the rules complicated. (AMHA currently hardships a stallion for $1,200 or a mare for $600; AMHR registers AMHA-registered stallions for $400 plus their age-appropriate registration fee and mares for $200 plus their age-appropriate registration fees.) And don't accept papers from smaller registries in lieu of AMHA or AMHR registration; most buyers don't consider them registered minis unless they're registered with at least one of the big two organizations.

Always use a measuring standard to verify any animal's true height. The recorded heights on registration certificates are notoriously inaccurate.

American Shetland Pony

Registered by: American Shetland Pony Club

Origin: Created by crossing imported British Shetlands with Hackney and Welsh ponies and selecting for refined conformation

Size: Not to exceed 46 inches (117 cm) (11.2 hands) in height, measured at the highest point on the withers. Shetlands average 42 inches (107 cm) (10.2 hands) tall.

Color: Any color but Appaloosa

Type: Two, Classic and Modern

Two Shetland Ponies, a Classic and a Modern, facing one another

The American Shetland Pony Club was founded in Morton, Illinois, in 1888; since then they've registered more than 150,000 ponies.

Today's leggy, ultra-refined American Shetlands no longer resemble their sturdy British Shetland ancestors. There are two types: Modern and Classic. The Modern is a lean, elegant, stylishly high-stepping pony suited to roadster and fine harness competition. Classics resemble American Shetlands of the 1950s. Some Classics also qualify as Foundation Shetlands. Foundation Shetlands must come from only "A" papered ponies on the pedigree printed on their registration papers and measure 42 inches (102 cm) or less; Foundation Shetlands are more conservative in type and have slightly more bone and substance.

The registry designates an animal's background by placing an "A" or "B" after its registration number. Shetlands with Hackney Ponies in their backgrounds have a "B" prefix and must show in Modern Shetland classes. Ponies with an "A" suffix do not have any Hackney heritage and can choose between Modern and Classic classes, depending on the pony's conformation and movement.

American Shetlands are very intelligent, hardy, active ponies. Classics are well suited to be children's riding ponies, and both Classics and Moderns make superb driving ponies for pleasure and shows.

LITTLE-KNOWN FACTS

- The Kennedys' Falabellas weren't the first miniature equines to grace the White House lawn. Teddy Roosevelt's sons, Quentin and Archie, owned a beautiful, 39-inch (99 cm) pinto Shetland named Algonquin that they kept at the White House during their father's time in office. Once, when Archie was ill, Quentin enlisted the aid of the family coachman and smuggled Algonquin to Archie's room by means of the White House elevator.
- The most famous Shetland pony of all time is arguably Merrylegs of *Black Beauty* fame.
- The most expensive Shetland pony in history was Supreme's Bit of Gold, who sold for $86,000 in 1958.
- American Shetland ponies are measured at the highest point of their withers, whereas American Miniature horses are measured at the last hairs of their manes. Therefore, many early-day Shetlands measuring 40 inches (102 cm) or less would have been small enough to register as American Miniature Horse Association Class B Miniature horses.

British Shetland

Registered by: Shetland Pony Society of North America, Shetland Pony Stud-Book Society (UK)

Origin: Shetland Islands (UK)

Size: Shetland Pony Society of North America: ideally no more than 44 inches with a maximum height of 46 inches (112–117 cm). Shetland Pony Stud-Book Society: no more than 42 inches (107 cm)

Color: Any except Appaloosa

Type: Compact, cobby, broad; small, elegant head with tiny ears; rounded barrel, short back, strong hindquarters; long, full mane, forelock, and tail

British Shetlands have remained unchanged since the breed evolved on the Shetland Islands hundreds, maybe thousands, of years ago. The island's harsh weather and a scarcity of feed made the British Shetland the compact, hardy animal it is today.

The Shetland Pony Stud-Book Society was founded in 1890 to maintain the breed's purity by encouraging the breeding of high-quality animals. Hallmarks of this ancient breed are short, muscular necks; compact, stocky bodies; short, strong legs with shorter than normal cannon bones in relation to their size; and intelligence mixed with character. British Shetlands have long, thick manes and tails and a dense double coat in winter to withstand harsh weather.

There are few purebred British Shetlands in North America at this time, though interest in the breed is growing. The Shetland Pony Society of North America registers British Shetlands, though not all SPSNA-registered Shetlands are of British origin. The organization also double registers animals with American Shetland Pony Society Foundation papers as well as Canadian Shetlands.

British Shetland Pony

Caspian

Caspian horse

Registered by: Caspian Horse Society of the Americas, Caspian Horse Society (UK), International Caspian Society
Origin: Iran
Size: From less than 10 hands (40 inches [102 cm]) to 12.2 hands (50 inches [127 cm])
Color: All colors except spotted (pinto or Appaloosa)
Type: According to the Caspian Horse Society of America and the UK, the Caspian is a horse, not a pony. Its limbs, body, and head are in proportion to each other. The overall impression is of a well-bred, elegant horse in miniature.

The Caspian, a breed of great antiquity and a precursor of the Arabian horse, descends from now-extinct miniature horses that lived in the region of Persia from 3,000 BC through the seventh century AD. The breed's direct connection to these ancient horses was verified by DNA testing performed in the 1990s by Gus Cothran at the University of Kentucky's Horse Genome Project.

In 1965, Louise Firouz, an American-born horsewoman living in Iran, discovered a remnant population of Caspians in the Elburz Mountains south of the Caspian Sea. She brought three Caspians back to her Norouzabad Equestrian Center in Tehran. A year later she returned to the mountains and obtained seven mares and six stallions, becoming the first serious breeder of Caspian horses in more than 1,000 years.

When Prince Philip of Great Britain visited Iran in 1971, the Shah presented him with a pair of Caspian horses. Intrigued, Prince Philip approached Ms. Firouz about exporting breeding stock to Great Britain. Over the next eight years, she exported 29 Caspians, thus providing foundation stock to fuel the rebirth of this ancient breed outside of the Middle East.

According to the American Livestock Breeds Conservancy, about 2,000 Caspians are now registered throughout the world, 600 of which are in North America. Caspians are listed as a Critically Endangered breed on the ALBC Conservation Priority List (see pages 16 and 17).

BREEDER'S STORY

Gary, Carlene, and Chase Norris
Briar Patch Miniature Overo Horses, Missouri

GARY, CARLENE, AND CHASE NORRIS breed show-stopping overo pinto American Miniature Horses on their farm near Winona, Missouri. Flashy, homebred Briar Patch horses have garnered top honors at major shows throughout the country.

When we asked Carlene to comment on the Briar Patch breeding program, here is what she had to say.

"We started in the miniature business in 1994, with just one for a pet, and then decided we wanted to start breeding them for show. We've made countless friendships over the years and learned so much from these little horses; we've never regretted it once.

"We were about five years into breeding when overos first started getting popular. We were fascinated with their color patterns and the genetics that go along with overo breedings, and we learned early on that most people love color in minis. Since our horses' bloodlines carry LWO (Lethal White Overo) and sabino overo genetics, we test all of our horses for the Lethal White Overo gene before breeding so we don't get lethal white foals.

"We like our breeding mares to be around 30 to 32 inches tall, and we like fine-boned horses for show. We try to breed for the conformation of a full-size horse, only in miniature.

"Some buyers want horses that are double registered. It depends on what registry they want to show in: American Miniature Horse Association, American Miniature Horse Registry, or Pinto Horse Association of America. For promotion apart from showing, we advertise in magazines, have our own Web site, and advertise some overseas.

"There is always room for new breeders. My advice to new breeders of miniatures is to do your research. Caring for horses takes up a lot of time. During foaling we stay home — we try not to miss any births in case there is a problem. You can't breed miniatures and then let them birth out in the pastures. Most births go smoothly, but sometimes with minis the foal's feet are turned in the wrong position or its head is back or the foal just doesn't get out of the sack after birth. We strongly recommend being there during foaling season."

Falabella

Registered by: Falabella Miniature Horse Association
Origin: Argentina
Size: Falabella Miniatures come in a variety of sizes with no height restriction; most are between 28 and 34 inches (71–86 cm) tall
Color: Any color is acceptable
Type: No specific type is preferred over another

Falabella horse

In 1845, Irish-born horse trader Patrick Newtall was traveling south of Buenos Aires in Argentina when he came upon a group of unusually small horses. He bought some and began breeding them on his Argentine ranch. Later his son-in-law, Juan Falabella, added European Thoroughbred genetics for refinement; Shetlands and miniature horses from Belgium were also added to the mix. By the early 1900s, the ranch consistently produced miniatures less than 33½ inches (85 cm) tall, and by 1927, Juan's grandson, Julio César Falabella, had amassed a herd of several hundred of these small horses. The breed came to the attention of North Americans when, in 1962, Julio sold two Falabellas to President John F. Kennedy, who gave them to his children, Caroline and John-John. Both *Time* and *Newsweek* magazines published images of them grazing the White House lawn.

Fewer than 1,000 Falabellas are registered with the Falabella Miniature Horse Association. Only small herds are known to exist in most countries, and the estimated worldwide population is only a few thousand. It is the Falabellas' prestige, rarity, and long history that set them apart in miniature horse circles. Many are double (and triple) registered as American Miniature Horses.

Miniature Donkeys

Cobby and cute, miniature donkeys belong to the subspecies *Equus asinus* of the broader family Equidae. The ones we know and love descend from pint-size donkeys brought to North America from the Mediterranean islands of Sicily and Sardinia, where they were originally used as working donkeys.

Miniature donkeys first came to America in 1929 when New York stockbroker Robert Green imported six jennies and a jack that he had purchased sight unseen during a trip to Europe. A year later marauding dogs attacked the herd, killing three jennies. The jack and the three remaining jennies comprised the first breeding herd of miniature donkeys in the United States. Green soon imported more tiny donkeys from the Mediterranean region, and by 1935 he'd amassed 52 head. Wealthy buyers such as Henry T. Morgan and August Busch Jr. also imported animals from the Mediterranean after beginning with Green-bred stock.

In the early 1950s, Daniel and Bea Langfeld of Danby Farm (already breeders of world-class Shetland ponies) purchased a miniature donkey for their daughter, who had cerebral palsy. They soon became large-scale breeders with as many as 225 donkeys in their herd. The Langfelds widely promoted their donkeys, charming the readers of national horse magazines with ads documenting the ongoing adventures of Parader's Seventy-Six Trombones (a Shetland colt) and Ricardo (his miniature donkey sidekick).

In 1958, Bea Langfeld established The Miniature Donkey Registry, which she turned over to the American Donkey and Mule Society in 1987. Today there are more than 50,000 donkeys registered in the Miniature Donkey Registry studbook, some of which have up to 500 recorded ancestors tracing all the way back to the first miniature donkeys in America. The height limit for MDR-registered donkeys is 36 inches (91 cm).

A second miniature donkey registry, the International Miniature Donkey Registry, was incorporated in 1992. It differs from the Miniature Donkey Registry in that it registers donkeys in two divisions (Class A donkeys are 36 inches [91 cm] and under, and Class B donkeys are 36.1 to 38 inches [92–97 cm] tall) and it awards registered stock two-, three-, and four-star ratings based on each donkey's conformation.

Miniature donkey

> **"Miniature donkeys possess the affectionate nature of a Newfoundland, the resignation of a cow, the durability of a mule, the courage of a tiger, and the intellectual capacity only slightly inferior to man's."**
>
> — Robert Green, America's first breeder of miniature donkeys

There's a lot to like about miniature donkeys. They're large enough for children to ride and strong enough to pull a cart carrying an adult and one or two children; they're easygoing and gentle to a fault; they're economical to keep and feed; and they're arguably the cutest creatures on God's green earth. They're readily available throughout North America, and there are also fair-sized populations in Britain, Australia, and parts of Europe. And they're the perfect choice for knowledgeable breeders who want to show a profit breeding miniature equines.

Miniatures can be purchased for next to nothing (unregistered geldings and jacks bring $100–200 in some locales) or for hefty five-figure prices. There is a strong market for high-end miniature donkeys, especially of the color du jour (at the moment, solid black with dark points).

Although miniature donkeys range in height from 26 to 36 inches (66–91 cm) (38 inches [97 cm] for International Miniature Donkey Registry stock), judges and breeders prefer donkeys in the 32- to 34-inch range.

Miniature donkeys come in all colors, including spotted. Spotted miniature donkeys (and mules!) can be double-registered with the American Council of Spotted Asses (see Resources).

Showing

Miniature donkey owners can show their animals at miniature donkey shows or open shows for donkeys and mules of all sizes and kinds. Showing donkeys is like showing horses but less stressful. Donkey and mule shows are more laid-back than most horse shows, and they're specifically designed so that competitors have a good time.

Miniatures are shown in halter classes (judges rate the donkey's conformation and movement), color classes (50 percent of the score is based on color and the other 50 percent on conformation), showmanship classes (judges rate the handler's ability to prepare and show a halter donkey, rather than the donkey itself), as well as a variety of performance classes including driving (pleasure driving, obstacle driving, team driving, reinsmanship, turnout, and harness races). There are also costume classes, trail and jumping classes (handlers lead the donkeys in these events), snigging classes (dragging a log through an obstacle course), and coon-jumping classes (in which the donkey jumps a hurdle from a standstill).

Entries can also play games such as musical tires (played like children's musical chairs), catch your ass (a timed event in which donkeys are released at the opposite end of the arena and competitors run, catch their own donkeys, and race them back to the starting line), diaper race (contestants vie to be the first to lead their donkey to the other end of the arena, diaper themselves, and race the donkey back to the starting line without the diaper falling off), and a slew of equally hilarious events.

Packing

Miniature donkeys are well suited for recreational backpacking. They're easy to handle and close to the ground, which makes loading them a breeze. And they're small (and tidy) enough to transport to the trailhead in a van or an SUV.

Recreational packing gear for miniatures falls into one of three classes: training packs, companion packs, and packsaddle and pannier (pack bags) combinations. Good packing gear of any type provides adequate padding, alleviates pressure on the spine, and is very stable, allowing for slightly different weights in each saddle bag without shifting to the side or moving while the donkey walks.

Training Packs

A training pack, also called a day pack, is a simple, soft-sided, saddlebag-like affair. It has a single girth and a built-in breastcollar and britchin (a harness that fastens around the donkey's chest and drapes across its hindquarters to keep things from sliding backward or forward when traveling uphill or downhill). The fabric connecting the two bags lies directly on the donkey's back; this is not a good thing — too much weight pressing directly on any animal's spine over a prolonged period of time can inflict permanent damage. Training packs should be used for carrying light loads on day hikes and nothing more.

Companion Packs

The companion pack is a nice choice for longer day trips or overnight camping (but not for extended trips). A good companion pack features a thick, divided pad that keeps the weight off its bearer's spine, detachable panniers (pack bags) like those used in full-scale packing gear, a single girth, and a built-in britchin and breastcollar to keep it in place. Companion packs designed for packgoats fit most miniature donkeys.

BREEDER'S STORY

Jon, Mary, and Jay Nissen of Nissen's Lazy N Ranch, Iowa

JON AND MARY NISSEN, along with their son Jay, maintain one of the oldest herds of quality registered miniature donkeys in North America. Their Lazy N Ranch in north-central Iowa has been home to world-class donkeys and mules for close to 35 years.

The awards won by Lazy N donkeys and mules would fill a small house; their LN prefix is a byword in donkey circles, and miniature donkeys from their herd consistently top the country's most prestigious sales — proof that the Nissens know their stuff. We asked them to share their thoughts with us.

"Describing the miniature donkey breeding program at the Lazy N Ranch is easy: miniature Mediterranean donkeys are our passion! We have owned or bred more than 50 State and National Champions. Our goal at the Lazy N is to improve and preserve the true miniature donkey. We breed for eye appeal through better symmetry and attractive colors. We also breed to produce a stronger, more useful donkey through better conformation and temperament selection. Our goals are influenced by the market, by the show ring, and, most important, by our own personal concept of the ideal original old-world Mediterranean Miniature Donkey.

"We've always loved working with longears. We broke and sold a lot of mules, but we wanted our business to pay us back a little more. We decided that driving the little ones and doing parades would be fun and still make money.

"We had a lot of fun with our project. We once drove a hitch of 24 miniature donkeys in the Milwaukee Circus Parade! We went a lot of places and met some great folks, all the while enjoying our beloved longears and slowly working into breeding and raising quality miniatures.

"We have always been obsessed with pedigrees. Even when we raised mules, our mares had nice backgrounds to go with their looks. Our fascination with history and our knowledge of the importance of pedigrees launched many long letters of correspondence with Bea Langfeld, the originator of the Miniature Donkey Registry. This information helped us lay the foundation of the breeding program here at the Lazy N. Most of our donkeys have extensive pedigrees, tracing many times to the original imports. They reproduce their good qualities and survive and thrive with few problems, just as their ancestors did.

"We also prefer bold, rich colors with no diluting genes — good reds, blacks, and contrasting spots — but we have super, quality individuals in gray dun and brown, too.

"That obsession with pedigree continued as we got deeper into the miniature donkey breed, and we found many had backgrounds that trace back to the original imports. These donkeys had a stocky, blocky look that was unlike the more leggy-type standard donkey. Thick donkeys with larger, rounder rumps are also better at keeping the harness in place when driving, so we breed for that type.

"The miniature donkey originated primarily in Sicily and Sardinia — two very small areas. It is very possible that these animals were inbred or at least linebred. The breed thrived and certain traits became fixed. Americans began importing miniature Mediterranean donkeys in 1929, and at the same time some larger donkeys known as 'Irish donkeys' were being imported. In some instances, these Irish donkeys may have been crossed with the miniatures; this is possibly how a taller, leggier miniature donkey type originated.

"Today we see more blocky, smaller-type miniature donkeys, probably because breeders are collecting donkeys of original type and concentrating on breeding those genetics.

"As with any animal and in any economy, there is always a demand for good stock. As a breeder, one must be careful to not get caught up in trends just to make a fast buck. This can be detrimental to the breed. One should instead continue to breed to improve the quality and consistency of each generation.

"There is room for new miniature donkey breeders, but they need to learn and implement the best breeding methods. Too many breeders are blindly breeding and the outcome is more pet-quality animals, which is an easy market to overpopulate.

"My advice to new breeders is to develop an eye for quality and balance. Study *genetics* if you are going to breed animals; don't just raise babies. Have a system of breeding, and learn how to apply good principles to concentrate desirable genes and achieve more consistency of type. In miniatures, smaller is better as long as quality is not sacrificed. If you are always outcrossing, a larger, stronger hybrid generally results; thus correctly implemented linebreeding is a valuable tool in a 'masters' breeding program.

"Certain animals and bloodlines have 'shrinking' genetics. Only outcross using the bloodlines of animals that have and express the shrinking factor; otherwise your miniatures are likely to increase in size. Quality is important, too, but realize that the 'perfect' animal does not exist. Find the strengths and weaknesses of your animals and then make crosses that will improve their offspring."

Sawbuck-style packsaddle

Packsaddle and Pannier Combinations

Packsaddle outfits include two large, detachable panniers and a sawbuck-style packsaddle (the sturdy type of packsaddle usually associated with packmules; see illustration) secured with one or two wide girths as well as a britchin and breastcollar. The sawbuck provides complete spinal relief, making it possible for the donkey to carry considerably heavier loads than what can be carried with the other packs. Unfortunately, packsaddles and panniers for miniature donkeys are hard to find (donkey entrepreneurs, take note), so miniature donkey packers must adapt goat-packing saddles (which fit most miniatures surprisingly well) or custom-build their own.

Driving

Miniature donkeys are the perfect driving animal for nervous or novice drivers, yet they're also a joy for anyone who loves driving but doesn't like the fuss of coping with a potentially fractious horse. While frightened horses and ponies spook and flee, donkeys freeze in their tracks when confused or afraid. We'll talk more about that in a bit.

Miniature Mules (and Hinnies)

Mules are the sterile offspring of a donkey jack and a mare. Hinnies are the less-common opposite sterile cross: they are sired by a stallion and out of a jenny. They come in a wide range of sizes, from massive draft mules to tiny miniatures, and it is hard to tell the two types apart by looking at them. Miniature mules (and hinnies — but to keep things simple, let's refer to them both as mules) are a growing sensation among miniature livestock enthusiasts; little wonder, as they're so darned cute!

Anything you can do with a miniature horse or a miniature donkey you can do with a mini mule, with the exception of breeding it (mules are hybrids and, with very few exceptions, sterile). Their rarity makes them stand out in virtually any venue, so if you're looking for an unusual miniature, think mules.

The American Donkey and Mule Society (see Resources) registers mules of all sizes, but the primary registering body for miniature mules and hinnies is the American Miniature Mule Society. The organization registers mules in two divisions: Class A (under 38 inches [97 cm]) and Class B (over 38 inches and under 48 inches [97–122 cm]).

Miniature mules can be shown at all-mule venues and at combination donkey and mule shows in miniature mule classes as well as classes written for full-size mules. Class topics are far-ranging and include halter, showmanship, riding, driving, and all the fun events like coon jumping, snigging, and costume classes. Mini mules make first-class recreational packers, too.

Larger miniature mules are sturdy enough to carry children and small adults.

BREEDING MINIATURE MULES

Potential breeders must realize that not all donkey jacks breed mares. Jacks prefer females of their own kind. By the same token, some mares refuse to be bred by a jack unless previously exposed to donkeys. For the best results, jack foals should be isolated from jennies at weaning time and kept with mares during their formative years if they're to be used in a mule breeding program. Not having jennies on the same farm (or at least within sight of young jacks) helps.

Stallions are usually willing to breed jennies to produce hinnies. The conception rate for this cross is extremely substandard, however, and relatively few hinnies are born.

BREEDER'S STORY

Jim and Nancy Eubanks of Summer Shade Farm, Ohio

JIM AND NANCY EUBANKS are longtime breeders of longeared equines. Active in the Midstates Mule and Donkey Show Society, they show and breed miniature mules and miniature and standard-size donkeys at their Summer Shade Farm, located near Bethel in southern Ohio.

We asked Nancy to tell us about their cute miniature mules that are registered with the American Miniature Mule Society. This is what she said.

"We started raising standard donkeys in the 1970s, then had Mammoths, and a decade later we decided to switch over to miniature donkeys. In the late 1990s I wanted to see if we could raise some miniature mules that had some color to them. We bought four registered miniature mares — two leopard Appaloosas and two pintos — to use with our spotted miniature jack. We turned them in with the jack, but we never saw him breed any of them and were disappointed. To our shock, one night the next spring we spotted a little mule standing beside one of the pintos! He was the first of three mini mule births we had that first year, and since then we've had a total of a dozen or so newborn mini mules. In 2007, we had our first miniature hinny. We have very much enjoyed our 35-plus years raising donkeys and miniature mules.

"Getting color on these little mules is more difficult than it first appeared. Our sorrel and white pinto mare always throws the typical mule color pattern of some white socks or stockings and a white rump splash, along with one blue-eyed white mule foal. One of the Appaloosa mares had a gorgeous leopard mini mule, TNT, but she hasn't had another loud-colored one since!

"There is a market for miniature mules, but demand seems to be affected by the economy, as it is with most other livestock. Our little mules are all handled from birth, but we try not to sell them to anyone who does not have some mule background or experience because we feel that they need that to be successful (both mules and buyers)!

"There is room for new miniature mule breeders but not on a large scale due to the economy. We do get calls for miniature mules from all over the country, and we have sold them as far away as California.

"I can only speak about miniature donkeys, horses, and mules, but they are much easier to work with than larger animals, simply because of their size. People always seem interested in animals that are smaller or larger than the norm. My advice to new breeders is to do your homework before going out to look, and buy the best animals you can reasonably afford based on your breeding goals."

Starting with Miniature Equines

The most important thing to remember when beginning with mini equines is to buy animals suited to your needs. If you want to breed, buy the best individuals you can afford, taking into consideration conformation, disposition, breeding records, registration status, pedigree, and size. If you want to show in halter classes, choose halter-quality animals. When showing in performance or driving for pleasure, select strong, sound animals with good minds. A show mare, stallion, jack, or jenny with many championships under its belt may cost $20,000; a pretty cob-type American Miniature mare or a well-trained miniature donkey gelding to drive costs closer to $500.

Buying Miniature Equines

Everything we talked about in chapter 4 applies to buying miniature horses, ponies, donkeys, and mules, but there are several additional points to consider.

Dwarfism

Dwarfism is rampant among miniature horses and, to a lesser degree, miniature donkeys. If you plan to breed, learn to recognize the subtle signs of dwarfism we discussed in chapter 4. Reread that information now; it's that important.

Teeth

Miniature equines, particularly American Miniature Horses and miniature donkeys, are prone to a number of dental abnormalities that may or may not be easily corrected. These include:

- **Poor occlusion.** A severe underbite is strongly associated with dwarfism, and poor occlusion of any type requires ongoing dental work in the form of floating (rasping) teeth to remove sharp edges that form when grinding surfaces don't line up, so look for miniatures with good bites.
- **Crowding.** Surprisingly, the teeth in the average mini's mouth are the same (or nearly the same) size as those as full-size horses and donkeys. Large teeth in a small jaw equate with painful crowding and even impacted teeth. Veterinary dental intervention may be necessary.
- **Retained caps.** Equines lose their baby teeth between two and three-and-a-half years of age. Normally, emerging permanent teeth push out caps (pieces of baby teeth remaining after their roots dissolve) as they come in. In minis, caps sometimes adhere to erupted permanent teeth. This may cause swelling below the eyes, localized sinus infections, and runny eyes. Fortunately, you can usually remove caps yourself.

Type- and Breed-Specific Problems

Learn about the maladies that affect the species and breed you're interested in owning. For instance, all of the animals described in this chapter, except the Caspian, are more prone to obesity and feed-related maladies such as laminitis (see box below), hyperlipemia (see box, page 297), and Cushing's disease (page 131) than are their full-size counterparts. These conditions can be prevented (or, in the case of Cushing's, treated), but every owner must be able to recognize early symptoms.

LAMINITIS

Laminitis is caused by inflammation inside the sensitive laminae (layers) of the hooves. Chronic laminitis is often referred to as *founder*. There are two sorts of laminae in an equine's hoof: sensitive laminae attached to the coffin bone and insensitive laminae attached to the inner hoof wall. The function of the laminae is to keep the coffin bone suspended within the hoof capsule and to allow for normal hoof wall growth over the coffin bone. *Founder* means to sink to the bottom (like a ship); this is precisely what happens when the coffin bone sinks down through the hoof capsule due to an afflicted animal's body weight tearing through weakened laminae. Obese equines are exceedingly prone to developing laminitis, as are small equines on overly rich diets or ones that have even temporarily gorged on lush spring grass or rich feed. Every owner of equines large or small *must* learn to recognize the earliest stages of laminitis.

Early symptoms: Heat in the hoof wall; increased digital pulses in the pastern; lying out flat, reluctant to rise; hesitant gait ("walking on eggshells"); standing in a sawhorse stance with the front feet stretched out in front to alleviate pressure on the toes and with the hind feet pushed farther forward than normal to bear more weight. If in doubt, call your vet. Unless quickly and aggressively treated, laminitis often cripples an equine for life.

A typical laminitic rocking-horse stance

Facilities and Feeding

Miniature equines can be housed in a variety of structures, such as existing horse facilities, three-sided field shelters, and moveable structures such as large Port-A-Huts (see Resources). The most important things are that their living quarters be kept draft-free and provide adequate cover from the weather.

Fences must be tall and stout enough to keep predators out and miniature equines in. Equines do well in cattle-panel, woven-wire, and electric fencing, and they quickly adapt to Electro-Net and similar portable fences. Barbed wire is unsuitable for any type of equine. High-tensile fencing is a poor choice for reactive horses (mules and donkeys don't tend to race blindly into hazards the way excited horses do), because on impact the fencing slices into the animal's skin.

It can't be said enough: don't overfeed easy-keeping miniature species. Miniature horses and small-breed ponies are apt to eat themselves into obesity and its associated dangers. High-quality grass hay should be the basis of a mini equine's diet, and most animals can have free access to it. Youngsters, mares in late gestation, and lactating mares need more protein. Limit treats; those calories add up.

HYPERLIPEMIA

Hyperlipemia is a potentially fatal condition triggered by starvation or a stressful event that causes an enormous mobilization of fat from the tissues to the liver. The liver is overcome by the load, fills with fat, and fails to function properly, usually resulting in death. Miniature horses, donkeys, and mules, as well as small-breed ponies — particularly obese ones — are all very prone to developing hyperlipemia if they stop eating for more than 24 hours. Mares, jennies, and mollies (female mules and hinnies) are more likely to develop hyperlipemia than are stallions, jacks, johns, or geldings. Never, for any reason, drastically cut back a small equine's feed in an effort to help it lose weight.

Symptoms: Anorexia, lethargy, weakness, and depression, followed by jaundice, ventral edema, head pressing and circling, and other indicators of liver and kidney failure.

Treatment: If a small equine stops eating and you suspect hyperlipemia, call your vet without delay. Early intervention is imperative; if untreated, hyperlipemic animals become increasingly depressed, uncoordinated, and usually die within 10 days.

Foaling Miniature Equines

Many people enter the miniature horse and donkey breeding business thinking it's exactly like birthing full-size horses. It's not.

For example, miniatures of all sizes, both large and small, are far more prone to experience serious foaling dystocia than are full-size mares. Because of the difficulty in manually repositioning foals inside such tiny reproductive tracts, Caesarians aren't terribly uncommon.

And miniature horse mares are notorious for experiencing premature placental separation, commonly known as red bag deliveries (see box below).

You *have* to be there when a miniature equine foals. If you (or better yet, a vet) are there to help, everything usually works out fine.

Birthing Miniature Equines

Let's assume you've purchased a pregnant miniature mare or jenny. Keeping in mind that no individual is likely to show every sign, these are some things you'll observe.

RED BAG DELIVERIES

Picture this: Your mini mare is in labor (lying down and straining), but her water never breaks. Instead of the milky white amnion appearing at her vulva, a purplish red, velvety-looking membrane appears. This is a serious emergency. The foal is being deprived of oxygen due to premature separation of the placenta. The red bag (chorioallantois) needs to be ruptured and the foal delivered as soon as possible — you don't have time to call your vet.

You must cut or tear the membrane and extract the foal as quickly as possible. The membrane is thick and slippery and not easily torn, so it's best to keep blunt-nosed surgical or leg-wrap removal scissors or a not-too-sharp knife in your birthing kit for just this purpose.

Inside the membrane you'll find the foal surrounded by a white bag — the amnion. Tear the amnion and help the mare deliver the foal as quickly as possible, pulling the foal out and down toward the mare's hocks in conjunction with her contractions. After a red bag delivery, closely monitor the foal for signs of oxygen deprivation, such as lethargy and disinterest in nursing or its dam.

A TIME FRAME FOR FOALING

1. In a normal foaling, the actual delivery (second-stage labor) takes approximately 20 to 40 minutes.
2. A mare or jenny should deliver the placenta (third-stage labor) within an hour after foaling; if she hasn't passed it in two hours, call the vet.
3. The foal should stand within about an hour and nurse within half an hour after that.

Most mares and jennies begin "bagging up" (developing enlarged mammary glands) four to six weeks prior to foaling. Some individuals show little development until days or even hours before foaling.

A hormone (relaxin) causes the muscles and ligaments in the pelvis to begin relaxing three or four weeks before foaling. This causes the rump to become increasingly steeper as labor approaches — the area along the spine seems to sink and the tailhead rises. This is noticed both from a hips-to-tail side view and from a side-to-side rear view.

The perineum, the hairless area around the vulva, sometimes bulges during the last month (this is more evident in miniature equines than in full-size mares and jennies). About 24 hours before foaling occurs, the bulge diminishes and the vulva becomes longer, flatter, and increasingly flaccid.

A few days before foaling, muscles in the floor and walls of the abdomen begin relaxing and its shape changes, making the mare's or jenny's belly seem to come to a point when viewed from the side and become narrower when viewed from the front or back.

As the cervix begins to dilate, usually a few days to a few hours before foaling, the cervical seal (wax plug) liquefies. When this occurs, females often discharge strings of mucus from their vulvas. This can be clear, thin goo; a thicker, opaque white substance; or a thick, amber-colored discharge tinged with amniotic fluid.

The following behaviors indicate that the mare or jenny is in first-stage labor; this generally lasts for 12 to 36 hours prior to actual foaling:

- Her udder will engorge with milk, to the point where the teats are filled to bursting. If her udder is pink, it will blush a rosier red and take on a shiny, moist look as foaling approaches. Wax plugs may form on the ends of her teats or she might drip or stream milk.

- She may drift away from the herd to seek a nesting spot, sometimes in the company of her dam, a daughter or sister, or another companion. She'll yawn and stretch (stretching helps put her foal into birthing position); urinate frequently and/or pass small amounts of manure; and she may go off her feed.
- When she begins pacing, pawing, kicking at or watching her belly; circling; or lying down and getting up again — especially with her tail cocked out behind or kinked to the side — it's time!

The foaling (second-stage labor) should progress as described in our Breeding chapter. Turn back to chapter 11 and read it now.

Typically, roughly an hour after the foal arrives, the mare or jenny goes into third-phase labor to deliver the foal's placenta — the tissues in which it developed inside of her uterus. With equines, it's very important that these tissues come out promptly and intact. A retained placenta, or even a retained piece of placenta, can lead to metritis (infection within the uterus) and subsequent laminitis. Follow these tips for aiding with third-phase labor:

- Leave the stall and reduce distractions to encourage the mare or jenny to lie down and expel the placenta.
- Don't pull exposed portions of the placenta. Pulling encourages the uterus to prolapse — something you never, ever want to experience.
- Wear protective gloves when handling placental tissue. In the unlikely event the mare or jenny is infected with a zoonotic disease such as brucellosis, it can be passed to humans who handle the placenta with bare hands.
- After it's expelled, spread out the placenta and check if pieces are missing. It should be T-shaped and inside-out (the purplish red part on the outside), with one torn spot from which the foal emerged. If you think something is missing, place it in a bucket of cool water, cover it, and call the vet.

Intact horse placenta

Foals are programmed by nature to seek sustenance in dark places, but that includes any dark area in the stall. Experienced mares and jennies circle and nudge their foals to put them in the correct position to nurse, though first-timers may have to be haltered and held until their foals make the right connection. Once you've dipped the foal's navel and you're sure he is nursing (watch his throat to see if he's swallowing), leave the mare or jenny and her foal alone to bond.

INVERTED TEATS

A commonly encountered complication unique to foaling miniatures is inverted teats. Minis have, at best, short teats, and as the udder swells, the teats are buried so deep that the foal can't suckle. The foal may appear to be nursing but grows increasingly weaker due to lack of fluids and nutrition. Always strip a mare or jenny's teats to make certain they're open and milk is flowing. If you can't reach mama's teat to strip it, neither can her foal!

16

Miniature Goats

IS A FARM A FARM WITHOUT GOATS? I don't think so, but I'm a goat enthusiast and hope that you are, too. Goats' irrepressible joie de vivre makes every farm a happier place, but only if you know what you're getting into (sometimes a sense of humor is an asset, too).

Goats are sweet, playful, intelligent, and personable; they are also the species most likely to escape inadequate fences and wreak havoc on gardens, orchards, and the hood of your next-door neighbor's new car. They don't mean to be destructive, but curiosity combined with supreme agility leads to problems unless your fences are goat-tight. Be prepared — really prepared — before adding goats to your farm.

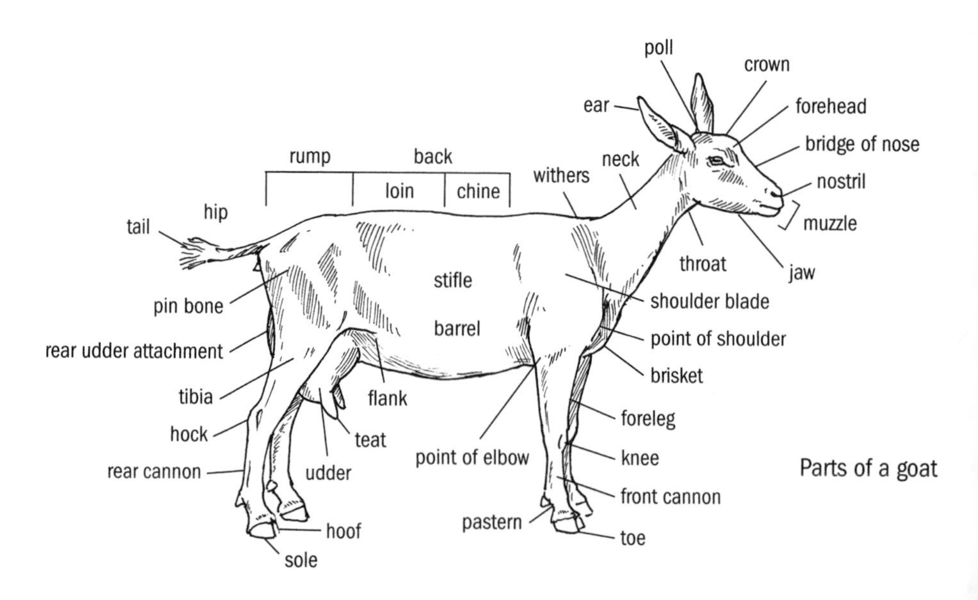

Parts of a goat

Getting into Goats

Much of what we talked about in section 1 applies to keeping goats. While well-tended goats tend to be healthy, happy goats, there are a few additional things worth mentioning.

Buying Goats

Before shopping for goats, learn what your chosen breed or type should ideally look like; a highly productive miniature dairy goat and a stocky pet breed like a Pygmy goat look very different indeed. And know what aesthetic features are important to you; for instance, if you want a milk goat but don't like most dairy goats' angular look, try a fleshier breed such as a Mini-Nubian or Kinder.

GOAT PHYSIOLOGY

Temperature: 101.5–104°F (38.5–40°C) for adult goats
Heart rate: 60–90 beats per minute
Respiration rate: 12–20 breaths per minute
Ruminal movements: 1–1.5 per minute
Life span: Usually 10–12 years (the known record is 23)

DOES
Age at puberty: 5–10 months
Breeding weight: 60–75 percent of adult weight
Heat cycle: every 18–23 days
Heat duration: 12–36 hours
Ovulate: 12–36 hours after onset of standing heat
Length of gestation: 144–157 days
Number of young: 1–5 (twins and triplets are the norm for most miniature breeds)
Breeding season (seasonal breeders): August–February
Breeding season (aseasonal breeders): Year-round

BUCKS
Age at puberty: Average 5–6 months (although some of the miniature breeds reach puberty much younger than this)
Primary rut: August–January
Breeding ratio: 1 adult buck to (up to) 30 does

DON'T CALL THEM BILLIES!

In an effort to counteract the ornery, tin can–eating billy goat image of yore, savvy goat owners call their goats bucks and does (in the manner of deer) instead of the old-fashioned terms, billies and nannies.

Look for a good rumen on all goats you buy; the larger the rumen (within reason) the better the goat can digest coarse fare. The rumen bulges out on the left side of a healthy, well-fed goat, especially one that has been feasting on hay, forbs, or pasture. This is true of either sex (all of my big Boer cart wethers look pregnant after eating).

Check the goat's bite — how its lower teeth meet up with its upper dental pad. It should be a perfect match. Goats need proper occlusion to browse and graze, and a bad bite is a serious fault in show goats and breeding stock. Be especially vigilant about checking the mouths of Roman-nosed animals, because that head configuration is the one most likely to result in "monkey mouth" or underbite.

Check an adult doe's udder very carefully (the udder is her complete mammary system; she has two teats, not two udders). Feel it. A dairy doe's udder feels soft and velvety; udders of other breeds might not be quite as silky but they should be pliable, not hard (hard spots indicate previous bouts with mastitis, and the mastitis is likely to reoccur). The udder should be well attached, not pendulous and dangling near the ground. Newborn kids can't latch on to a sagging udder, and it's hard to fit a milk pail under one. A dairy doe's udder should be held up close to her body and it should be *capacious* — a word goat owners use to mean that it should have space to hold a lot of milk. For hand milking, you want teats that are big enough to comfortably milk (comfortable for you and the doe). In miniature breeds, that means a teat that can fit your thumb and three fingers. Each teat should have a single orifice (the hole through which milk flows), and the orifice should be large enough to generate a decent stream of milk. It takes forever to milk a doe with tiny orifices, so if you'll be hand milking, ask to milk a doe before you buy her.

And when purchasing breeding stock, by all means check teats on both sexes. A buck with poor teat structure is likely to pass the trait along to his offspring, and he will be heavily penalized in the show ring. (A buck's rudimentary teats are located in front of his scrotum.) Goats are prone to a host of teat anomalies that make it difficult to milk them and sometimes even for their kids to nurse.

These anomalies also disqualify goats for show competition, so even in non-dairy breeds, check those teats!

Ask about CAE

The big three health threats to ask about when buying goats are caseous lymphadenitis (CL), hoof rot (we discussed both of these in chapter 8), and caprine arthritic encephalitis (CAE).

CAE is a goat-specific disease related to ovine progressive pneumonia (OPP) in sheep. First identified in 1974 and initially called viral leukoencephalomyelitis of goats (VLG), it's a progressively crippling disease caused by a retrovirus (a type of RNA virus such as HIV in humans that reproduces by transcribing itself into DNA; resultant DNA inserts itself into a cell's DNA and is reproduced by the cell). Unfortunately, there is no vaccine against CAE, and there is no cure. There are tests, however, to identify infected animals, so it's best to buy from annually tested, CAE-free herds or to have goats you plan to purchase tested at your own expense.

The virus that causes CAE is responsible for two separate syndromes in goats: a neurological disease of the spinal cord and brain in young kids and a joint affliction of older goats.

Kids affected with CAE first show signs of the disease between one and four months of age. The virus causes progressive weakness of the hind legs leading to eventual paralysis in a few days up to several weeks. Kids remain in good spirits and continue to eat and drink.

The arthritic form of CAE usually surfaces between one and two years of age. Some affected goats are badly crippled within a few months of the onset; in others, the disease progresses slowly over a span of years.

The virus that causes CAE is transmitted from dam to kid through colostrum and milk. To prevent transmission of the virus, offspring of CAE-positive dams must be taken away at birth, before they suckle. They need to be bottle raised on either properly heat-treated colostrum and milk from a CAE-free doe, ewe, or cow or on IgG supplement followed by CAE-free milk or milk replacer. To heat-treat colostrum, heat the colostrum to 135°F (57°C) in a double boiler or water bath and maintain temperature for one hour.

Fencing for Goats

Carefully review chapter 6 before buying goats and keep this in mind: If there is a weakness in your fences, goats will find it and exploit it.

Some people find that multiple strands of electric wire are effective fences for goats, but if you go this route, opt for a first-class fence charger and make certain it's working at all times. Unless plank fences are lined with woven wire, forget about using them. We think the best all-around goat fences are built of sturdy, woven-wire field fencing reinforced with a strand of electric wire that has been placed inside at shoulder height to keep inhabitants from leaning sideways into the fence and scratching their sides and backs.

For smaller enclosures and where money is no object, we recommend welded-wire cattle panels. They're easier to work with than woven wire, more durable, and less likely to be bowed by itchy goats.

Keep in mind that horned bucks, even miniature bucks, can blitz the toughest fencing in record time. I once knew a Savanna buck whose claim to fame was his ability to completely obliterate a cattle panel buck pen overnight. He weighed 200 pounds (90 kg). Your Pygmy buck may weigh 50 pounds (23 kg), but if he has horns, he can eventually take down a pen, too!

Housing for Goats

Miniature goats are incredibly easy to house. A dry, draft-free place to hang out is often enough. Four or five Pygmies can find happiness in a single Port-A-Hut if they have an exercise yard or pasture beside it. I know two Nigerian Dwarfs who live in a super-sized Dogloo. The key words are dry and draft-free. Above that, improvise.

Feeding Goats

Clean, high-quality grass hay should be the backbone of every goat's diet. Wethers, open does, and bucks outside of breeding season need nothing else. Youngsters, late gestation and lactating does, and hardworking bucks in rut, however, also need concentrates. Feed a properly formulated commercial mix or work with your county extension agent and your local feed mill to formulate one based on locally available ingredients and your needs.

All goats need access to minerals in loose form (preferred) or as a lick formulated for goats. Goats think tub licks are lip-licking good because they usually contain a lot of molasses; however, beware of tub licks in the sizzling summertime. A Texas-based Boer breeder friend and her husband had to work hard to free a doeling who climbed into a hot-lick tub full of sun-softened lick mix late one summer day, where she fell asleep. As evening approached and the day cooled down, the lick mix hardened and fastened her firmly on her side in the tub!

Think like a Goat

The trick to staying ahead of goats is knowing how they think. Goats are intelligent and curious. Things must be investigated, climbed upon. If you know what they'll likely do before they do it, you can prevent the common problems associated with keeping goats.

Harmony in the Herd

A well-defined pecking order exists within every herd, be it 2 or 102 goats. Where a goat stands in this hierarchy depends on its age, sex, personality, aggressiveness toward other goats, and the size of (or lack of) its horns. Unweaned kids assume their dam's place in the order and often rank immediately below her after weaning.

Newcomers fight to establish a place in the herd. Fighting for social position is conducted one-on-one; established herd members don't gang up on a new one.

When fighting, males shove one another, butt, and side-rake opponents with their horns; does usually butt, jostle, and push. To butt one another, goats position

OOOO, THAT SMELL!

The first thing many prospective goat owners want to know is: "Do bucks reek the way people say they do?" The answer is: It depends.

As breeding season approaches, most bucks' scent glands shift into overdrive, producing a plentiful supply of strong-scented musk to attract a mate. Add to this the fact that bucks spray urine on themselves, and you can see how they've gotten the reputation of being quite smelly.

That said, age, breed, and the company he keeps affect the amount of musky scent a buck exudes. Yearling bucks are rarely as stinky as older males. People who keep several breeds claim that meat bucks are smellier than fiber bucks, and dairy bucks are worse than both combined. A single buck with or without a doe or wether for companionship is rarely as odiferous as one housed with a horde of other bucks. And some bucks simply smell better or worse than the norm.

What to do when your friendly, smelly friend rubs up against you to mark you with his scent? Haul out a can of mechanic's hand cleaner like GoJo or Fast Orange, and scrub the stink away. Try it. It really works!

> **WATCH THOSE HORNS!**
>
> It's easy to get hurt by horns, even when your horned goat doesn't mean to hurt you. When working closely with a horned goat, make a small hole in each of two old tennis balls and push one onto the tip of each horn. When you're finished, pull them off and save them to use another day.

themselves a few feet apart, facing each other. They rear on their hind legs and swoop forward, down, and to the side to smash their horns against those of their opponent. Bucks do not back up and charge in the manner of rams.

Other forms of aggression include staring, making threats with horns (chin down, horns jutting forward), pressing horns or forehead against another goat, rearing without actually butting, and ramming an opponent's rear end or side. However, there is very little infighting once each herd member knows and accepts its place.

The herd is led by its herd queen, usually an old doe that has head-butted, shoved, and threatened her way to the top of the ladder. She's rarely (if ever) challenged, and she remains herd queen until she's removed from the herd or she becomes too feeble to lead. When that happens, the position is often assumed by one of the old queen's daughters.

Most of the year, any bucks in the group, including the herd king (the alpha male), defer to the herd queen. During rut, however, the herd king assumes leadership of the group. The herd king breeds all of the does — underling bucks don't breed. Bold bucks, however, constantly challenge the herd king, so herd queens generally outlast many kings.

Since the herd queen leads the herd to food, if you lead your herd with a pail of feed (instead of trying to drive them), your goats will treat you like the ultimate herd queen. However, if you insist on driving goats from the rear (bringing up the rear to protect the group is the herd king's job), they will perceive you as a two-legged herd king — not a good position to be in when there are four-legged contenders in the herd.

The Buck Stops Here

Bucks, seasonal and aseasonal breeds alike, enter rut as autumn approaches and stay in rut through the first months of winter. The volume and motility of a buck's semen is greatest at this time, even among aseasonal breeders.

Bucks that would live peaceably with other bucks the rest of the year become testy toward one another during rut. Each considers himself herd king, so if bucks from different herds get together, watch out.

During rut, scent glands located near a buck's horns (or where his horns used to be) secrete strong-scented musk. When bucks rub their foreheads on a person or object, they're spreading their scent. Bucks also twist themselves and grasp their penises in their mouths. They sometimes masturbate on their bellies and front legs and then sniff themselves and flehmen.

Bucks "display" when courting does in heat. They enurinate (spray urine on themselves — their bellies, front legs and chest, and into their mouths and beards), paw the earth, and emit an amazing series of vocalizations called "blubbering." If the object of his affection doesn't flee, a courting buck assumes a slight crouch and sashays closer to the doe with his head slightly extended and his tail curled up across his back. He gobbles (just like a turkey!), flicks his tongue, and if his ladylove allows it, he'll sniff and nuzzle her sides and vulva. If she urinates, he'll catch urine in his mouth and flehmen. He obviously thinks he's quite the lady's man.

If the doe allows him to breed, he'll make a number of false starts before ejaculating. When he ejaculates, he thrusts his hips forward and leaps off the ground while twisting his head back and to one side. Small, young, or otherwise inexperienced bucks sometimes fall off the doe and onto the ground. When he's finished with his first try, he'll lick his penis and flehmen, rest a short while, and start again.

Because highly socialized bucks view humans as herd members, women and children working around bucks in rut should be aware of and discourage courtship behavior. A buck that perches his forefeet on his caretaker's chest and gobbles in her face means no harm, but being bowled over by an amorous, smelly buck is just not fun. Likewise, when in rut, some otherwise easygoing bucks perceive human male caretakers as competition and react accordingly. No matter how small the buck or the size, age, or level of experience of his human handlers, people who work around him should never take a buck for granted.

Those Does

From early autumn through midwinter, seasonal breeders (such as dairy breed does) cycle, or come in heat, every 18 to 21 days; aseasonal breeders (such as Pygmy and Nigerian Dwarfs) cycle year-round. Ovulation occurs 12 to 36 hours after the onset of standing heat.

SIGNS OF HEAT

- Interest in nearby bucks
- Increased activity level (especially fence walking)
- Loud and strident vocalizing
- Tail wagging (also referred to as flagging)
- Frequent urination
- Mounting other does and being mounted by them
- Decreased appetite
- Lower milk production

Does are stimulated by the appearance and scent of a buck. They rub their necks and bodies against him. A fully receptive doe stands with her head slightly lowered, her legs braced, and her tail to one side. She may urinate when he sniffs her.

Once impregnated, a doe enters anestrous and most does stop coming into heat. A pregnant doe gives birth to one to five offspring 144 to 157 days after the onset of her last standing heat.

A day or so before kidding, the average doe becomes fretful and anxious. A few hours before giving birth, she'll usually leave the herd and seek a secluded birthing place. She may take along an "auntie" (usually a doe's grown daughter or her own dam) to keep her company and cheer her on.

As hard labor begins, the doe rolls onto her side to push. She may struggle to her feet, turn, and reposition herself; this helps put her kids into correct birthing positions.

After each birth, she rises and licks her newest arrival clean. Does recognize their newborns by scent, so cleaning is an important element in the bonding process.

In a best-case scenario, the doe's afterbirth passes a short time after her final kid is delivered. Does frequently want to eat the afterbirth so as not to attract predators, but because they may choke on the membranes, most owners remove it as soon as it's delivered.

Neonatal kids are wired to seek darkness (in places such as armpits or groin) and warm, bare skin. Most kids struggle to their shaky legs 10 to 30 minutes after birth and begin actively seeking a teat. Encourage this; kids should ingest colostrum within the first hour or two after their birth. Help weak or disoriented kids by holding them near their dam's udder, although most resist a teat placed directly into their mouths.

Kids kneel to nurse, and they bunt their dam's udder to facilitate milk letdown. A rapidly wagging tail means a kid is suckling milk. After feeding, contented kids take naps. A kid that cries, seems to be doing a lot of noisy suckling, or constantly probing at his dam's udder isn't finding enough to eat. Without intervention, this kid could die.

Goats are a lying-out species like cattle and deer. Some does leave their very young kids in what they deem safe spots while they go off to feed, returning four to six times a day to suckle them. When kids begin nibbling at pasture and browsing, they start following their dam; this can occur anywhere from one to seven days after birth. When scientists conducted sheep and goat cross-fostering studies, lambs raised by does grew faster than kids raised by ewes or even kids raised by their own dams because the lambs trailed their surrogate mothers all

SIGNS OF IMPENDING KIDDING

- **A filled udder.** It may be stretched so tightly that it's shiny-looking and teats may jut out to the sides (called "strutted udder").
- **Mucus.** A glob of mucus may cling to the doe's vulva or hang in strings. With some does, a string of mucus appears a few days prior to kidding.
- **Restlessness.** An hour or so before kidding, most does begin circling and pawing the ground or bedding to hollow out a soft spot on which to give birth (called "nesting"). They lie down, get up to reposition themselves, flop down, rise — repeating this process many times.

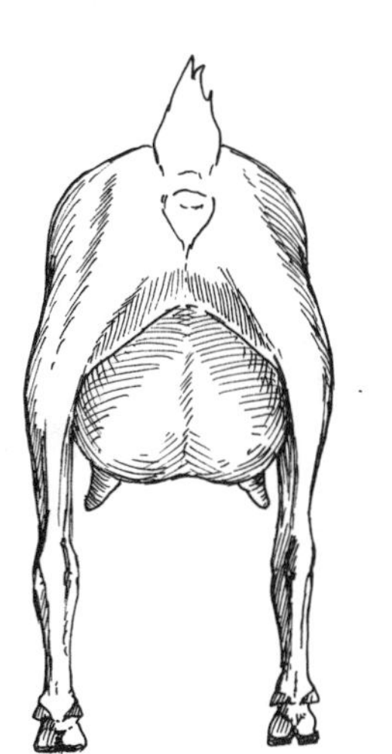

Strutted (filled) udder

- **Introspection.** Once nesting begins, many does' attention turns inward. They murmur softly to their unborn kids in a "mama voice" (low and sweet) used only prior to kidding and during the first few days of the new kids' lives.

day and thereby suckled more often. Conversely, ewes were constantly upset by their foster offspring when the kids refused to shadow their woolly moms.

When neonatal kids, and even older weaned ones, are handled for what they perceive as unpleasant procedures, such as getting shots or being treated for coccidiosis, they shriek, a high-pitched distress call that's sure to bring the nearby neighbors out in droves. You won't miss this the first time you hear it.

The Breed You Need

Miniature goats fall into one of four classifications: pets, dairy goats, fiber goats, and heritage breeds, though some categories overlap. Miniature dairy and fiber goats, for instance, make perfect pets. One predominately dairy breed (Kinder goats) and two pet breeds (Pygmy goats and Myotonic goats [also known as fainting goats or fainters]) are sometimes classified as meat goats, too. In fact, Pygmy does give a surprising amount of yummy, high-butterfat milk, making the Pygmy a bona fide triple-purpose miniature goat breed.

Miniature goat enthusiasts have a fine lineup of breeds to choose from. And as experimental breeders use Pygmy and Nigerian Dwarf genetics to downsize larger breeds, new ones are being developed all the time. Miniature and naturally small breeds range in size from tiny Pygmy goats and Nigerian Dwarfs to San Clemente Island goats of next-to-"normal" size. Whatever size you want, it's out there!

And if you're looking to help conserve a rare breed, there are several miniature goats to choose from. The small- to medium-size endangered San Clemente

WHY GOATS CLIMB ON CARS

Goats climb on cars (and tractors and haystacks and the roof of your house or anything else that intrigues them) because they're there. The many hundreds of breeds and types of goats in the world descend from the bezoar goat (*Capra aegagrus*), with the possible exception of a few breeds such as Angora and Cashmere goats that may in part descend from the markhor (*Capra falconeri*). Bezoar goats and markhors are mountain goats. Enough said.

The only reliable way to keep goats from climbing on something you'd rather they leave alone is to fence them out of the area. Barring that, arm waving and blasts from a high-pressure water gun like a Supershooter might help. But don't count on it.

MEASURING A GOAT

Measure height at the withers using a measuring standard, not a flexible tape. The goat should be standing on a firm, flat surface and holding his head in a typical position, not pulled up, down, or out — any of which may significantly alter height measurement. Imagine a line dropping straight down through the goat's shoulders to the floor. The goat's front legs should be set evenly under his shoulders, not forward or backward. His rear legs should set squarely under him, neither drawn too far back nor too far forward.

The Pygmy Goat Registry requires that the goats' cannon bones be measured. This type of measurement is taken at the outer edge of sharply bent knee and pastern joints using calipers.

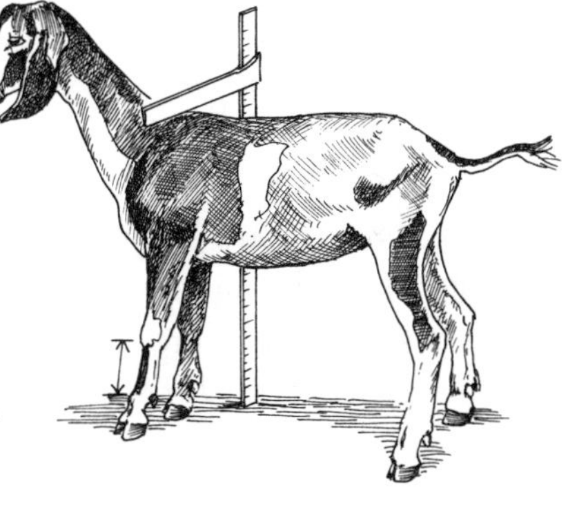

Measure goats from their withers to the floor.

Island goat is listed as Critical on the American Livestock Breeds Conservancy's 2009 Conservation Priority List, which means there are fewer than 200 annual registrations in the United States (see pages 16 and 17). Also listed are Myotonic goats (classified as Watch) and Nigerian Dwarfs (classified as Recovering).

Learn more about the following breeds by contacting their breed associations; you'll find them listed at the back of this book in the Resources section.

Australian Miniature Goat

Registered by: Miniature Goat Breeders Association of Australia (MGBA), Australian Miniature Goat Association (AMGA)
Use: Pet, dairy, and fiber
Origin: Australia
Size: Miniature Goat Breeders Association of Australia: adults up to 21 inches (53 cm). Australian Miniature Goat Association (at three years of age): Grade D, 25 inches maximum height (53 cm); Grade C, 23 inches maximum (58 cm);

Grade B, 22 inches maximum (56 cm); Grade A, 21 inches maximum (53 cm); Purebred, 20 inches maximum (51 cm)
Color: Any combination of colors is acceptable
Ears: Vary by type
Facial profile: Varies by type
Coat: Varies by type
Horns: Usually horned

While they are not yet available in North America, no discussion of miniature goats would be complete without mentioning these interesting small goats from Down Under.

The Australian Miniature Goat Association calls the Australian Miniature "a genetically small animal. Its conformation is similar to the larger breeds, with all parts of the body in balanced proportion relative to its size. They are hardy, alert animals with impeccable temperaments that come in all colours and colour combinations." Each registry records goats of several different types and handles classification in a slightly different manner.

The Australian Miniature Goat Association recognizes three breed types based on coat and ear length: Minikin (short, shedding cashmere coats and any ear type that is shorter than muzzle length), Sheltie (long, nonshedding coats and any ear type), and Nuwby (short, shedding cashmere coats and longer than muzzle-length ears).

The Miniature Goat Breeders Association of Australia recognizes four types based on ear style determined at three years of age: Elf (short ears), Pixie (upright ears), Munchkin (folded ears), and Nuwby (pendulous ears).

Australian Miniature Goats come in a number of interesting types and coat varieties.

MINIATURE DAIRY GOATS

The Miniature Dairy Goat Association (MDGA) maintains registries for scaled-down versions of the main full-size dairy goat breeds: Mini-Alpines, Mini-LaManchas, Mini-Nubians, Mini-Oberhaslis, Mini-Saanen/Sables, and Mini-Toggenburgs; the specifics for each breed are discussed in its breed entry.

To create the F1 (first) generation of any miniature dairy goat breed, breeders mate a Nigerian Dwarf buck to a registered, full-size doe of the desired breed; the offspring are registered as Experimentals. Through continual upgrading, breeders create registered "Americans" at the third generation and "Purebreds" at the sixth generation.

To be registered in the MDGA, mature does must be at least 23 inches (58 cm) tall and mature bucks 24 inches (61 cm). Maximum heights vary by breed, but the preferred maximum height is 1 inch (2.54 cm) shorter than the minimum height for the corresponding full-size breed.

The organization's goal is to produce compact, high-production does suitable for hand milking on the small farm. Miniature dairy goats produce between 2 pounds (1 pint; 0.9 kg)) and 10 pounds (well over 1 gallon; 4.5 kg) of milk per day. And thanks to the influx of Nigerian Dwarf genetics, many mini dairy goats are year-round breeders.

Kinder

Registered by: Kinder Goat Breeders Association (KGBA)
Use: Dairy, meat, pet
Origin: Snohomish, Washington, USA
Size: Adult does, 20 to 26 inches (51–66 cm); a maximum of 28 inches (71 cm) for adult bucks
Color: Any combination of colors is acceptable
Ears: Long and wide, resting below horizontal and extending to the end of the muzzle or beyond when held flat against the jaw line
Facial profile: Straight or dished

Kinder

Coat: Short, fine-textured
Horns: Horned, but to show at sanctioned shows they must be disbudded or dehorned

Kinder goats are a combination of full-size Nubian and Pygmy genetics. A registered Pygmy buck bred to a registered Nubian doe produces first-generation Kinder offspring; after that Kinders must be bred to other Kinders, though backcrossing to either parent breed is permissible.

Kinders are dual-purpose meat and dairy goats. Some Kinder does give an impressive amount of milk, such as Zederkamm Daffodil, who gave (on official milk test) 2,290 pounds of 5.5 percent butterfat milk in a single lactation. Butterfat content for the breed ranges from 5.5 to 7.5 percent.

Miniature Myotonic (or Fainting Goat)

Registered by: Myotonic Goat Registry (MGR), International Fainting Goat Association (IFGA)
Use: Pet, meat
Origin: Tennessee, USA
Size: Does three years of age or older can be no more than 22 inches (56 cm) tall for registration in the International Fainting Goat Association's Mini-Myotonic herd book; bucks up to 23 inches (58 cm) tall may be registered. The Myotonic Goat Registry doesn't maintain a separate mini registry; minis are registered in the main herd book. Some stand as little as 17 inches (43 cm) tall and weigh no more than 50 pounds (15 kg).
Color: All colors, combinations, patterns, and marking are acceptable; pied black and white is the most common color

DO FAINTING GOATS REALLY FAINT?

In a word: No. They are affected by a genetic disorder called "myotonia congenita," an inherited, neuromuscular disorder caused by mutations in the CLCN1 gene and characterized by an inability to relax muscles after contraction. When Myotonic goats are startled or scared, their skeletal muscles, especially in their hindquarters, contract, hold, and then slowly release (amount of time varies according to individual). "Fainting" episodes are painless, and the goats remain awake (they often continue chewing food they happened to have in their mouths at the time) until the stiffness passes.

Ears: Usually medium in width and length, most often held horizontally from the side of the head and facing slightly forward. Some Myotonics have a characteristic and noticeable ripple halfway down the ear.

Facial profile: Usually straight with a slight stop below the eyes separating the forehead from the lower face.

Coat: Most are short-haired but some have longer, thicker coats; the coat should be straight, not wavy

Horns: Most are horned, though polled genetics exist

Miniature Myotonic

Myotonic goats, also known as Tennessee Fainting Goats, Nervous Goats, Stiff-leg Goats, Scare Goats, and a dozen or so additional colorful names, descend from four "fainting" goats brought to Tennessee in the 1880s by an itinerant farmworker named John Tinsley.

Myotonic goats are stocky, muscular, and wide in proportion to their height. They come in a vast array of sizes, colors, coat styles, and horn shapes but all share a common characteristic: they "faint" (see Do Fainting Goats Really Faint? page 316). Other breed characteristics include prominent bony eye orbits that make them appear somewhat pop-eyed, a distinctive ripple partway down the ear, and massively muscled hindquarters caused by the contraction and release of these muscles during fainting episodes.

Because their condition precludes a lot of climbing, Myotonic goats stay put in fences better than many other breeds. They're unusually good mothers, most breed year-round, and twins or triplets are the norm.

Myotonics are an American heritage breed still being monitored by the American Livestock Breeds Conservancy; additional breeders are needed.

Miniature Silky Fainting Goat

Registered by: Miniature Silky Fainting Goat Association (MSFGA), American Silky Fainting Goat Association (ASFGA)

Use: Pet

Origin: Virginia, USA

Size: Adult does, 23½ inches (60 cm) maximum height; adult bucks, 25 inches (64 cm) maximum height

Color: Any combination of colors is acceptable
Ears: Ideally erect; horizontal placement permitted
Facial profile: Dished
Coat: Long, straight, and flowing, reaching nearly to the ground; lustrous, smooth, and silky to the touch. An abundance of chest, neck, and facial hair including long bangs and cheek muffs. The ideal coat resembles and feels like that of a Silky Terrier dog.
Horns: Yes

Miniature Silky Fainting Goat

Arguably the most unique miniature goat of all is the cute, perky Miniature Silky Fainting Goat developed by Renee Orr in Lignum, Virginia. The foundation sires for Orr's herd were two long-coated Myotonic bucks named Bayshore's Rogues Pierre and Bayshore's Napoleon, which she bred to long-coated Nigerian Dwarf does with Myotonic goats in their background. This background notwithstanding, not all Miniature Silky Fainting Goats faint, though many do.

Mini-Alpine

Registered by: Miniature Dairy Goat Association (MDGA)
Use: Dairy, pet
Origin: Full-size Alpines were developed in the French Alps but are considered a Swiss breed
Size: Minimum 23 inches and maximum 29 inches for mature does (58–74 cm); minimum 24 inches and maximum 31 inches for mature bucks (61–79 cm)
Color: See Alpine Colors page 319
Ears: Erect
Facial profile: Straight or slightly dished
Coat: Short to medium
Horns: Yes (and unless disbudded, they grow impressive horns!)

Mini-Alpine

ALPINE COLORS

Alpine goats, both full-size and miniature, come in an array of attractive colors with interesting French names:

- Cou blanc (coo blanc), or "white neck." White front quarters and black hindquarters with black or gray markings on the head.
- Cou clair (coo clair), or "clear neck." Front quarters are tan, yellow, off-white, or shading to gray; hindquarters are black.
- Cou noir (coo nwah), or "black neck." Black front quarters and white hindquarters.
- Sundgau (sundgow). Black with white markings such as on underbody and facial stripes.
- Pied. Spotted or mottled.
- Chamoisee (shamwahzay). Brown or bay; characteristic markings are black face, dorsal stripe, feet, and legs; some have a martingale running over the withers and down to the chest. Spelling for male is *chamoise*.
- Two-tone chamoisee. Light front quarters with brown or gray hindquarters. Differs from cou blanc or cou clair by not having black hindquarters.
- Broken chamoisee (sundgau, cou blanc, etc.). A solid chamoisee (sundgau, cou blanc, for example) broken with another color by being banded or splashed with white.

The Miniature Dairy Goat Association describes the Miniature Alpine as "an alert, gracefully hardy animal that adapts and thrives in any climate while maintaining health and excellent production." Inquisitive, intelligent Alpines are one of my favorite breeds in both full and miniature sizes.

Mini-LaMancha (Mini-Mancha)

Registered by: Miniature Dairy Goat Association (MDGA)
Use: Dairy, pet
Origin: California, USA
Size: Minimum 23 inches, maximum 27 inches for mature does (58–69 cm); minimum 24 inches, maximum 29 inches for mature bucks (61–74 cm)
Color: Any combination of colors is acceptable

Ears: Two types: gopher and elf. Gopher ears lack cartilage but have a ring of skin around the ear opening. Elf ears are triangular external ear flaps up to 1 inch long. Does may have either type, but only gopher-eared bucks can be registered.
Facial profile: Straight
Coat: Short and glossy
Horns: Yes

LaManchas are the only major dairy goat breed developed in North America. They likely descend from short-eared Spanish goats that accompanied Spanish padres to California. The name, it is said, may have been given to LaManchas when a crate of these unusual goats arrived at the Paris World's Fair for exhibition in 1904; the crate bore the inscription, "LaMancha, Cordoba, Spain."

LaManchas are noted for their peerless personalities, docile natures, and a steady production of milk moderately high in fat. Everyone who has LaManchas loves them.

Mini-LaMancha

Mini-Nubian

Registered by: Miniature Dairy Goat Association (MDGA); promoted by the National MiniNubian Breeders Club (NMBC)
Use: Dairy, pet
Origin: Full-size Nubians were developed in England, where the breed is called the Anglo-Nubian
Size: Minimum 23 inches and maximum 29 inches for mature does (58–74 cm); minimum 24 inches and maximum 31 inches for mature bucks (61–79 cm)
Color: Any combination of colors is acceptable
Ears: Long and pendulous
Facial profile: Straight to strongly Roman-nosed (the latter is preferable)
Coat: Short, fine, and silky
Horns: Yes

The English developed Anglo-Nubian goats by crossing native does with Jumna Pura, Zaraibi, and Chitral bucks from Africa and

Mini-Nubian

India. A British breed registry formed in 1919. Goats imported by Mr. J. R. Gregg of California in 1909 and 1913 formed the nucleus of the breed in North America, where it's now called simply the Nubian.

Nubian goats, full-size or mini, are elegant and graceful. They are unusually intelligent and inquisitive, with endearing, quirky personalities. Their one drawback, some say, is their voice, which is loud and somewhat strident. Some, however, like talkative Nubians (I do). This is hands-down my favorite dairy breed!

Mini-Oberhasli

Registered by: Miniature Dairy Goat Association (MDGA)
Use: Dairy, pet
Origin: Full-size Oberhaslis were developed in Switzerland
Size: Minimum 23 inches and maximum 27 inches for mature does (58–69 cm); minimum 24 inches and maximum 29 inches for mature bucks (61–74 cm)
Color: Chamois (bay-colored, sometimes described as being colored "like the wood on the back of a violin") with a black dorsal stripe, udder, belly, and lower legs; the head is nearly black with two white stripes on its sides. Black does, but not black bucks, can be registered.
Ears: Erect
Facial profile: Straight or dished
Coat: Short and silky
Horns: Yes

Full-size Oberhaslis are listed as Watch on the American Livestock Breeds Conservancy List. Once known as Swiss Alpines, they are more alert than some other breeds. They are noted for their sweet, tasty milk, their intelligence, and their great dispositions.

Mini-Oberhasli

Mini-Saanen (pronounced SAH-nen or SAW-nen)

Registered by: Miniature Dairy Goat Association (MDGA)
Use: Dairy, pet
Origin: Full-size Saanens were developed in Switzerland
Size: Minimum 23 inches and maximum 29 inches for mature does (58–74 cm); minimum 24 inches and maximum 31 inches for mature bucks (61–79 cm)
Color: Saanens are white or cream with pink or olive-colored skin, sometimes lightly speckled with black; all other colors are Sables

Ears: Upright, alertly carried
Facial profile: Straight or slightly dished
Coat: Short
Horns: Yes

Full-size Saanen goats originated in the Saanen Valley of the canton of Bern in Switzerland, where they were selected for milking ability, hardiness, and color. In 1893 several thousand head of Saanens were taken from the valley and dispersed throughout Europe; they came to the United States in 1904 and became the first breed registered in North America.

Saanens (and Sables) are heavy milkers and, large or small, they are strong, vigorous goats with rugged bones and friendly dispositions. Similar to the other Swiss breeds, however, they don't fare well in hot, humid parts of the world, and their light-colored skin predisposes Saanens to skin cancer. In southern climates, Sables are a better choice.

Mini-Saanen and sable-colored Mini-Saanen

SABLE OR SAANEN?

Saanens are solid white or cream-colored goats with light-colored skin. Sables are Saanens with pigmented, colored skin, though full-size Sables are now considered a separate breed. Sables come in an array of handsome colors and have been a part of the Saanen heritage for as long as there have been Saanens. The first Sables in the United States arrived on the same ship with the first Saanens and have been here ever since. Sables are the result of the pairing of two recessive genes, one from an animal's sire and one from her dam. If an animal has only one of these genes, she is white or cream, but if the animal has two, she is colored. Since colored genetics were brought in with the original Saanen imports, this will continue to happen as long as people breed Saanens.

WHAT ARE THOSE FUNNY THINGS ON MY GOAT'S NECK?

The dangly bits of skin suspended from some goats' necks are called "wattles" in North America and "tassels" in Britain, Australia, and New Zealand. They serve no known function.

Mini-Toggenburg

Mini-Toggenburg

Registered by: Miniature Dairy Goat Association (MDGA)
Use: Dairy, pet
Origin: Full-size Toggenburgs were developed in Switzerland
Size: Minimum 23 inches and maximum 25 inches for mature does (58–64 cm); minimum 24 inches and maximum 27 inches for mature bucks (61–69 cm)
Color: Toggenburg base colors range from light fawn to darkest chocolate, but all have the same markings: white ears with a dark spot in the middle of each ear; two white stripes down their faces from above each eye to their muzzles; white hind legs from hocks to hooves; white forelegs from knees downward (a dark band below each knee is acceptable); a white triangle on both sides of their tails; and a white spot at the root of their wattles or in that area if no wattles are present. Varying degrees of cream markings instead of pure white are acceptable.
Ears: Erect
Facial profile: Straight or dished
Coat: Short to medium length
Horns: Yes

The Swiss developed Toggenburg goats about 300 years ago in the Toggenburg Valley of the canton of St. Gallen in northeast Switzerland. Its supporters call the Toggenburg the oldest and purest of Swiss goat breeds.

They are marvelous dairy goats; a full-size Toggenburg, GCH Western-Acres Zephyr Rosemary, currently holds the Guinness Book of World Records title for most milk ever given by a dairy goat in a single lactation. She gave 9,110 pounds (4,132 kg) of milk amounting to nearly 1,140 gallons in 365 days. Many

Toggenburg does "milk through" without rebreeding each year, and lactations of 18 to 20 months are common.

Toggenburgs do best in cooler climates; they're easy kidders and good mothers, fine foragers, alert, and affectionate.

Nigerian Dwarf

Registered by: Nigerian Dwarf Goat Association (NDGA), American Dairy Goat Association (ADGA), International Dairy Goat Registry (IDGR), American Goat Society (AGS)
Use: Dairy, pet
Origin: Nigeria, Africa
Size: Ideal size of mature does is 17 to 19 inches (43–48 cm); acceptable includes does up to 21 inches (53 cm) tall. Ideal size of mature bucks is 19 to 21 inches (48–53 cm); acceptable includes bucks up to 23 inches (58 cm) tall. Ideal weight for does and bucks is around 75 pounds (34 kg).
Color: Any combination of colors is acceptable, although Pygmy-specific markings are penalized
Ears: Erect
Facial profile: Straight
Coat: Soft, with short- to medium-length hair
Horns: Yes

Nigerian Dwarfs are perfectly scaled-down miniature dairy goats: their body parts are in balanced proportion. Does give up to 2 quarts of 6 to 10 percent butterfat milk per day.

Nigerian Dwarf kids weigh about 2 pounds (0.9 kg) at birth, but they grow quickly. They are precocious breeders and can be fertile at seven weeks of age. Most does can be bred at seven to eight months of age, although it's better to wait until they're at least one year old. Bucks can be used for service as young as three to six months of age. Nigerian Dwarfs breed out of season, and litters of three and four kids are common.

Nigerian Dwarf

Nigora

Nigora

Registered by: American Nigora Goat Breeders Association (ANBA)
Use: Fiber, pet
Origin: USA
Size: Undefined, but as a cross between Nigerian Dwarfs and full-size Angora goats, most are mini-size
Color: Any combination of colors is acceptable
Ears: Erect like a Nigerian's, droopy like an Angora's, or somewhere between
Facial profile: Straight to slightly dished
Coat: Individual Nigoras produce one of three types of fiber: Angora, cashmere, or a combination of the two
Horns: Yes

First-generation Nigora goats are created by breeding registered Nigerian Dwarf bucks to registered, full-size Angora does (the reverse breeding is acceptable but, due to potential problems at kidding time, rarely advised). Thereafter, offspring (from any matings) that are no more than 75 percent of one breed or 25 percent of the other are acceptable. Nigora goats that are predominantly Angora in breeding are called "heavy Nigoras" (for their larger size and heavier fleece), Nigoras from predominantly Nigerian breeding are called "light Nigoras" (for their smaller size and lighter fleece production), and goats that are roughly half-Angora and half-Nigerian are known as "standard Nigoras."

Nigora goats are easygoing, friendly, and intelligent. Many are kept by handspinners for their fiber and by others as pets.

Pygmy Goat

Registered by: National Pygmy Goat Association (NPGA)
Use: Pet
Origin: West Africa
Size: Does one year old and older: minimum 16 inches (41 cm) tall with 2⅞-inch (7.5 cm) cannon bones; maximum 22¾ inches (58 cm) tall with 4½-inch (11.4 cm) cannons. Bucks one year old and older: minimum 16 inches (41 cm) tall and 3-inch (7.6 cm) cannons; maximum 23⅝ inches (61 cm) tall with 4⅝-inch (12.5 cm) cannons.

Pygmy

Color: Carmel, agouti, black, or spotted
Ears: Medium-size, firm, erect
Facial profile: Somewhat dished
Coat: A full coat of straight, medium-long hair. Does may or may not be bearded. On bucks, abundant hair growth is desirable; the beard should be long and flowing and the copious mane draping capelike across the shoulders.
Horns: Yes

The Pygmy goat is genetically small, cobby, and compact; its limbs and head are short relative to its body length. It is full-barreled and muscular; its body circumference in relation to height and weight is proportionally greater than that of dairy breeds. It is also hardy, agile, alert, and animated, good-natured, and gregarious.

The Pygmy is indeed the quintessential pet goat, though it was raised for meat in its African homeland. Does give from 1 to 2 quarts of milk (which is from 5 to more than 11 percent butterfat) that the NPGA says is higher in calcium, phosphorus, potassium, and iron than milk from full-size dairy breeds, and it's lower in sodium, too.

Pygmy goats are in fact *achondroplastic* dwarfs. Achondroplasia is also found in dachshunds, corgis, bassett hounds, and munchkin cats. Pygmies, like these other animals, are designed to be this way and unless they're obese (don't overfeed pregnant does!) or bred to bucks of larger breeds, Pygmy does experience no more kidding dystocias than goats of other breeds.

Pygora

Registered by: Pygora Breeders Association (PBA)
Use: Fiber, pet
Origin: Oregon, USA
Size: Does average 22 inches tall (56 cm); minimum size is 18 inches (46 cm) at two years of age. Bucks average 27 inches (69 cm); minimum size is 23 inches (58 cm) at two-and-one-half years of age. There is no maximum size limit.

Pygora

Color: All Pygmy colors and their dilutions are acceptable, plus white
Ears: Medium-long and drooping
Facial profile: Straight or (preferably) dished
Coat: Individual Pygoras produce one of three types of fiber: Angora, cashmere, or a combination of the two
Horns: Yes

The Pygora breed was developed in the early 1970s when handspinner Katherine Jorgensen began breeding Pygmies with Angoras to recreate the fiber she saw growing on Navajo goats living on an Arizona reservation. The Pygora Breeders Association was formed in 1987 to register the offspring of registered Pygmy goat bucks and full-size, registered Angora does.

Pygoras, according to breed literature, "have the docility of the Angora and the spunk and playfulness of the Pygmy." They are alert, curious, friendly, cooperative, and easy to handle.

San Clemente Island

Registered by: San Clemente Island Goat Association
Use: Heritage
Origin: San Clemente Island off the coast of southern California, USA
Size: 26 inches (66 cm) or more for does and 27 inches (69 cm) or more for bucks
Color: Light brown to dark red or amber with markings as follows: black head with two brown stripes down the face from above or around the eyes to the muzzle; a black patch on the cheek or jaw and a small black spot on the chin; a black cape over the shoulders, up the top of the neck, and down the front legs. The underside of the neck is brown. Ears are black outside and brown inside. A black dorsal stripe runs down the back, and there is some black on the back legs and flanks.
Ears: Narrow with a distinctive crimp in the middle; carried horizontally
Facial profile: Dished
Coat: Variable
Horns: Yes

San Clemente Island goat

Slightly larger than most of the other breeds we've discussed, though still significantly smaller than most

(continued on page 330)

BREEDER'S STORY

Jeanne DuBois
One Goat Shy Farm, Wisconsin

JEANNE DUBOIS and her husband, Steve Voss, live on One Goat Shy Farm (so named, Jeanne says, "because a goat breeder is always one goat shy of having enough goats") in southwestern Wisconsin, where they breed top-flight Mini-Nubians. Here's her story.

"My husband and I live on a 20-acre farm in southwestern Wisconsin, among the rugged and scenic ridges of the Driftless Region where the terrain consists of high ridges alternating with deep valleys, or coulees, as we call them.

"We moved here from Minneapolis, soon after I became very ill and needed to escape the city. Immediately, I started looking for goats. Our property included a couple of barns, a tiny hayfield, and 13 acres of woods, so it was obvious to me that we needed goats. I thought that with fresh goat milk, chicken eggs, and loads of our garden vegetables, my health would improve a little.

"After months of Web research, borrowing every book the local library had on goats, I decided on Mini-Nubians. I knew that I needed a small animal, and I decided against a miniature cow, because I knew I couldn't handle something that large. At first I didn't think I could even handle goats, because the full-size Nubians I'd looked at were too big. After discovering the mini breeds, though, I knew I could do this. Why Mini-Nubians? I loved their rich milk, and my husband liked their floppy ears.

"Then I discovered it wasn't easy to buy even a few Mini-Nubian does. These girls are few and far between, and as for the mature ones already milking, forget it! But I kept looking and visiting farms, and finally discovered Green Gables Mini-Nubians, near Eau Claire, Wisconsin. I bought two milkers from Eliya Forrester, plus two young bucklings to go with them — I already had major breeding plans.

"Milking? I learned the hard way that I needed a milking stand. Hay? I learned that it wouldn't be piled neatly in one place, but pulled onto the floor by the goats, as if to say 'I'm a goat, that's what I do.' Fences? Actually we lucked out with the fencing thing. We found Mini-Nubians easy to fence because they're a little too big to simply climb the fence, like Nigerians do, but not so heavy that they bend the fence down by leaning against it.

"We've been very happy that we chose mini-goats. Our Mini-Nubians give us everything we were looking for from goats: the best milk and butter I've ever had (not to mention the best ice cream!), a manageable amount of milk, and a good market for the kids. They are also easy to manage and relatively inexpensive to raise.

"This spring was our first experience with kidding. We were worried that we wouldn't be able to sell all those kids, but we already have prospective buyers for all of them. They sell themselves; who could resist that cute little furball or its greeting 'maaah' every time it hears the kitchen door open?

"My breeding program emphasizes the production of goats that are about 24 inches high (tall enough to milk without lying underneath the goat's belly) and milk at least two quarts a day. I maintain their health without the use of complicated feeds and medications. My goats are raised as naturally as possible and on organic feed. This goes a long way toward preventing health problems.

"I want my goats to be the perfect goat for a small homesteading family; no muss, no fuss. Given our current economic conditions, a lot more of these goats are going to be needed by families who hope to cut their food and medical expenses by being more self-sufficient and independent. More mini-goat breeders are needed to supply the need of these families and to help the current breeders breed even better mini-goats.

"If you're interested in breeding mini-goats, I suggest you think about what you bring to the table: you need to like to be around animals and willing to put the goats' welfare ahead of watching your favorite TV show. Do you have any experience or skills with animal care, farming, or gardening? That helps. Enjoying being outdoors even when it's 30 below, or at least being able to bear it, helps too. Read everything. The books may not help when it's 3 a.m. and your goat is kidding and you can't find the flashlight, but at least the things you read will be in the back of your mind. Talk and ask questions about goats with everyone who will listen. People who breed goats are always willing to help, even when you ask the same 'stupid' question for the fourth time. Join the online Web communities; there are thousands of them just for goats.

After all of that consideration, if you still want to have a goat — and milk it too — go for it! You'll have a lot of work ahead of you, but you'll love every minute of it. Goats are downright fun and wonderful to know. They're fascinating, complex, rewarding animals and well worth sharing your life with."

full-size breeds, slender, deerlike San Clemente Island goats are the only Critically Endangered small goats on the American Livestock Breeds Conservancy's Conservation Priority List (see pages 16 and 17).

A good deal of colorful folklore surrounds these goats' origin; however, it appears they came to San Clemente Island in 1875, carried there by Salvador Ramirez, who claimed to have transferred them from nearby Santa Catalina Island. They thrived on San Clemente Island as feral goats until the mid-1980s, when the U.S. Navy, which took possession of the island in 1934 and now maintains a naval base there, ordered their extermination or removal. The Fund for Animals saved the breed from extinction by removing more than 6,000 goats, but it soon lost track of most of them.

Now, the breed is watched and protected. The new and very active San Clemente Island Goat Association registered 375 goats (254 does, 77 bucks, and 44 nonbreeders) by fall of 2008, and the estimated global population stands at 425 goats. Thanks to the efforts of a few dedicated individuals who sought out remaining purebred animals on the mainland, there is still seedstock to work with. To keep the breed alive, more conservation breeders need to raise the San Clemente. If you want to save a piece of history, there is no better candidate than the San Clemente Island goat.

17

Miniature Llamas

LLAMAS TOOK NORTH AMERICA BY STORM in the 1970s as the exotic critter du jour. It wasn't until the early 1990s, however, that breeders began developing llamas smaller than their standard-size cousins. In 1999 these breeders banded together to form the American Miniature Llama Association or AMLA (see Resources). Currently about 1,000 miniatures have been dually registered with the International Lama Registry (see Resources) and the AMLA.

Llama, Alpaca, or What?

The first thing prospective miniature llama purchasers should know is that not all sellers know the difference between llamas and alpacas.

To confound the issue, lamas (a catch-all term referring to all four South American camelid species: llamas, alpacas, guanacos [wild ancestors of llamas], and vicuñas [wild ancestors of alpacas]) are interfertile. Bred together, different species produce viable offspring, most of which fall into acceptable miniature llama parameters. These offspring include *huarizos* (llama sires and alpaca dams), *mistis* (alpaca sires and llama dams), *paco-vicuñas* (vicuña sires and alpaca dams), *llamo-vicuñas* (vicuña sires and llama dams), *llamo-guanacoes* or *llanacos* (guanaco sires and llama dams), and *paco-guanacoes* (alpaca sires and guanaco dams).

The good news is guanacos and vicuñas are still very rare in North America, so you're unlikely to encounter any of their crosses. Alpaca-llama crosses are another story. As alpaca prices fall, especially for excess male alpacas, more backyard breeders are adding alpaca males to their llama herds. The resulting crias (baby llamas) are adorable and the adults nice indeed, but they aren't miniature llamas.

LLAMA PHYSIOLOGY

Typical adult height measured at the shoulder	Standard llamas: 4–4.5 feet (1.2–1.4 m) Miniature llamas: 3 feet (0.9 m) Alpacas: 3 feet (0.9 m)
Typical adult weight	Standard llamas: 250–500 pounds (113–227 kg) Miniature llamas: 120–200 pounds (54–91 kg) Alpacas: 120–225 pounds (54–102 kg)
Temperature	99.5–102 degrees
Heart rate	60–90 beats per minute
Respiration	15–30 breaths per minute
Typical length of gestation	350–355 days
Llama colors	White to black through many shades of brown and gray in solids, spots, and patterns
Primary reason for development	Llamas: Pack animals; also widely used as religious sacrifices and for meat Alpacas: Fiber
Fiber measurement in microns	20–40 microns
Life span	15–25 years
Shape of ears	Llamas: long and banana-shaped Alpacas: shorter and spear-shaped
Shape of back in profile	Llamas: straight with high-set tail Alpacas: somewhat rounded with lower-set tail

Llamas and alpacas have noticeably different physiques. Llamas' body parts (head, back, legs, neck, and ears) are longer and straighter than those of alpacas. Llamas are noted for long, narrow ears that curve in at the tips ("banana ears" in llama terminology), whereas alpacas have shorter, more triangular, pointed ears ("spear ears"). Alpacas' faces are considerably shorter, too.

Miniature llamas (front) are much smaller than their full-size (rear) kin.

Llamas were developed in South America thousands of years ago as working animals, specifically to carry packs at high elevations, which is why their backs are straight and strong. Alpacas were developed alongside llamas but primarily for fiber, so straight backs weren't important. Alpacas' backs are gently rounded, and correspondingly, their tails are set lower than those of llamas.

Alpacas have a single coat, without the longer, coarser guard hair that overlays most llamas' fleece; most (but not all) llamas are double-coated with varying amounts of guard hair over their soft, inner fleece.

To sum it up, llamas are streamlined, elegant beasts, while alpacas look cuddly and cute. Mistis (and much more rarely, huarizos) combine characteristics of both parents and vary widely in shape and size.

To avoid identification problems, don't buy unregistered miniature llamas; registration papers that match the llama in question ensure you're getting the real thing.

Llamas at a Glance

All South American camelids have certain features. They are all, for example, modified ruminants. They chew cud but unlike true ruminants, such as cattle, sheep, and goats, their stomachs have three compartments instead of four.

Like true ruminants, they have a hard upper dental palate but no upper teeth in front, and a set of upper and lower molars in back. All have a split, prehensile (adapted for grasping) upper lip that helps them grip forage in unison with their lower incisors. Mature males have curved, fanglike "fighting teeth," too.

Bare, oblong patches on the side of each rear leg in both sexes are scent glands associated with the production of alarm pheromones. Llamas have scent glands between their toes as well.

Each foot consists of two digits with broad, doglike, leathery pads on the bottom and thick, down-curved toenails in front.

Llamas establish a hierarchy or "pecking order" in every group, be it 2 llamas or 60. Within this group they communicate through body language and through vocalizations ranging from gentle humming to loud shrieks.

Llamas usually rest in the *kush* (or *cush*) position (lying on their bellies with their legs tucked under their bodies) but sometimes lie on their sides or backs as well. They eliminate on community dung piles, and crias learn to use community potties within days after birth. When one llama eliminates, most of the rest of a group will line up to use the dung pile, too.

Except for females interacting with their own young crias, llamas don't initiate close physical contact with one another in the manner of many other species. Unless habituated to it, they prefer that humans not touch them either. They look cuddly but aren't, much to the disappointment of many first-time owners. Llamas can be likened to slightly wary cats. Most enjoy interacting with humans — but on their own terms.

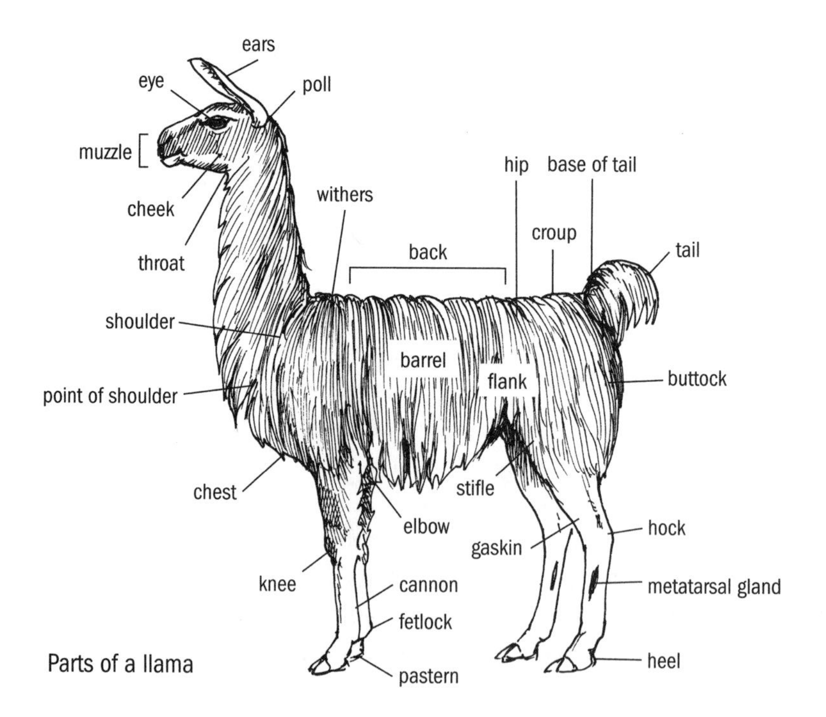

Parts of a llama

BODY LANGUAGE

When a llama stands rigidly erect, ears pricked forward, tail raised, it is focusing on a distant scene or object of concern.

When a llama strikes a broadside pose with its head thrown back, ears pinned, and nose and tail held high, it's saying, "This is my territory; shove off!"

When two llamas of either sex face one another, bodies rigid, ears pinned back, heads up, chins elevated, and tails held high, you're witnessing a standoff between fairly equal-ranked members of a herd. Eventually, one will usually turn its head or walk away, but if not, further animosities ensue, such as spitting, jostling, and verbal threats.

Low-ranking herd members, particularly crias, indicate submission by lowering their heads, with necks curved, and flipping their tails across their backs.

Llama-Speak

The South American native tribes who developed them call llamas their "silent brothers." In fact, llamas aren't silent at all — they make a wide variety of vocalizations. These are the ones you'll often hear:

Humming. Llamas don't hum because they're happy or content; they hum when they're feeling lonesome, bored, overheated, cold, in pain, frightened, worried, curious, cautious, or distressed. A more peaceful hum is the gentle, low-pitched sound mothers hum to their crias (and their crias softly hum it back).

Clicking. Llamas click to intimidate other llamas, such as when telling a herd mate to "leave me alone." They also sometimes click if they're worried or concerned. Females click when their babies wander too far away or if another mom or cria ventures too near. Llamas snort in response to the same stimuli; clicks are sometimes interspaced with snorts.

Growling. Llamas growl (a rumbling sound made deep in their throats) to communicate a warning when they're mildly annoyed.

Alarm calling. When a llama perceives danger, it issues a distinctive, shrill, undulating whinny to alert the rest of its herd. Some give an alarm call only occasionally; others give them many times a day.

Screaming. Llamas scream when infuriated. Males scream when battling one another or when warning another male away from their territory. Some llamas also scream when they're deeply frightened or stressed.

Orgling. Males orgle while breeding. *Orgling* is a throaty, gargling sound that varies greatly from individual to individual but it always indicates mating activity. If one male orgles, all the other intact males in the area may orgle, too. Males orgle throughout the breeding process — from courtship through the final act — sometimes for more than an hour at a time.

Spit Happens

Llamas spit at one another for numerous reasons: to indicate displeasure or intense fear, to establish dominance over a herd mate, or to discipline her own or another female's cria. The average llama, however, rarely (if ever) spits at human beings.

When a llama tenses, pins back its ears, and elevates its chin, it's annoyed and could spit very soon. The mildest type of spitting is an "air spit": the llama emits an explosion of air. Air spits mean, "Shove off!" or "I'm frightened (or annoyed)."

Then comes the "My mouth is full, so take this!" spit. This spit happens if a llama air spits while it has something in its mouth. The llama spews saliva, chewed grass, or dry feed. It means, "Keep annoying or scaring me and you'll be sorry!"

Finally there is the true, hurled-from-the-depths-of-the-stomach spit that no human or animal wants to encounter — ever. Regurgitated stomach contents are slimy, stinky, and disgusting beyond words. After a regurgitated stomach contents spitting occurs, both the llama that spits and the one spat upon typically spend 15 to 30 minutes with their mouths hanging open in response to the terrible taste and smell. Try to avoid this spit if you possibly can.

Llamas come in several coat styles.

WHY MINI LLAMAS FOR HANDSPINNERS?

The same reasons for raising miniatures over standard-size creatures apply for handspinners (and those who market fleece to handspinners) as for everyone else: miniatures require less feed and space than full-size llamas; they're easier to handle; there is a strong market for their offspring; and they're just so darned cute. But there is a huge bonus for handspinners raising miniature llamas: because two or three miniatures can be kept in place of a single, large llama, it's easy to include a wider variety of fleece types and colors in a typical herd.

Fiber Types, Not Breeds

There are no separate breeds among llamas, be they standard-size or miniature; they're categorized according to fiber type and sometimes by their ancestors' country of origin (Argentine, Chilean, Bolivian llamas).

Although some call llama fiber "wool," it's hollow (true wool has a solid core), so it's actually hair. Llama fiber is lighter and warmer than sheep's wool, and it contains no oily lanolin, so it produces greater yields (more yards of yarn per ounce of fiber). Because it's warmer, lighter-weight llama wool garments produce the same heat retention as considerably bulkier sheep's wool items.

Like other fibers, llama fiber is measured in microns. A micron is 1/1,000 of a millimeter in diameter or 1/25,000 of an inch. Generally, alpaca fiber is finer than llama fiber (the lower the count, the finer the fiber), but there are exceptions. Alpaca fiber measures less than 20 microns (the standard grading system calls this "royal alpaca") to more than 35 microns (classification: "very coarse"), while most llama undercoats measure from 20 to 40 microns. Yarn composed of more than 5 percent fiber measuring 22 microns or greater is too coarse and itchy to wear next to human skin.

Most llamas are double-coated. Fiber sheared from double-coated llamas contains up to 20 percent guard hair that must be picked out (usually done by hand) before the undercoat is processed into yarn. Removing guard hair is a nitpicky process so spinners should avoid llamas with an abundance of guard hair.

All llamas fall into one of several basic fiber types: classic, woolly, and suri.

Classic Llamas

Classic llamas are also known as ccara (pronounced CAR-uh) or ccara sullo (CAR-uh SOO-yoh) llamas. They have short, double coats. Guard hair usually accounts for 15 percent or more of a classic llama's overall fleece. Classics have soft, semi-crimpy (minutely wavy) undercoats topped by guard hair, and short hair on their heads and legs, especially below the knees. These llamas shed their undercoats, so they needn't be sheared or clipped; it's easy to remove shed fiber from the tools used to groom them. Because classic llamas' coats pick up fewer sticks and less debris than woolly llamas, they're easier to keep clean, making them the preferred type for packing (even a miniature llama is useful on the trail) and public relations work.

Woolly Llamas

Most breeders simply refer to them as medium- or long-wool llamas, but there are actually several types of wooly llamas.

Curaca (cur-AH-cah) llamas have less guard hair (3 to 15 percent on average) than classic llamas have and longer wool on their bodies, necks, and legs. Like classics, they have hair instead of wool below their knees and hocks. Curaca llamas partially shed, but many require clipping or shearing every few years.

Tapada (tah-PA-dah) llamas have dense, sometimes silky, sometimes crimpy, sometimes loosely wavy coats, as well as wool (not hair) on their heads and below their knees. They have less than 1 percent guard hair and are considered to be single-coated.

Longer-coated tapadas are frequently confused with lanudas (la-NOO-dahs). Lanudas are single-coated, silky-woolly, long-coated llamas. They have wool-fringed ears and tails, woolly faces, and abundant fiber all the way down their legs to their feet.

Some tapadas and lanudas are referred to as silky llamas. They don't shed, so they require full body shearing every year (and considerable grooming in between). Tapadas and lanudas yield the most, and usually the best, fiber.

Suri Llamas

A suri llama's single coat of fiber grows parallel to its body and hangs in long locks of high-luster fiber. Fiber drapes down the sides of the suri llama's body in twisted or flat locks that resemble dreadlocks, though the fiber isn't actually matted. Locks are round, form close to the skin, and twist uniformly the length of each lock. Suri fiber is slick, soft, and shiny (used to make worsted cloth), and when the llama moves, its fiber moves freely. Suri fiber gives suri llamas a distinctive, flat-sided appearance. Standard-size suri llamas are comparatively rare; miniature suri llamas are rarer still.

SURI AND SILKY

To create suri and silky llamas, ancient breeders selected for superfine guard hair to minimize the differences between it and the animals' undercoats, eventually producing a single-layer, lustrous fleece.

Conversely, to produce low-micron, woolly llama fleece, they selectively bred to minimize the percentage of guard hairs to produce a fine and uniform fleece with a lot of crimp. Woolly llama coats have a matte appearance due to the lack of guard hair (only guard hair is lustrous).

Getting into Miniature Llamas

Not all small llamas are truly miniature llamas that have the potential to sire or produce miniature stock. Small llamas may actually be alpaca hybrids, and small may equate with lack of early care and proper feeding. The best way to check the authentication of the miniature llamas you're buying is to look for the American Miniature Llama Association seal on a potential purchase's International Lama Association registration papers. And even then, realize that miniature llamas are still in the early stages of development and they could produce offspring that are oversize but not as large as full-size llamas.

Aberrant Behavior Syndrome

Before buying llamas, in general, and breeding males, in particular, be certain you recognize aberrant behavior syndrome (ABS). Originally called berserk male syndrome (BMS), this dangerous behavior pattern is the result of mishandling young camelids of *both* sexes.

Overhandling, such as bottle feeding, causes baby llamas to imprint on humans; this is what triggers ABS. Llamas become aggressive and hostile toward people, spitting at and charging anyone who comes near them. Males are more likely to develop ABS, but females can be dangerous, too. ABS llamas and alpacas share some or all of this background:

- They were separated from their mothers at an early age and bottle-raised
- They were raised apart from other llamas (particularly adult role models)
- They were handled by people unaware of normal camelid behavior (they were raised by inexperienced owners or displayed in petting zoos)

THE AMERICAN MINIATURE LLAMA ASSOCIATION DIVISION

Only International Lama Registry papered llamas are eligible for dual registration with the American Miniature Llama Association. International Lama Registry handles the paperwork, affixing the AMLA seal to each animal's ILR certification. This seal indicates which division the animal is registered in:

- Miniature Llama (llamas of any sex, three years of age or older, that measure no more than 38 inches [97 cm] at the shoulders)
- Miniature Foundation Llama (females used in breeding programs for producing miniature llamas; females measure 38.1 to 40 inches [97.1–102 cm] at three years of age)
- Immature Status (any llama less than three years of age whose dam is a fully registered Miniature Llama or whose mother is a registered Miniature Foundation Llama and whose sire is a registered Miniature Llama; these papers must be updated at three years of age)

Thankfully, however, not every obnoxious llama is an ABS llama. Some are simply frightened or spoiled. Previously unhandled llamas are afraid of human contact and react, sometimes spectacularly, when subjected to scary procedures like being cornered and caught at toenail trimming time and when shorn. There is nothing wrong with these animals that patience and training won't cure; they react the way they do out of fear. The experts at Southeast Llama Rescue (see Resources) say to expect untrained llamas to:

- Swing their rumps or shoulders into people while trying to avoid being touched
- Run into people with their chests when trying to evade capture
- Swing their necks and heads around, especially when being haltered or touched on the head or neck
- Kick
- Spit
- Fall on a person who is trying to pick up a foot
- "Spook" at new stimuli while being led, dragging anyone holding onto the lead rope

When truly aggressive llamas hurt you, they mean it — and when crias are handled incorrectly, aggression can surface at an early age. Mild indications of ABS include:

- Crias that greet people by flipping their tails across their backs and curving their necks into a U-configuration. This is submissive behavior to be shown to another llama; llamas should *not* consider humans part of their herd.
- Any llama that positions herself to prevent people from passing by
- Any llama that follows people within a 2-foot (0.6 m) distance without enticement such as food
- Males that orgle at humans

If aggression isn't nipped in the bud, behaviors intensify and may escalate to:

- Screaming at or spitting on people without provocation
- Charging at, jumping over, or trying to crawl through fences to reach bystanders
- Charging into humans — ramming them with knees and chest and possibly stomping them after they've been knocked down
- Rearing
- Biting

Can these llamas be saved? Sometimes, but to do it you'll need help to safely handle the task. Contact Southeast Llama Rescue (see Resources); they've successfully rehabilitated hundreds of aggressive llamas and can help you formulate a plan.

Keep in mind that the typical aggressive llama is an intact, hormone-crazed male, so the first thing to do is have him gelded. Then give him time for his hormone levels to subside. According to Southeast Llama Rescue, most of the time, males that are gelded and then rehabilitated are successfully retrained.

Nip aggression in the bud, however, by discouraging overfamiliarity in young llamas. It's all right to interact with youngsters but don't allow them to mouth your clothing, nibble your shoes or shoelaces, sniff your crotch, rub up against you, shoulder in to you, or wrap their necks around you — all of which are precursors to later aggression. Adult llamas don't allow crias to behave in this manner, and neither should you.

LLAMAS LOVE DUST!

Llamas create dust bowls, where they can roll to fluff up their fiber and maintain its insulative qualities. Sometimes after or during rolling they lounge in their dust bowls, flat out on their sides or on their backs with their tummies turned to the sun.

Facilities for Llamas

Llamas, except for a few breeding males, are easy to contain. They are, however, extremely curious animals and notorious for threading their long, fragile necks through any opening their heads fit through. For this reason, the best fencing for llamas is woven-wire field fencing or cattle panels with *small* openings; this is especially important if you'll be raising inquisitive crias.

Unless sheared late in the summer, llamas kept in field shelters or barns rarely have a problem staying warm. Staying cool in the summer is another story indeed, as they're very prone to deadly heat stress. Pastured llamas need lots of shade where they can hang out during the heat of the day. Field shelters and natural cover such as trees are equally effective at shading and cooling them. Ventilation is as important as shade, so llamas kept indoors need fans.

One of the beauties of miniature llamas is that if you're in reasonable shape, you probably won't need a handling chute for routine procedures (a must when raising full-size llamas). Unless you can walk up to each of your llamas and slip on a halter without fuss, however, you will need a catch pen. This is a small, sturdy enclosure where you can herd a single llama to be caught. Four 8-foot long, 4½-foot tall pipe panels that are chained together (the sort used for making equine round pens) work well.

Feeding Llamas

Except for young stock and late gestation or lactating females, your llamas will probably only need quality hay and a good mineral mix formulated for llamas. Llamas wrest the maximum amount of nourishment from the feed they consume and, consequently, easily get fat. Don't overfeed; fat llamas are much more prone to heat stress than trim, healthy llamas.

Keeping Llamas Healthy

Llamas, as a whole, are hardy, easy-care livestock. They're susceptible to the same sort of parasites (coccidia, stomach worms, lice) and illnesses (enterotoxemia, bloat, colic) other barnyard species are prone to but usually to a lesser extent. And they require the same dewormers, vaccines, and medicines to prevent or cure them. There are, however, a few problems you are more likely to encounter when raising llamas, such as high stress levels, teeth issues, and high susceptibility to meningeal worm.

Avoid Stress

Llamas suffer terribly when stressed, and stress leads to serious medical problems. Signs of stress include humming, open-mouthed breathing, pacing, tense lips, "worry wrinkles" below the eyes, tail swishing, and refusing food or water. A solitary llama is invariably stressed: llamas are herd animals and they never like to be alone, so it's kindest (and from a medical standpoint, best) to keep more than one. Other stressful situations include being caught when they aren't used to it, toenail trimming, shearing, being separated from a favorite friend, being subjected to unpleasant veterinary procedures, and being hauled.

Weather extremes, especially blazing heat and humidity, head the list of long-term stressors. It's imperative that you cool any llama suffering from heat stress as quickly as you can. A good way is to hose it down, saturating it all the way to the skin, using lots of cold water, and then place it in front of a fan to finish cooling off.

In hot, steamy climates, keep plenty of plastic milk jugs of ice-cold water and bagged ice cubes in the freezer at all times. Pack these under an overheated, kushed llama or hold them against a standing llama's underbelly, armpits, or groin.

Indicators of heat stress include lethargy, drooping lower lip, staggering, open-mouth breathing, elevated temperature and respiration, and drooling.

TWO EASY PLOYS TO COOL YOUR LLAMAS

Llamas love sprinklers as much as children do. Set up a sprinkler hose or oscillating sprinkler in their loafing areas, and they'll have a great time!

And just as children love to frolic in a wading pool, so do our long-necked, woolly friends. Some llamas kush in wading pools, while others simply stand in the pool and cool their feet. Keep in mind that llamas who spend a lot of time immersed in water may damage their fiber, but better to have damaged fiber than a dead beast.

THE LLAMA TEMPERATURE INDEX

When is hot *too* hot? Add the current Fahrenheit temperature and humidity rating. If the combined total is less than 110, things are fine. If it equals 150, consider this a yellow flag — don't do anything to further stress your animals such as hauling, weaning, breeding, or trimming toenails. A reading of 180 or greater means trouble. Haul out the barn fans and encourage your llamas to loaf in the shade.

Prevention is better than a cure. Llamas always need access to plenty of clean, cool water. Place extra containers of water in shaded areas, and when it's really sizzling, periodically drop ice cubes in popular containers to cool the water. Use fans; even cheap box fans work for a few head of llamas. Don't let llamas get fat; overweight animals are especially prone to overheating. Shear every llama, every year.

Trimming Teeth

At 18 to 24 months of age, intact male llamas begin growing six sharp fighting teeth or fangs, two on the top jaw and one on the bottom, on each side of the mouth; a few females and geldings get them, too. Unless these are blunted or removed, males use them when fighting one another, inflicting serious damage to their opponents.

Pulling teeth often results in serious jaw injuries, so fighting teeth are usually sawn off at gum level. Your vet can do it using a Dremel-type tool or flexible cutting wire designed to slice hard surfaces without damaging soft tissue. (It may be sold as OB wire.) You can also do it yourself, but first have a vet or experienced person show you how.

Adult male llamas grow impressive fighting teeth.

Aged llamas and younger ones with poor occlusion sometimes need to have protruding front teeth trimmed. A battery-powered Dremel tool with a diamond wheel does the trick. You can remove a lot of tooth without injuring the llama, but never grind all the

SHOWING MINIATURE LLAMAS

The Alpaca Llama Show Association (ALSA) sanctions hundreds of camelid shows across the United States each year. It now recognizes a full range of halter classes for registered miniature llamas. It also allows miniatures to be shown in performance classes such as packing to carry lighter weights and to jump the smaller jumps designed for alpacas participating in the same events.

way down to root level because exposed roots are prone to dental infections. Whether using OB wire to trim fighting teeth or a Dremel tool to grind back incisors, stop frequently to let the tooth you're working on cool down. Never cut llama teeth with side-cutters or nippers — they will shatter!

Watch Out for Meningeal Worm

Llamas that share grazing and watering areas with white-tailed deer are at risk for meningeal worm, also called "deer worm" or "meningeal deer worm." White-tailed deer are the natural hosts for this parasite. Ground-dwelling slugs and snails serve as intermediate hosts between deer and other species. Although this parasite doesn't bother deer to any extent, in other species, such as llamas, alpacas, goats, and sheep, larvae migrate to the host's spinal cord and brain causing rear leg weakness, staggering gait, hypermetria (exaggerated stepping motions), circling, gradual weight loss, and paralysis leading to death.

In endemic areas, many vets prescribe off-label, preventive ivermectin injections at 30-day intervals throughout the spring and summer months. Treatment is difficult and generally unsuccessful, so if you live where white-tailed deer are present, it's important to discuss this serious problem with your vet.

Suspect meningeal worm any time a llama staggers or becomes weak in the hindquarters. If it happens, call your vet without delay because untreated infestation leads to paralysis and usually death.

Trimming Toenails

Trimming a llama's toenails can be an interesting endeavor, because most llamas violently resist having their legs handled. They are prey animals and instinct tells them that if they can't run when danger beckons, they're toast. And male llamas, even geldings, fight (and play-fight) by biting at one another's lower legs

HAULING LLAMAS

Llamas kush as soon as the conveyance they're riding in begins to roll down the road. This, coupled with the fact that llamas won't poop except on designated dung piles, makes hauling miniature llamas in goat totes, SUVs, vans, and even cars (with the backseat removed) a breeze.

No matter what you haul your llamas in or where you take them, pack along a bag of llama droppings from home so you can start new dung piles where you stop en route and also at your final destination.

until one or both fall to their knees. Is it any wonder they resist at nail-trimming time?

You'll need a good deal of patience, and unless you're agile and strong, possibly a handling chute, to trim llamas' toenails. It's wise to wear spit-resistant clothing, too, including goggles.

Start with your hand on the llama's shoulder or haunch and run it down the leg till you reach the foot. Give the llama a verbal signal ("Foot," "Leg," "Up"), grasp the lower pastern, and pick up the foot. If you practice before nail trimming day and reward your llamas for picking up their feet (hand over treats while the foot is in the air, not after you've placed it back on the ground), your llamas won't object so much.

Use spring-handled hoof cutters of the sort used to trim the hooves of sheep and goats. Most farm stores carry them. Carefully trim the sides of each nail, avoiding soft tissue. Be conservative; if you cut into soft tissue, the nail will bleed. Then snip straight across the tip and shape the nail by rounding any sharp edges with a file.

Breeding Llamas

Females don't come into heat; they are induced ovulators: the act of breeding causes the female to ovulate. In South American camelids, ovulation occurs about a day and a half after mating.

Females typically produce a single cria. Twinning is a very rare occurrence, which is surprising since nature provided llamas with four teats like a deer or a cow. Crias are nearly always born during daylight hours and delivered from a standing or squatting position. Llamas have attached tongues, so they can't clean their newborns in the manner of other species.

Before Breeding

Don't breed females until they're physically mature — usually around 14 to 18 months of age. Llama gestations typically last 350 to 355 days, but normal crias have been born as early as 315 days and as late as 375 days after conception. Crias' bodies, like those of most newborns, don't thermoregulate very well, so you don't want them born in the dead of winter nor in sultry summer's heat.

During Breeding

A typical mating session begins with the amorous, orgling male pursuing the object of his affection, trying to convince her to stop running and lie down. If she's willing, she does; if she's already pregnant, she'll "spit him off," meaning she'll repulse his advances.

If she's interested, she kushes and he mounts, gradually inching his body closer to hers. When he's close enough, he clutches her sides with his forelegs and snakes his long, thin penis through her vagina and all the way into her uterus, where he dribbles sperm until the mating ends. A typical mating takes 20 minutes or longer.

Sound (orgling) and stimulation (penetration and the tight grasp of the male's front legs) causes the female to release hormones that cause her to ovulate approximately 30 to 40 hours after mating.

Most owners test for possible pregnancy after seven days by reintroducing the couple. If she spits him off, she's probably pregnant. Further testing should occur at weekly intervals up to day 28 after breeding. At this point, if she's still spitting the male off, she's probably pregnant.

Two llamas mating

LLAMA MILK

When compared with milk from cows, goats, and sheep, llama milk is higher in sugar (6.5%) and lower in fat (2.7%) and energy content or calories (70.0 kcal/100 g).

Ultrasound imaging is a more reliable method of establishing pregnancy. Experienced llama vets ultrasound between 45 and 60 days postmating, or a vet can run a blood test to check the female's progesterone levels. Both tests work well.

After Breeding

Don't let your pregnant female get fat. Almost 85 percent of fetal growth occurs during the last trimester of pregnancy, so females don't need concentrates in their diets until then. Feed high-quality grass hay during early gestation, making certain the llama has a plentiful supply of clean water and a properly balanced mineral supplement available at all times. Unless she's grossly overweight, begin feeding a small amount of grain about six weeks before her first due date.

Continue deworming, using safe chemicals such as fenbendazole and ivermectin. Don't use Valbazen to deworm pregnant females; though it's an excellent dewormer for non-pregnant llamas, it's been strongly implicated in spontaneous abortions and birth defects.

Four to six weeks before her first estimated delivery date, give your female a CD/T booster, so she's sure to pass antibodies against these diseases to her cria through her colostrum (first milk). A cria's immune system doesn't start functioning until it's three or four months old, so it needs passive transfer of antibodies from its dam to survive.

Remove males from the herd during your female's last trimester. As birth approaches, the developing cria's placenta produces more estrogen, which sometimes makes the female smell as though she's receptive. If mated (and even some geldings are very insistent), she could lose her pregnancy. It is especially important that no intact males be present for the birth. Males sometimes attempt to breed females in the act of giving birth and if that happens, her cria could be killed.

Before the Big Event

Although it's not the norm, llamas sometimes give birth weeks before their expected due date, so during your female's last trimester, stay alert for these signs:

- **Udder enlargement.** Most females "make bag" two or three weeks before giving birth, although some hold out until the last few days.
- **Nesting behavior.** A day or two before her cria is born, the typical female seeks out a place to give birth. Once she chooses it, she'll return to that spot time and time again. She may roll more than usual, and when resting, she'll probably kush.

- **Last-minute physical changes.** As her body prepares to give birth, a female's vulva becomes relaxed and swollen, with or without a bit of stringy discharge.
- **Isolation.** A few hours before the big event, she'll leave the herd, with or without a companion, and go to her nesting spot.

Keep in mind that nearly all crias are born between 7 a.m. and 2 p.m., but birthing can occur night or day.

It's Time!

When the female goes into labor she'll repeatedly lie down, rise, and lie down again. She'll hum or even moan and make many trips to the dung pile where she'll strain but make little if any dung.

As contractions come closer together, she'll hum more stridently. She'll continually drop, roll, and get up again, but while she's up, her back remains arched and her tail up.

Most females give birth standing or squatting but a few prefer to deliver lying down. Crias dangle from their dams for what sometimes seems like an eternity (it's usually 5 or 10 minutes, in fact) before plopping to the ground. Don't panic; this is nature's way of draining fluids from the cria's airways and is an essential, albeit scary (for the owner!), part of birthing llamas. Apart from this,

Most llamas give birth while standing.

PREEMIE CRIAS

If a newborn cria seems unusually weak (it can't hold its neck up shortly after birth or stand by two hours of age), has a weak or missing suck reflex (put your finger in its mouth to find out), has very short fleece or no teeth (the tips of its central incisors should be breaking through the gums), or has quite floppy ears, it may be premature. Keep this baby warm, and call your vet!

birthing llamas go through the same process as other birthing livestock; refer to chapter 11 for particulars.

Some llamas construe eye contact as aggression, so don't make eye contact with a female who has a newborn cria. Many females that would never otherwise spit at humans will do so to defend their babies.

Some authorities think handling suckling crias, especially males, leads to aberrant behavior syndrome, but judicious handling is probably all right. How-

BREEDER'S STORY

Richard and Gayle Dumas of the Fuzzy Farm, Virginia

ONE DAY GAYLE DUMAS was peeling potatoes and half listening to the television when an "I Love Alpacas" commercial happened to air. She stopped what she was doing, logged on to the Internet, and instantly fell in love with alpacas. Soon she owned a select, small group of alpacas and full-size llamas. In November of 2002, the couple added miniature llamas to the growing Camelid family at their Gloucester, Virginia, farm. Nowadays, in addition to raising some of the best miniature llamas in the world, Richard and Gayle offer an agisting (boarding) service for llamas and alpacas. Gayle also markets top-of-the-line raw and carded alpaca and llama fiber, along with natural and hand-dyed yarns.

Gayle's love of these exquisite animals and the couple's commitment to them is evident when she speaks of her camelid friends.

"Initially we purchased land and built a home in the country on 10 acres, then after a couple of years we wanted to do something constructive with our property, but I am not willing to raise animals whose destiny is the dinner table. We started with alpacas but soon discovered the delightful personalities and beauty of llamas, so we purchased a couple of young llamas, not realizing how large they would be at maturity. I started researching miniatures and learned that they are perfectly proportioned llamas, only smaller. They have the same delightful personality, intelligence, grace, and curiosity as big llamas, and if carefully selected, they have wonderfully soft, light, and warm fiber that rivals alpaca fiber for fineness and quality.

"Our only regret is that we didn't start sooner and that it took us awhile to fine-tune our breeding program to incorporate Argentine llamas. Rare Argentine llamas are noted for their fine fiber, their mellow dispositions, and their strong robust bone. They usually have heavy fiber all the way down their legs,

ever, don't let the baby push you around, mouth your clothing, or otherwise become a pest.

Healthy, well-grown crias can be weaned between four and six months of age. Remember that a solitary llama is a stressed one, so don't neglect to give the new weanling a friend. Another cria or a friendly older llama of the same sex is best, but an amiable sheep or goat makes a fine pal, too.

even between their toes. We'd noticed that many miniature llamas, while small in stature, seemed to lack substance. So we hope that the select small Argentines we have added to the mix will remedy that.

"The markets for Argentine and miniature llamas are strong partly because of these animals' rarity, and miniatures appeal to people with limited space. Their small stature, their grace, and their intelligence make them perfect for public relations or as therapy animals. No one can look into their huge, expressive eyes without falling in love. When you put Argentine and miniature llamas together, the results are exquisite!

There is certainly room for additional ethical breeders. This portion of the industry is in its infancy. Few people are aware that miniature llamas exist, but once they meet one, they're hooked.

"My best advice to someone considering breeding miniature llamas is to get into the business for the right reason: the love of the animals. Raising any animal is a deep commitment; they must be taken care of every single day, no matter how rotten you feel or how bad your day has been. But the rewards are beyond words.

"On the practical side, be sure to ask a lot of questions. Ask other breeders and owners who they trust. Visit many farms, and see how the animals interact with their owners: is it clear that they are loved and well cared for? What about mentoring after the check clears — is the seller willing to help? Don't be in a hurry; you are dealing with real lives. Be sure of what you are getting into before making the commitment. Check into the availability of an experienced camelid vet in your area or a vet with a willingness to learn. Help your vet learn by gifting her with reference books, or sponsor her attendance at camelid seminars. The llama she saves could be yours!"

18

Miniature Pigs

PIGS AREN'T BUILT FOR PULLING CARTS or packing loads, nor are they easy to milk. Few jobs exist for swine (though French truffle-sniffing pigs spring to mind), so historically they've been raised for a single purpose: to eat.

Be that as it may, in the late 1980s when small swine in the form of Vietnamese Potbellied pigs became available in North America, pigs became pets as well. This hasn't always been in the pigs' best interest.

PIG PHYSIOLOGY (ADULT PIGS)

Temperature: 102°F (38.9°C)
Heart rate: 60–80 beats per minute
Respiration: 12–30 breaths per minute
Natural life span of domestic pigs: 10–20 years
Age at puberty:
Boars: 3–8 months
Gilts: 3–7 months
Estrous cycle: 17–25 days
Heat duration: 24–72 hours
Ovulate: 36–42 hours after onset of heat
Length of gestation: 114–115 days (3 months, 3 weeks, and 3 days)
Number of young: Breed dependent but averages 4–10 (the largest litter we know of was 18 piglets produced by a Hungarian sow named Nyanyóca)

The Pet Pig Boom of the 1980s and 1990s

In 1985, Canadian zoo director Keith Connell traveled to Vietnam and brought back 18 pigs of the Í breed, planning to breed and to sell them to other zoos in North America. These black, wrinkly pigs with sagging backs were large by modern potbellied pig standards — some weighed up to 300 pounds (136 kg) — but considerably smaller than North America's standard domestic pig. In 1986, he sold some stock to American pet breeders and the Potbellied pig craze was on. These pigs and their descendants became known as the Con or Connell line.

At first, pigs sold for thousands of dollars each (one brought $37,000), prompting early breeder Keith Leavitt of Texas to import Mong Cai pigs from northeastern Vietnam near China and the Gulf of Tonkin. His stock became known as the Lea or Leavitt line.

Most of today's purebred potbellied pigs descend fully or in part from Con and Lea bloodlines, though additional importations, including potbellied pigs from Britain and Sweden, soon followed.

America's enchantment with potbellied pigs grew by leaps and bounds. Early promoters crowed that potbellied pigs were "the size of a cocker spaniel," easily potty trained, and flawless house pets. The same promoters cranked out piglets at the rate of three litters per sow per year, with up to six or seven piglets per litter.

At the same time, buyers watched their "35-pound pigs" grow up and up and out. Potbellied pigs don't stop growing until they're five or six years old. Teensy $1,500 piglets were maturing into 150- to 200-pound hogs.

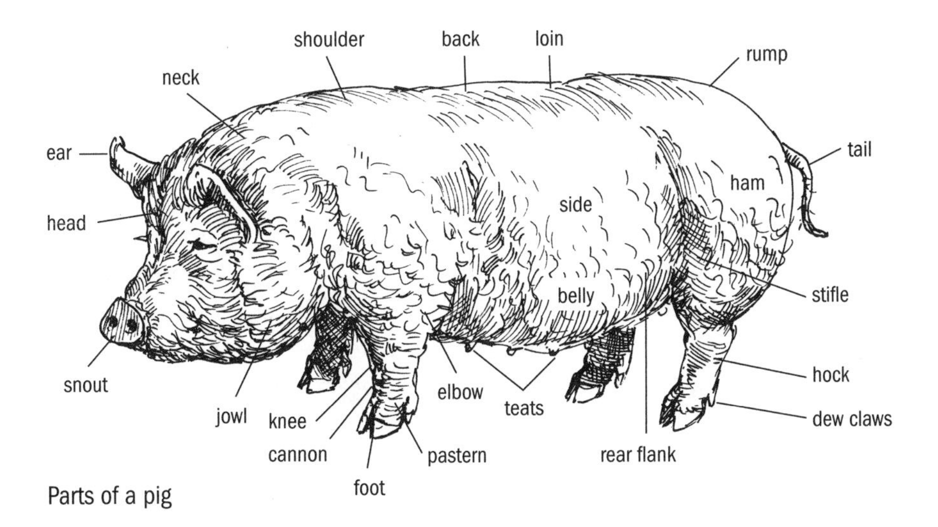

Parts of a pig

Worse, owners quickly discovered that pigs are smart and strong-willed, and they live their lives from a pig's viewpoint. Behaviors that seem perfectly normal to a pig, like rooting (all of the linoleum from the kitchen floor), chewing (interior walls), and nipping to establish a place in the herd (which comprised the humans in their household) upset pig owners very much indeed. Owners who worked through problems with their house pigs loved them; the ones who didn't began giving them away. Pig sanctuaries sprang up across the country, and humane groups became inundated with unwanted pet pigs, yet promoters kept breeding more piglets because there were still interested new buyers.

"[Pigs] have the cognitive ability to be quite sophisticated. Even more so than dogs and certainly three-year-olds."

— Professor Donald Broom, Cambridge University Veterinary School

"[Pigs] are able to focus with an intensity I have never seen in a chimp."

— Dr. Sarah Boysen, Penn State University

According to Dr. Linda Kay Lord in "Survey of Humane Organizations and Slaughter Plants Regarding Experiences with Vietnamese Potbellied Pigs," published in 1997 in the *Journal of the American Veterinary Medical Association* (see Resources), at the time there were between 250,000 and 1 million pet pigs in the United States, most of them Vietnamese Potbellied pigs or crosses between these and domestic swine. In a survey of humane organizations in seven states (responders represented 68 percent of the total number of humane groups in the seven-state region), the author was dismayed to discover that over an 18-month period, fully 55 percent were asked to take Potbellied pigs — to the tune of 4,380 requests. Only 72 percent of pigs were accepted, and 21 percent were ultimately euthanized. Furthermore, 485 slaughterhouses responded to her questionnaire and indicated that more than 4,000 potbellied pigs were slaughtered for meat during the same period.

The moral of this story: If you love potbellied pigs, please don't breed them. And if you breed Kunekunes or any other type of mini pigs, think twice before promoting them as pets. In the right hands, pet pigs shine; in others, they're a disaster waiting to happen. If you do market piglets to the public, don't fudge the facts. Make sure buyers know what they're getting into, and screen every buyer before you sell. Piglets are smart, affectionate creatures that bond deeply with the humans they live and interact with; they deserve better than to be discarded later on.

VIETNAMESE PIGS IN VIETNAM

Ironically, the swaybacked, pug-nosed Vietnamese pig is a vanishing breed in its Southeast Asian homeland. In the early 1990s, while the American Potbellied pig craze was gaining full momentum, the Swedish government provided Vietnamese farmers with large swine to cross with native breeds to boost pork production. At the time there were four breed types developed by natural selection in localized regions of Vietnam.

- **Í** (with arguably the shortest breed name in the world) are the most common unimproved swine in Vietnam. They are fairly large by miniature standards, often tipping the scale at up to 200 pounds (91 kg). Í have little hair and wrinkly black skin with narrow foreheads, dished faces, and small upright ears. They originated in the western part of North Vietnam near the Red River Delta.
- **The Mong Cai pig,** also from North Vietnam, is slightly larger (220 pounds [100 kg]) and hairier than the Í. Mong Cai are usually white with black heads and white snouts and black markings elsewhere on their coats.
- **The Heo Mai** hails from the mountainous regions of South Vietnam and is sometimes called the Vietnamese primitive pig. Heo Mai are smaller than the North Vietnam breeds (90 to 100 pounds [41–45 kg] on average). Heo Mai are exceptionally swaybacked and have longer snouts than the Í.
- **The Co,** once commonly encountered throughout central Vietnam, is another small pig, the largest tipping the scales at 90 pounds (41 kg). Its body is shorter and its back is less swayed than the other types of potbellied pigs.

The Co breed type has largely disappeared since the introduction of production swine from Sweden, and the other types are largely restricted to out-of-the-way regions of Vietnam. The good news: In an effort to prevent their extinction, Vietnam's government now issues subsidies to farmers raising indigenous pigs.

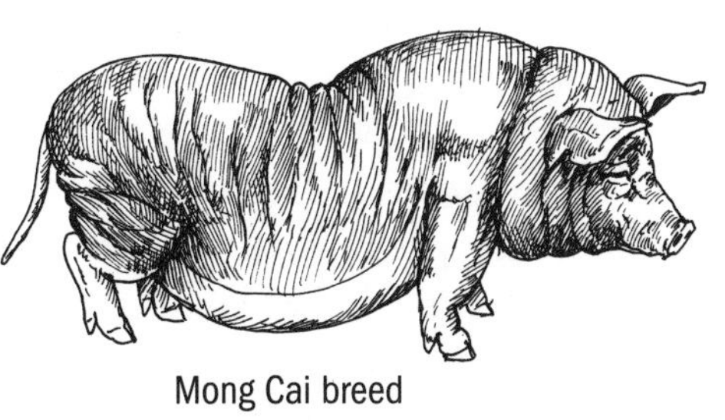

Mong Cai breed

Think like a Pig

Before getting pigs it's best to learn to understand them. You can't outmuscle a pig (even miniature pigs are incredibly strong), so it behooves you to learn how to outthink them. Pigs are intelligent creatures; some studies place their IQ immediately below non-human primates and well above dogs. Researchers teach pigs to play computer games, manipulate food and water devices, and turn fans and radiant heaters on and off. They're more easily housetrained than many puppies and readily learn to walk on a leash.

How Pigs Perceive Their World

Pigs see in color — likely at least green and blue. We know this because their eyes consist of rods and cones with two distinct wavelength sensitivities. They have wide-angle vision and can see roughly 310 degrees around them but they probably cannot focus their eyes. Because of this, pigs often refuse to enter dark places or to cross shadows, door thresholds, or areas where floor colors or textures change. Directing a light to these areas may help. Also, when frightened, they may not watch where they're going and race full-tilt in to unwitting humans.

Their sense of smell is unusually well developed, as evidenced by the truffle-sniffing pigs of Europe. Pigs detect scent in the wind and probably recognize one another by sight and smell.

Pigs also have an acute sense of hearing, and they hone in on sounds by moving their heads. Loud or unusual sounds tend to frighten them.

They are quite sensitive to touch. Scratch a pig's back or tummy and it all but melts.

Pigs communicate through a wide range of snorts, grunts, and barks indicating hunger, thirst, alarm, fear, affection, courtship, and more. In scary situations they also emit pheromones to communicate fear. Squealing is a porcine distress signal.

Pigs are omnivores and relish a wide variety of vegetable and animal foods, including carrion. They have been known to snack on neonatal lambs and kids, so they shouldn't be kept with other stock at birthing time. They find certain tastes in particular appealing — in one university study their food of choice was apples.

Pigs are gregarious and, unless raised in a single-pig household, prefer the company of other pigs. Apart from dominance issues, they are gentle and affectionate toward others in their group (including humans).

Normal pigs alternate between wakefulness and sleeping for about 12 out of 24 hours and are drowsy or completely asleep for the other 12 hours.

PORCINE TRIVIA

- Pigs more readily approach a person who is sitting, squatting, or kneeling than one who is standing up.
- When holding piglets, restrain them as you would a dog; never hold them upside down by their hind legs — this can hurt them.
- Pigs have very little herding instinct. It's infinitely easier to lead them with a bucket of food than to drive them.
- Pigs foam at the mouth when challenging another pig or a human.

Social Hierarchy

In wild pigs, a family group (known as a sounder instead of a herd) consists of several females and their offspring. Males live a solitary existence or in bachelor groups.

Pigs, both wild and domestic, are organized by one of two social orders: teat order and dominance hierarchy or pecking order.

- **Teat Order.** Within a week after birth, piglets select and begin defending their position at their dam's udder. Stronger, more aggressive piglets claim teats closer to their mother's forequarters, as these teats secrete the most milk.
- **Pecking Order.** Once weaned, littermates squabble to establish a dominance hierarchy among themselves and with any other piglets placed in their midst. Fighting consists of head butting, neck biting, shoulder slamming, and foaming at the mouth. Within 24 hours, most battling ceases and piglets accept their place in the pecking order — at least until newcomers are added to the mix. Once dominance is established between two individuals, subsequent encounters usually consist of only grunts and threatening postures.

Nesting Behavior

Piglets are born unable to regulate their own temperature, so they huddle close to their mother and littermates to help them survive. Pigs that are kept in housing outdoors burrow into their bedding for the same reason.

Pigs eliminate up to 5 pounds (2.3 kg) of soft but formed feces per day, but they will neither urinate nor defecate near their sleeping and eating areas. Piglets learn potty manners from their dams. Pigs also avoid toys, feeders, and other items soiled with manure. They want to be clean.

Staying Cool

Because they lack adequate sweat glands to release warm moisture and cool themselves, pigs seek shade and wallow in water or mud to cool down. They instinctively dig wallows; if you don't want yours to dig one, provide a children's wading pool in its stead.

Pigs Root

Pigs are hardwired to root — that's what their nose disks are for. Some breeds root more than others, but all pigs root to some degree. Unsupervised house pet pigs destroy carpeting and linoleum; outdoor pigs dig in the lawn. Pig producers sometimes "ring" pigs by inserting metal rings in the sides or centers of their nose disks, but rings hurt and frustrate the pig, so this is not recommended.

In addition to rooting in the earth, pigs quickly raise their heads in a rooting motion guaranteed to tip over feeders, water receptacles, and sometimes human bystanders. Be forewarned.

BEFORE YOU GET A PIG

Check into local zoning ordinances before buying or adopting a pig, especially if you live in town. Most municipalities — large and small — have Potbellied pig ordinances of one kind or another. In some cases, pet pigs are classified as livestock and simply can't be kept; in others, you may have to comply with a long list of regulations. These regulations from Chapter 8.19 of the Riverside Municipal Code in California are typical.

According to the municipal code, a Potbellied pig is defined as "a domesticated miniature Vietnamese, Chinese or Asian pot-bellied or pot-belly pig not exceeding one hundred twenty-five pounds in weight and eighteen inches measured at the shoulder." Only two such animals can be kept on any single family residentially zoned property. All pigs must be:

- Licensed on an annual basis
- Spayed or neutered
- Detusked
- Provided with a fenced yard and maintained in a safe, clean, odor-free manner when kept outdoors
- Restrained by a harness and leash or similar restraint not longer than six feet and held by a competent person while on a street, sidewalk, or other public place

Getting into Pigs

Guinea Hogs and Ossabaw Island Hogs are ideal breeds for niche-marketed pork production, and at this writing, there's a fine market for Kunekune pet pigs. Opportunities are out there. If you'd like to raise pigs, here's what you need to know.

Buying Pigs

Before buying miniature pigs, home in on your needs. What a pork producer requires may be irrelevant to pet producers and vice versa.

Pork producers require stock adapted to their rearing systems. For instance, someone who plans to pasture raise Guinea Hogs should try to buy stock from someone who is already doing that. Pork producers need productive, meaty breeding stock that gains well on the type of feeds available in their locale. Pet breeders need stock with good dispositions and that breed true; no one wants to buy a prospective house piglet and have it grow to immense size.

Color and breed type aren't important when lean pork is the end product, but they're very important to someone marketing pet pigs. And if you plan to sell pigs as breeding stock, begin with registered pigs; without registration papers, many potential buyers will pass.

No matter what type of pigs you raise, always start out with healthy animals and keep them that way. Whether raised for meat or as pets, no one will buy diseased pigs.

A Pig in the House

Can miniature pigs make great house pets? Yes, indeed. Join a YahooGroup like Pig Info and Chat or Pot Spot to see how many families have miniature pigs living in their homes. That said, keeping a pig in the house is not for the faint of heart. Before deciding to keep one, we suggest you download a free copy of *The Shelter Worker's Guide to Pigs* from the PIGS, A Sanctuary Web site (see Resources), then decide if a house pig sounds right for you. If it does, I recommend reading *The Complete Guide for the Care and Training of Pet Potbellied Pigs* by Kathleen Myers (see Resources). Even if your piglet is a Kunekune or any other mini pig breed, read it — you'll be glad you did.

According to *The Shelter Worker's Guide to Pigs*, common problems associated with keeping pigs in the house full time include their rooting up floor coverings, destroying walls (pigs love drywall), overturning furniture and potted plants, opening refrigerators and cabinets, and urinating or defecating outside the litter box. These problems can be surmounted by pig-proofing the home (removing items the pig might destroy) and supervising the pig when it's loose in the house.

However, the publication continues, aggressive behavior is intensified when a pig is housed indoors full time, especially in single-pig households, because it's difficult for the pig to establish a hierarchal structure in a one-pig home. It may head-swipe, charge, and bite humans the way it would engage with a herd of its peers. This, too, is surmountable but requires understanding and diligence on the part of everyone who comes in contact with the pig. Remember: Once dominance is established, hostilities cease — but the humans must emerge as top pigs.

A better solution is to allow your house pig to spend part of its time outdoors, or keep it outdoors the entire time. Having two pigs also helps (they can establish hierarchy between them). If keeping a full-time outdoor pig, you must prevent loneliness.

Outdoor Housing

Outdoor pets can be kept in existing buildings, horse stalls, Port-A-Huts (see Resources), or medium-size or larger dog houses and plastic Dogloos. Port-A-Huts and dog houses should face away from the wind, and in cold climates, they should be fitted with a flap to keep out inclement weather. Housing must be draft-free and should be filled with a deep bedding of old blankets, hay, or straw.

Fencing for pig pens must be stout and firmly secured to fence posts set deep in the ground. Hog panels are a good solution, and when used in conjunction with wooden fence posts and a nice gate, they are neat enough for most any setting.

Feeding

Don't feed farm pig feed, dog kibble, or table scraps to pet pig breeds like Potbellied pigs or Kunekunes. Buy feed formulated specifically for miniatures (most

Pigs don't have to live in a mansion. Give them a dry, draft-free shed, and they'll be fine.

large pet stores carry it). It's not terribly expensive, and an adult Potbellied pig requires only two to three cups per day.

Pigs naturally love treats, but don't overdo it. A small handful of fruit or Cheerios works well. And though they repeatedly tip their water containers over until it drives you silly, make certain pigs have access to clean water at all times.

Health Care

Ask your vet which vaccinations are necessary in your locale and how frequently your pigs should be wormed.

Pigs have naturally runny eyes, and a waxy buildup tends to form in their ears. Clean runny eyes with warm water and a washcloth and waxy ears with a dry washcloth or paper towels.

All pet pigs, with no exceptions, should be neutered or spayed. Boars emit a strong, offensive odor, and they're generally only interested in eating and breeding. If they can't find a female in heat, they'll attempt to mate with rocks, logs, blankets, and even human legs. Unless detusked, they can do considerable damage with their strong jaws. You definitely don't want a pet boar.

And the typical unspayed female becomes quite erratic during her heat cycle, which occurs every 17 to 25 days year-round. Some become more loving, but others become aggressive and tend to bite.

Breeding

The breeding of pets such as Kunekunes and of pork-producing pigs such as Guinea Hogs and Ossabaw Island Hogs is in many ways radically different, yet some things are much the same. Here are some of the things to consider when producing either type.

BASIC BREEDING TERMS YOU SHOULD KNOW

Farrow. The act of a sow or gilt giving birth

Gestation. The 113- to 116-day period when a sow or gilt is pregnant; from breeding until farrowing

Gilt. A young female pig that has not yet farrowed

Sow. A mature female pig

Standing heat. When a sow or gilt assumes a rigid stance and maintains it while a boar courts her or a human presses down on her back. This indicates she is ready to be bred.

PORCINE BREEDING TRIVIA

A boar's penis has counterclockwise, corkscrewlike ridges near the tip that function like threads in a screw. During breeding, these ridges are grasped by several sets of interlocking rings within the female's cervix. This causes the male to ejaculate.

Sperm is ejaculated in three distinct factions. The first is watery and contains few sperm cells, the second is composed of mainly sperm cells, and the third is a gel-like substance that temporarily plugs the female's cervix to prevent sperm from trickling out.

The Porcine Birds and Bees

A sow's estrous cycle averages 21 days but can range from 18 to 23 days. During most of this cycle, follicles are developing on the sow's ovaries; each ovulates a single egg. As the time of standing heat approaches, these follicles grow larger and begin releasing estrogen; this causes the sow to display signs of heat.

When she is fully receptive, a sow stands still when pressure is applied to her back. Other signs of heat include riding other pigs, fence walking, vocalizing, and increased swelling and redness of her vulva. Standing heat lasts three to four days for sows but only one or two days for gilts. Gilts begin cycling around five to six months of age, but it's best to delay breeding until their second or third heat.

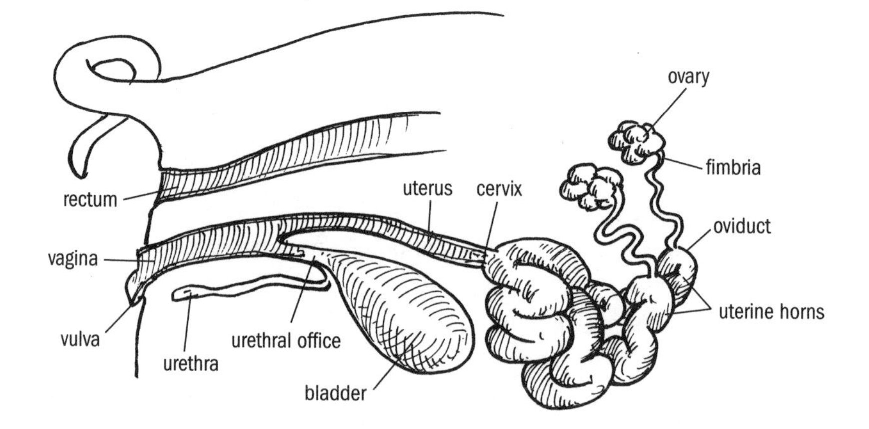

Sow's reproductive system

Boars are attentive toward sows and gilts in heat. They sniff females' vulvas, nuzzle their flanks, and attempt to mount. Matings take 5 to 10 minutes from start to finish. If both the female and the male are healthy and the male isn't overused (a boar should generally breed no more than five sows per week), conception is likely to occur.

Sow Care

As mentioned above, gestation lasts 113 to 116 days. Reducing stress is important during the first two months of this gestation time frame because stress has been implicated in early fetal death; non-stressed sows and gilts deliver more pigs.

Sows and gilts can be housed individually or in groups. They require roughly 14 percent crude protein feed throughout gestation, and the amounts should be adjusted so that the pigs are neither thin nor obese when they farrow. Beginning about a week prior to farrowing, their diet should incorporate bulky feed such as oats, barley, or hay to prevent them from becoming constipated. Clean, fresh water should be available at all times.

About two weeks before farrowing, deworm each sow and treat her for external parasites. A week before her due date, move her to her farrowing quarters (perhaps a farrowing pen for production hogs or an empty stall in the barn for pet pigs). Reduce her feed for about two days prior to farrowing.

PRE-FARROWING BEHAVIOR

Watch for these signs as your sow's farrowing date approaches:

- Teats enlarge and become firmer up to 10 days prior to farrowing. During the same time period the sow's vulva will begin to swell.
- About 2 days before her delivery, the sow's teats become turgid and, if checked, they secrete clear fluid.
- Twelve to 24 hours before farrowing, the sow will begin to secrete milk. At the same time she's likely to become restless — frequently getting up and lying down, pawing, rooting the floor — and she will attempt to build a nest using any available materials.
- About 6 hours before farrowing she will come into full milk.
- Between 30 minutes and 4 hours before giving birth, the sow will begin breathing heavily. Fifteen minutes to an hour before farrowing she'll quiet down and lie on her side, then begin straining. She may pass blood-tinged, oily fluid and meconium (fecal material from her unborn pigs) before the first piglet appears.

After furrowing she will be famished. Switch her to a 16 percent protein diet, remove the bulky items from her feed, and provide copious amounts of clean drinking water.

Farrowing

In a normal farrowing, about half of the piglets are born headfirst and the other half hind feet first. They are usually born about 15 minutes apart. Total delivery time varies according to litter size, but most deliveries of miniature pigs require less than 2½ hours. The placenta should be delivered within two hours of the final piglet's birth.

If gestation exceeds 116 days, and if the sow is drastically off-feed prior to farrowing, anticipate problems. If the sow strains hard without delivering pigs, or if the time between the delivery of two piglets exceeds one hour and the sow still has a full abdomen, something is wrong.

Lending a Helping Hand

If you have to assist, follow the protocols in chapter 11, keeping these extra points in mind: a sow's uterus slopes downward and then divides into two horns, both of which might contain piglets. To pull a forward-facing piglet, grasp both front legs and pull, or place your hand in its mouth and pull on its lower jaw. To pull a pig in hind-feet-first position, grasp both hind legs and pull. Pull the piglet all the way out, and make certain it's breathing before going back in for another pig.

As each piglet is born, make sure its nostrils and mouth are free of mucus so it can breathe. If the broken umbilical cord is more than 2 inches long (measured from the piglet's abdomen), trim it to 2 inches in length. Dip the cord in iodine, dry the piglet with a towel, and place it in a bedded basket under a heat lamp where the sow can see it.

Newborn Behavior

After farrowing is completed and piglets are returned to the sow, the sow will grunt softly and offer her udder to the piglets to nurse. Within the first day or two, each piglet will select its own nipple and return to it whenever feeding. A sow quickly settles into a nursing routine. She feeds her piglets at roughly two-hour intervals, inviting them to nurse by rolling onto her side and grunting softly. Each piglet rushes to its feeding station and nuzzles the sow's udder, which triggers milk letdown, and the piglets nurse.

Neonatal piglets are extremely cold sensitive and require temperatures near 90°F (32°C) to survive, whereas sows prefer temperatures in the 65 to 70°F (18–21°C) range. To provide heat for the newborns (and a place where they can snooze without being accidentally crushed by their mother), block off a section of the farrowing pen so that only the piglets can enter, and hang a heat lamp over it so they can stay warm. *A fallen heat lamp can burn down your barn*; make certain you hang yours correctly. Hang it very securely and at a distance from any combustible material (don't place bedding directly below or near the heat lamp). Ideally, piglets will lie scattered around or under the lamp; if they pile up under it, they are too cold and you should move the lamp a little closer to them.

Piglets need the warmth of a heat lamp to survive, but hang it correctly so it doesn't fall and burn down your barn.

Piglet Care

Sometimes sow's milk doesn't contain enough iron for baby pigs, so three or four days after piglets are born, many producers routinely inject them with supplemental iron. Discuss this with your vet.

CONCORDE SQUEALS

Any time you pick up piglets, expect ear-splitting squeals. Pigs hate to be picked up. (Think about it: In the wild the only time a piglet's feet leave the ground, it's being carried off by a predator.)

A piglet's squeal can range from 110 to 115 decibels. Compare that with the Concorde jet, which is usually less than 112 decibels.

Plan ahead. Wear earplugs.

Many pork producers use side cutters to clip off piglets' needle teeth to prevent them from injuring one another or tearing their mother's teats; others feel it's totally unnecessary. If you do it, it should be done when the piglets are a day old.

Pork producers generally introduce solids at about one week of age, beginning with 20 percent protein pig prestarter and then switching to 18 percent protein starter once they're readily eating. This should be placed in a shallow pan in the piglets' sleeping area where their mother can't reach it. Discard old feed once a day and replace it with new stuff.

Piglets are prone to scours, which can quickly lead to dehydration and death. If you don't know the cause (and there are several), consult your vet without delay so you can treat it correctly.

Piglets are usually weaned at four to six weeks of age. The best way to accomplish this is to remove the sow from the farrowing pen, leaving the piglets in familiar quarters. Fresh feed and clean water should be available at all times. Optimal sleeping area temperature for four-week-old pigs is 80 to 92°F (27–33°C).

LET THE BUYER BEWARE!

When seeking miniature pigs, take height and weight claims with a grain of salt. Many breeders tout pictures of adolescent pigs, but remember: Pigs grow until they're five or six years old.

And be especially wary of ads or Web sites hawking "teacup" or "micro" pigs. Yes, they are tiny piglets, but how big will they be when they mature? Ask for verification, preferably from a vet. Breeders who have their pigs' interests at heart won't be offended.

And be aware that, by far, the most often quoted reason people give for surrendering a pig to a rescue or sanctuary is, "I never dreamed he'd get so big!" Don't fall for the "size of a cocker spaniel" line; most of the time it simply isn't true.

The Breed You Need

Miniature pigs fall into one of three categories: pet pigs, laboratory mini pigs, and heritage hogs for pastured pork production. Potbellies and Kunekunes are pet pigs; Ossabaw Island and Guinea Hogs are production pigs. However, some people do keep Guinea Hogs as pets because they are so docile. All four breeds are naturally small pigs; none are man-made miniatures.

Guinea Hog

Registered by: American Guinea Hog Association
Use: Meat and occasionally pets
Origin: Some authorities believe they descend from bristly haired, sandy-red swine called Red Guineas that slave traders carried from West Africa and the Canary Islands to America
Size: 150–300 pounds (68–136 kg); 22–27 inches (56–69 cm) tall; 46–56 inches 117–142 cm) long, measured from between the ears to the base of the tail
Color: Black or bluish black, sometimes with a touch of white on the nose and feet
Ears: Medium-size and erect
Facial characteristics: Short to medium in length, with a broad snout and slightly dished profile
Body characteristics: Straight to slightly arched back; from the side they present a long, rectangular appearance
Tail: One curl
Coat: Medium to long, coarse, and bristly

Today's Guinea Hogs descend from small, hairy, black hogs kept on homesteads throughout the southeastern United States in the 1800s and well into the early twentieth century. Historical names for these ancestors included Acorn Eaters, Yard Pigs, Pineywoods Guineas, and Guinea Forest Hogs. Some had long noses, others short; they were variously described as small-, medium-, or large-framed; and they ranged in size from 100 to 300 pounds. Other traits they had in common with today's Guinea Hogs included the ability to fend for themselves, producing meat and lard with minimal human input.

Guinea Hogs have calm, friendly temperaments. They are minimal rooters and do well on grass; they tend toward obesity if overgrained. The Guinea Hog is an ideal heritage breed for pastured pork production.

There are fewer than 200 purebred Guinea Hogs in the world today, and they're listed as Critically Endangered on the Conservation Priority List of the American Livestock Breeds Conservancy (ALBC). They excel at homestead chores such as tilling the garden and ridding the place of pests such as snakes and rodents; they produce tasty cuts of succulent pork; and because of their size and easygoing nature, they make fine pets as well.

Guinea Hog

MINIATURE PIGS FOR SPECIALTY PORK

The whys and hows of heritage pastured pork production are beyond the scope of this chapter, but there are excellent books on the subject. I recommend *Storey's Guide to Raising Pigs* and *Dirt Hog: A Hands-On Guide to Raising Pigs Outdoors . . . Naturally*, both by Kelly Klober (see Resources).

For additional information about using Guinea Hogs or Ossabaw Island Hogs in pastured or organic pork production programs, contact the American Guinea Hog Association, the National Sustainable Agriculture Information Service (ATTRA), or the American Livestock Breeds Conservancy (ALBC) (see Resources).

Kunekune

Registered by: American Kunekune Breeders Association
Use: Pet
Origin: New Zealand
Size: According to the British breed standard, 120–240 pounds (54–109 kg), 24–30 inches (61–76 cm) tall

Kunekune

Color: Cream, ginger, brown, black, and spotted
Ears: Erect or flopped and inclining forward
Facial characteristics: Broad face, short to medium-length snout, slightly dished profile
Body characteristics: Most Kunekunes have a pair of wattles, also called *tassels* or *piri piri*, under their chin. They have level or slightly arched backs, short legs, and a nicely rounded body.
Tail: One curl
Coat: Completely covered with hair ranging from short and straight to long and curly

Kunekune (pronounced *cooney cooney*) roughly translated means "chubby" in Maori. Swine are not indigenous to New Zealand, and how they arrived there is open to conjecture, but they were an integral part of native Maori culture for many years. Kunekunes were nearly extinct by the 1970s when two wildlife

JULIANI PAINTED MINIATURE PIG

There is presently no breed organization or registry for the Juliani Painted Miniature Pig, a very rare pet pig that its few American breeders claim came to North America from Europe, where several breeds were used to develop a small, intelligent pig. Julianis have small- to medium-size erect ears, a slight potbelly, short hair, a slightly swayed back, and a short, straight tail. They come in red, red and black, red and white, white and black, black, silver, and silver and white. Their legs are slightly longer than those of a Potbellied pig.

park owners, Michael Willis and John Simster, heard about them and set out across New Zealand to buy some. They were only able to find 18 pigs in all of New Zealand. From these pigs, which were later mixed with additional stock, the New Zealand studbook was formed.

Kunekunes are grazers and easy keepers, requiring little or no grain in their diets. Though the Maori certainly ate them, today's Kunekunes are primarily pet pigs. They are friendly, intelligent, and easygoing.

Ossabaw Island Hog

Registered by: Ossabaw Island Hog Registry (maintained by the American Livestock Breeds Conservancy)
Use: Meat
Origin: Descended from Spanish pigs placed on Ossabaw Island by Spanish sailors, possibly as early as the 1500s
Size: 100–250 pounds (54–113 kg)
Color: Black, black with white spots, white with black spots; often has a reddish tinge
Ears: Small, upright
Facial characteristics: Long snout, slightly dished
Body characteristics: Heavy shoulders and head, relatively lightweight hindquarters
Tail: One curl
Coat: Very hairy with heavy bristles on the head, neck, and topline

Ossabaw Island Hog

BREEDER'S STORY

Jim and Lori Enright of USA Kunekunes, California

LONG-TIME PIG ADMIRERS and keepers Jim and Lori Enright live near Mira Loma, California, where they breed rare Kunekune pigs.

Lori became interested in Kunekunes after keeping a friend's pig. The Enrights are tireless promoters of these interesting small pigs. Their Kunekunes take part in a 4-H project called Rare Livestock Breeds Conservation, in which the Enrights allow 4-H youth to display and care for the pigs at fairs, shows, and various educational venues, where they educate the public about genetic diversity and the adaptability and historic value of heritage breeds.

The Enrights also founded the American Kunekune Breeders' Association and have been actively registering pigs since 2006.

It's obvious that Lori loves her sweet Maori pigs. Here is what she had to say about getting into the business and her current approach to breeding.

"While investigating rare breed swine, I contacted Tim Harris of Harris Associates, Ltd., an expert in animal transport. He suggested I contact a group that was planning a pig buying trip to Great Britain. So I e-mailed Professor Steven Moeller, Ohio State Swine Extension Specialist, and the next thing I knew, I was invited to travel with a group leaving for England to purchase breeding pigs — in less than two weeks! Jim suggested that 'they might have your little pigs over there' and added, 'Why don't you go?'

"In England, I met Andy and Maureen Case, breeders of rare swine, including Kunekunes. They are the largest breeders of Kunekune pigs in the world. With Andy's help and that of Professor Moeller, we purchased the little group of Kunes that became our foundation breeding stock. Our imported pigs arrived in 2005, and our first litter was farrowed in June of 2006.

"We are very happy that we chose this breed. Kunekune pigs are small, colorful, wattled, and extremely friendly. They have very balanced conformation, wide 'teddy bear' heads, big ears that can be erect or semi-lop, unusual wattles or tassels that hang from the lower jaw, and short up-turned snouts. The thing that does it for me, however, is their temperament. Kunekunes are very brave pigs, and this makes them friendlier than most. The sows are sweet and docile; I have no problem processing their piglets while they are present. Boars are sweet, too, although I always tell folks to be careful around breeding boars. Ours have their tusks because there is no reason to remove them.

"Due to the rarity of the breed, its small size, and especially its temperament, there is a very strong market for these pigs. Most of our piglets are pre-sold before they are even on the ground. As soon as we knew that our imported pigs were on United States soil, we started to advertise them, resulting in a waiting list with deposits in hand; in fact, we couldn't supply the initial demand until our second litter was farrowed.

"Advertising is the key to our business. You'd be surprised how many people are really fond of pigs. We have received tens of thousands of inquiries from all across the United States, Canada, and the world.

"I don't think there is a very big chance of the Kunekune market becoming oversaturated, like the Potbellied pig market. While the Kunekune pig and the Potbellied pig are both miniature pigs, they are very different from one other. Buyers of Kunekunes are, therefore, different as well. Most people who contact us and purchase a Kunekune or two already own pigs or other livestock and do not live in urban settings. One of the first things I tell people concerned about size is that the Kunekune is larger than most Potbellied pigs. I tell them that 'you are not going to carry a Kunekune pig around in your purse.'

"We target folks who own other pigs or livestock, have acreage or farmland, are hobby farmers, and wish to breed their pigs, but we do sell pigs to folks who just want to keep a pig in the yard as a pet. Kunekunes make wonderful pets.

"Kunes serve a multitude of purposes. The breed was developed by the Maori people of New Zealand and kept for their meat. We've had calls from people wanting Kunes to glean orchards, graze vineyards, search for truffles, and the like. They don't root the way long-snouted breeds do, so they're easy on pastures; this makes them useful for keeping grass and weeds down. I've read that they are using pigs in some apple orchards to break the life cycle of a beetle that is ruining the fruit. Kunekunes would be great for that purpose.

"To someone thinking about breeding Kunekunes, I would say that it's important to learn about the behavior of pigs in general so you understand that pigs do not like to be restrained in any way. Due to the smaller size and placid temperament of the Kunekune, this behavior is much more manageable than with larger breeds."

WILD BOARS ARE A VIABLE OPTION

Pork raisers who are looking for a unique pig-raising experience and who live in states that allow the keeping of feral hogs should consider the European Wild Boar. Raised mainly for niche marketed meat and trophy hunts, if kindly raised from infancy onward they can also make fine, unusual pets. My own pet pig, Carlotta, is a crossbred from a European Wild Boar and an Ossabaw Island Hog. She's a gentle, good-natured outdoor pet pig.

The term *wild boar* applies to the entire species, not just the males. Their average weight is 200–500 pounds (91–227 kg), though individuals, especially males, can be considerably heavier. They have compact bodies with very long snouts, upright ears, large heads, small hips, and fairly short legs. Colors range from dark gray to black or brown, all with grizzled overtones. Both sexes have impressive tusks; boars' tusks may be up to 7 inches (18 cm) long with 4 inches (10 cm) or more protruding from their mouths. Wild Boar sows produce small litters of three or four pigs on average; the piglets are soft brown with darker horizontal stripes. Wild Boars are the ancestors of today's domestic pigs.

Wild Boar

Ossabaw Island pigs are classified as feral hogs, having evolved as an isolated population on Ossabaw Island off the coast of Georgia. They are hardy, efficient grazers that are ideal for pastured pork production. Beautifully fat-marbled Ossabaw Island pork is firm-textured, succulent, and tasty. Ossabaw hogs are larger in the shoulder area than conventional pigs, so they yield more roasts and chops.

The population of pigs today on Ossabaw Island has been pared to a bare minimum to avoid erosion of the island's natural resources. And because of quarantine restrictions, it's impossible to import stock directly from the island. Although scarce (like the Guinea Hog, there are only about 200 purebred domesticated Ossabaw Island pigs in the world and they are listed as Critically Endangered on the ALBC Conservation Priority List), stock descended from pigs taken from the island in the 1970s can be found. The breed needs dedicated conservators, but be sure to check your state's feral pig statutes before buying Ossabaw Island Hogs, as many states prohibit the keeping of feral pigs.

Potbellied Pig

Potbellied pig

Registered by: North American Potbellied Pig Association, National Potbellied Pig Registry
Use: Pet
Origin: Vietnam
Size (ideal according to the National Potbellied Pig Registry): 100–125 pounds (45–57 kg); 16 inches (41 cm) tall at three years of age
Color: Black, white, black and white, silver, red, or spotted
Ears: Small, erect
Facial characteristics: Short to medium in length; dished; often deeply wrinkled
Body characteristics: Pronounced potbelly, swayed back, short legs; skin may or may not be wrinkled
Tail: Straight
Coat: Sparse

Potbellies are the quintessential pet pig, and they make very good ones. They're intelligent, affectionate, and cute. They have, however, been bred indiscriminately

POTBELLIED PIG RESCUE

If you'd love to have a pet pig (or two or three or four), why not adopt your pet through a pig rescue or an animal welfare organization? Keep in mind that you aren't *buying* these pigs (adoption fees are used to maintain the group's other animals until they leave for new homes). Reputable rescue groups are concerned about the ongoing welfare of the animals they place; therefore, you must meet certain criteria to adopt, and in most cases, adoptive organizations retain ownership and the right to do farm checks on demand. The beauty of these adoptions, apart from the satisfaction you'll derive from helping a pig in need, is that rescue groups carefully prescreen both the animals they place and the homes in which they place them; thus, they're likely to find you a perfect match. And if the adoption doesn't work, the animal can be returned and re-homed. It's a win-win situation.

To locate a pig that needs your help, scope out the pig rescues and sanctuaries listed at the back of this book in Resources. If there aren't any in your locale, chances are the ones listed can direct you to animal shelters in your area that occasionally handle pigs.

for so long that buyers are often disappointed when they mature into much larger sizes than expected.

The best solution for all concerned is to adopt from a bona fide Potbellied pig rescue that cares about its pigs and works to make placements that last. The people there will help you find a pig or piglet perfect for your purposes, and you'll help save a life by clearing a space at the rescue for another pig that might otherwise not have had the chance to find a good home.

LABORATORY MINI PIGS

Because human and porcine physiology is very similar, researchers have developed an array of miniature breeds specifically for lab research. Except for Göttingen and Yucatan mini pigs, they are rarely available for sale to the public but well worth the search if you want one — they are extremely cute, and some types stay very small.

GÖTTINGEN

Developed during the 1960s at the University of Göttingen in Germany, these are the smallest of the laboratory mini pigs, weighing only 25–30 pounds (11–14 kg) at sexual maturity and 90 pounds (41 kg) when fully grown. Scientists developed them by combining Vietnamese Potbellied pig, German Landrace, and Hormel mini pig genetics. There are two varieties, white and colored, and both types have a rounded appearance and a short snout. Widely used in labs throughout Europe and in North America, they are used mainly in pharmaceutical and toxicology research.

HANFORD

The Hanford laboratory pig was developed in 1958 at the Hanford Laboratory in Richland, Washington, by crossing Palouse gilts with a Pitman-Moore boar; Yucatan and feral genetics were added later. The largest of the laboratory mini pigs, it weighs 150–200 pounds (68–91 kg) when fully mature. Hanford pigs have long snouts and are generally white. They have less subcutaneous fat and the largest heart and blood vessels of all laboratory mini pigs. They are used in the testing of implanted devices sized for humans.

HORMEL

Also called the Minnesota Miniature pig, this pig was developed at the Hormel Institute of the University of Minnesota in the 1950s

If you'd prefer to buy a Potbellied piglet (and these tips apply when buying Kunekunes and pet Guinea Hogs, too), choose wisely.

Buy a registered pig from a professional breeder. Unregistered pigs from backyard litters may not be purebred Potbellied pigs, and piglets with full-size pigs in their background aren't likely to stay small. If you buy an unregistered piglet, look at its tail. Potbellies have straight tails. A curved or curly tail indicates non-Potbellied breeding.

using Guinea Hog and Catalina feral pig genetics. Feral Pineywoods pigs from Alabama and Ras-n-Lansa pigs from Guam were later added to the mix. In the 1960s, Yorkshire genetics were added to create a white pig, which is sometimes called a Sinclar. All Hormels have short snouts and reach 35–50 pounds (16–23 kg) at sexual maturity. Hormel mini pigs come in dark, white, and spotted varieties.

OHMINI

Ohmini mini pigs were developed in Japan in 1945 by Hioshi Ohmi, who crossed Manchurian pig genetics with Hampshires, Durocs, and Hormel mini pigs. Ohmini mini pigs are black with a coarse coat, exceptionally long ears, and wrinkled skin. They weigh about 75 pounds (34 kg) at full maturity.

PITMAN-MOORE

Researchers at the University of Minnesota developed Pitman-Moore mini pigs using feral hog stock from Florida. Pitman-Moores tip the scale at 150 pounds (68 kg) when fully grown. They are white with dark spots.

MINI AND MICRO YUCATANS

Yucatan pigs are native to southern Mexico, Costa Rica, and a few other areas of Latin America. Also known as the Mexican Hairless, Labco, Maya (in Britain), Pelon de Cartigo (in Costa Rica), and Cuino (in Latin America), most Yucatans are black or slate gray, though a white variety exists. The pigs are relatively hairless with a short profile and short snout, and some individuals have wattles. Miniatures weigh about 150 pounds (68 kg) at full maturity, and micros (developed in 1978 at the Colorado State University Swine Laboratory) average 100 pounds (45 kg).

Ask for references *before* you view potential piglets.

Don't buy a pig in a poke. Visit the breeder and look at his or her breeding stock. The height of the parents will give you an indication of the height of the purebred piglets. Make sure the parents are at least four years old if you're using them to calculate their offspring's size.

When size matters, get a written guarantee. Make certain it spells out precisely what the breeder will do if the piglet grows too tall. Refund your money? Replace it with another piglet? Will you have to return your pig? Watch out for clauses that void the guarantee if you "overfeed" your pig. Excess feed triggers obesity but won't alter an animal's genetic background, and that's what determines height.

Look over the piglets very carefully. They should be bright-eyed and lively and have soft, supple skin. Properly fed Potbellied piglets have nicely rounded hindquarters and bellies, with a modicum of jowl at the neckline. Some sellers starve piglets to make them seem smaller to potential buyers. This is cruel and unnecessary. When these piglets begin eating a proper diet, they'll mature to their true genetic size.

Ask what the piglet and its parents have been tested for and expect written documentation. Be aware that it's illegal to cross state lines with pigs that haven't been tested for pseudorabies and brucellosis. Don't risk having your piglet confiscated!

19

Miniature Sheep

SHEEP ARE MARVELOUS CREATURES. But then, I'm prejudiced—sheep have been part of my life for years. And most of my sheep are natural miniatures, the kind of sheep you probably would like to have on your farm, too.

Sheep have a lot of virtues. They are relatively inexpensive to buy and keep, easy to care for, and long-lived, making them great investments. Given predator-proof fencing, minimal housing, good feed, and a modicum of daily attention, sheep will thrive. It's fairly easy to learn to care for sheep in a short period of time, making them a good choice for first-time farmers.

Most breeds produce wool and meat (young miniatures yield tender, tasty lamb) and since sheep droppings are firm and small, sheep make admirable lawn ornaments and mowers. They can even clear your land in the bargain.

REAL LAWN MOWERS RUN ON GRASS

Before lawn mowers were the norm, the rich employed groundskeepers with scythes to cut their grass — or they grazed sheep on their lawns. Sheep clipped grass on golf courses and in prestigious parks throughout the world, including New York's Central Park. And Thomas Jefferson, James Madison, and Woodrow Wilson all grazed sheep on the White House lawn.

Today, Turin, Italy, is following their lead by leasing 700 ovine lawn mowers to trim municipal parks and lawns at a savings of 36,250 euros (approximately $48,000) a year.

SHEEP PHYSIOLOGY (ADULTS)

Temperature: 101.5–103.5°F (38.5–39.5°C)
Heart rate: 70–80 beats per minute
Respiration: 12–20 breaths per minute
Ruminal movements: 1–2 per minute
Natural life span: 10–15 years (however, the oldest sheep in the *Guinness Book of World Records*, an Australian ewe aptly named Lucky, is 22-plus years of age and still going strong)

EWES

Age at puberty: 5–10 months (singles cycle younger than lambs from multiple births)
Breeding weight: 60–75 percent of adult weight
Heat cycle: every 13–19 days
Heat duration: 24–48 hours
Ovulate: 20–30 hours after onset of estrus
Length of gestation: Average 147 days; normal range 138–159 days (averages vary slightly by breed; early-maturing breeds have shorter gestation periods)
Number of young: 1–3 (singles or twins are the norm for most miniature breeds)
Breeding season: August–February

RAMS

Age at puberty: Average 5–7 months or 50–60 percent of mature weight (some breeds reach puberty much younger than this)
Primary rut: August–January
Breeding ratio: 1 adult ram to 35–50 ewes; one ram lamb to 15–30 ewes

While all sheep graze, primitive and hill breeds like Classic and Miniature Cheviots, Soay, and Shetlands browse as well as graze. Brambles, tree sprouts, kudzu, and briar roses are all fair game, and most sheep cheerfully devour cheatgrass, poison ivy, dandelion, purple knapweed, and a host of other undesirable forbs and grasses.

Besides, sheep are simply cute and fun to have around. When handled with kindness from lambhood on, sheep make peerless pets.

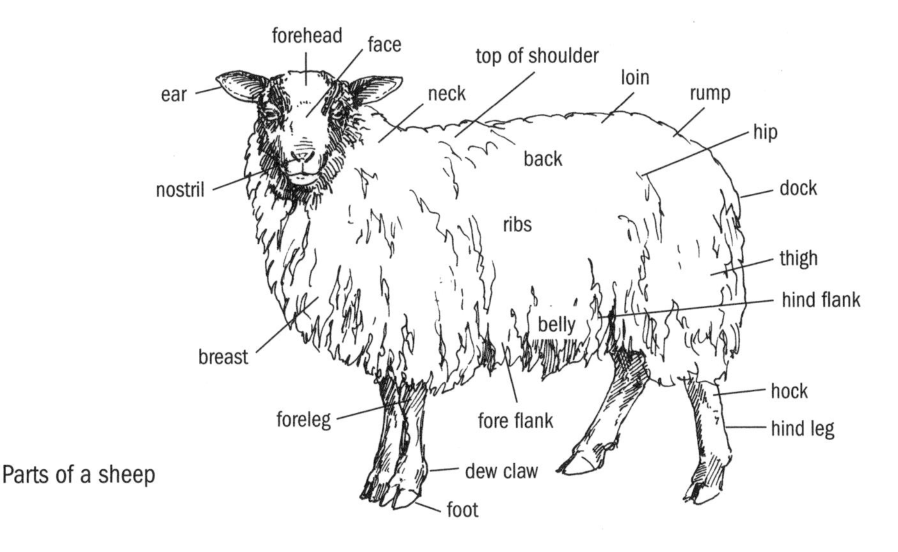

Parts of a sheep

Getting into Sheep

Sheep are basically low-key, easy-care animals, and the compact size of miniature breeds makes them even easier to handle than conventional sheep.

Buying Sheep

You should consider many of the same things when buying sheep as you would consider when buying goats.

Before shopping for sheep, learn what your chosen breed or type should ideally look like and know what aesthetic features are important to you. If you're raising sheep for wool, for instance, you'll want to think about which type of fiber you prefer.

Look for a good rumen on all sheep you buy (a large rumen equates with greater ability to take in and process feed). The rumen bulges out on the left side of a healthy, well-fed sheep, especially one that has been feasting on hay, forbs, or pasture. This is true of either sex.

Check the sheep's bite—how its lower teeth meet up with its upper dental pad. It should be a perfect match. Sheep need proper occlusion to browse and graze, and a bad bite is a serious fault in show sheep and breeding stock. Be especially vigilant about checking the mouths of Roman-nosed Miniature/Classic Cheviots since that head configuration is the one most likely to result in "monkey mouth," or underbite.

OVINE PROGRESSIVE PNEUMONIA

Sheep don't get caprine arthritic encephalitis (CAE), but they are prone to a similar, serious affliction called ovine progressive pneumonia (OPP). The retrovirus that causes OPP is closely related to maedi-visna, a disease commonly encountered in other parts of the world; only Iceland, Australia, and New Zealand are completely maedi-visna and OPP free. As with CAE, transmission is through infected colostrum and milk. Symptoms include weight loss, sluggishness, lameness, and a fibrous udder condition known as hardbag. Most OPP-infected sheep eventually develop secondary bacterial pneumonia, from which they die. There is a test but no vaccine or treatment for OPP.

Check an adult ewe's udder very carefully (the udder is her complete mammary system; she has two teats, not two udders). Feel it. As with does, a dairy ewe's udder feels soft and velvety.

Don't forget to check for suspicious lumps that might be CL (*caseous lymphadenitis*). It's a proven fact that testicle size in rams correlates with fertility, so choose rams with large, symmetrical testicles.

Fencing for Sheep

Carefully review chapter 6 before buying sheep. Sheep are not as difficult to keep confined as goats; nevertheless you should still employ similar fencing. Some people find that multiple strands of electric wire are effective fences for sheep. If you decide on this, buy a premium fence charger and make sure it works all the time. If you use plank fences, be sure they are lined with woven wire. We think the best all-around sheep and goat fences are built of sturdy, woven-wire field fencing reinforced with a strand of electric wire that has been placed inside at shoulder height to keep inhabitants from leaning into the fence to scratch their sides and backs.

If you have a smaller enclosure or do not have money constraints, opt for welded-wire cattle panels. They're easier to work with than woven wire, more durable, and less likely to be bent by itchy sheep.

No matter what you choose, rams will test your fences. Horned rams, like horned bucks, even miniatures, can wear down even the toughest fencing.

Housing for Sheep

Miniature sheep, like miniature goats, are incredibly easy to house. They often just need a dry, draft-free place to use for shelter. Several sheep can be housed in a

Port-A-Hut (see Resources), if given some pasture to roam around. There is wide room to improvise with your housing; just make sure it is dry and free of drafts.

Feeding Sheep

Just like goats, sheep should receive clean, high-quality grass hay. Wethers, open ewes, and rams outside of breeding season need nothing else. Feed concentrates to youngsters, late gestation and lactating ewes, and hardworking rams in rut. Use a properly formulated commercial mix or work with your county extension agent and your local feed mill to formulate one based on your needs and what is locally available.

One difference between sheep and goats that is especially troubling for people who keep both species is that copper is toxic to sheep in the quantities that goats (and horses, cattle, pigs, and most other barnyard species) require for optimal health. Copper toxicity results in liver disease, jaundice, and death. Sheep cannot tolerate a copper concentration any higher than 15 ppm (parts per million). Therefore, it's imperative that sheep owners examine nutritional information on

SHEEP TIPPING

Once a sheep's feet are off the ground and it can't touch anything with its hooves, it will remain still (see illustration on page 384). That's why shearers tip sheep to immobilize them for shearing. Tipping also works admirably well for routine tasks such as hoof trimming and doctoring minor wounds. Tipping a sheep seems daunting at first, but it's easy once you know how.

Standing on the left side of the sheep, reach across and grasp its right rear flank (don't pull its wool — that hurts). Bend its head away from you, back against its right shoulder. Lift its flank and pull the sheep back toward you. This puts it off balance, and it will roll gently toward you onto the ground.

Hold the sheep on its side, then quickly grasp both front legs and set it up on its rump so it's slightly off center, resting on one hip. If it struggles, place one hand on its chest for support and inch backward until it's more comfortable.

Don't tip a sheep right after it's eaten, as being tipped puts considerable strain on a full rumen. And don't let anyone try to help by holding the sheep's legs; if its hooves encounter anything, even your helper's hands, the sheep will struggle to break free.

feed packaging before giving commercial feed to sheep. Various studies indicate that mature ewes of British breed origin appear to be the most vulnerable to copper toxicity — and that encompasses all four miniature breeds.

Sheep still need access to other minerals, however, in loose form (preferred) or as a lick.

Docking Tails

If you raise Babydoll Southdowns or Miniature or Classic Cheviots, you'll have to dock your lambs' tails. Some people prefer long-tailed wool sheep, but this isn't a wise option; docking is done for a very good reason.

Wild sheep such as the Mouflon, ancestor of our modern sheep breeds, have short, hair-covered tails. European short-tailed breeds have short tails with little or no wool covering. Centuries of selection for wool production, however, has resulted in sheep with longer, thicker, woollier tails that can become encrusted with manure if the sheep has diarrhea or is housed in wet, dirty conditions.

Enter the blowfly, a particularly nasty external parasite that lays its eggs in wounds or in damp, goopy fleece (like long, wet tail wool). Blowfly eggs hatch in about a day. The resulting larvae (also known as maggots) lacerate and tunnel into their host's flesh, creating lesions. After a few days, secondary infection occurs and, if untreated, leads to toxemia or septicemia; eventually the sheep dies.

Docked tails also make shearing easier. Shearers generally charge a premium for handling sheep with filthy, full-length tails.

Lambs' tails should be shortened between one and seven days of age; the older the lamb, the more painful the operation. Tails are usually docked by banding, a process in which a tool called an elastrator applies a thick rubber band around the tail. This cuts off circulation to the tail and in about two weeks the tail falls off. Few neonatal lambs appear to feel extreme pain when being banded, although there are exceptions. In either case, the tail becomes numb in a few minutes' time. After that, it's business as usual for the newly banded lamb. Lambs whose dams were boosted with CD/T toxoid a few weeks prior

OFF WITH THEIR TAILS

- According to a 2002 Animal Health Survey, 91.7 percent of U.S. lambs are docked.
- In Australia alone, flystrike accounts for more than $170 million a year in losses.

to lambing need no further protection. Barring that, an injection of tetanus antitoxin at banding time is a very good thing.

A good deal of controversy surrounds the appropriate tail length of docked lambs. At one time, show sheep were docked so short that virtually no tail was left behind. However, ultra-short docking affects the muscles and nerves of a sheep's anus and prevents the sheep from lifting its tail to direct fecal pellets away from its hindquarters. Short docking may also contribute to rectal and vaginal prolapses. Tails should be left long enough to cover a ewe's vulva and a ram or wether's anus. The American Veterinary Medical Association, American Association of Small Ruminant Practitioners, and American Sheep Industry Association all agree that tails should be no shorter than the distal end of the caudal tail fold. Slightly longer is even better. To find this point, turn the tail over and band at or up to three inches below where the bare patch of skin close to the lamb's body ends in a point.

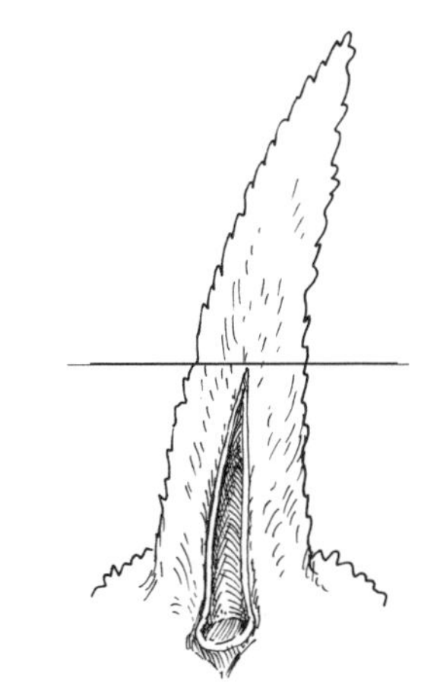

Band lamb's tails 1 to 3 inches below the point where the bare patch of skin close to the lamb's body ends in a point.

Shearing

Nonshedding wool breed sheep must be shorn on an annual basis, no exceptions. Not shearing leads to heat exhaustion; dirty, matted fleece; and great discomfort.

If you're fortunate, you can hire a professional sheep shearer to do the deed. Unfortunately, shearing is a vanishing trade and shearers are in very great demand. Ask around before you buy woolled sheep; if you can't find a reliable shearer, resign yourself to doing your own shearing, which is no easy feat. (Hint: You'll need to have plenty of time and a fitting or milking stand to restrain the sheep or else lots of patience and a strong back.) If you can't find a good shearer and aren't willing to take on the task yourself, buy Soay or shedding Shetlands that you can *roo* (comb or pluck instead of shear; see box on page 399).

To locate a shearer, ask other shepherds whom they hire to shear their sheep, search for business cards and flyers on the bulletin boards of feed stores and large-animal veterinary practices, peruse online sheep directories, Google *sheep shearers*, consult your county extension agent, or contact a nearby university with a vet school — they may have grad students who are experienced shearers.

WHICH CAME FIRST: DOMESTIC SHEEP OR YARN?

Ten thousand years ago, even before they domesticated sheep, New Stone Age humans gathered tufts of wild sheep fleece and twisted them into yarn.

Check shearers' references and explain your expectations up front. Shearers are paid by the head, so they want to work quickly. Sheep skin is thin and tender, and even the best shearers sometimes nick sheep, but multiple gashes, deep cuts, and teat and penis injuries are unacceptable, as is dragging sheep around by their wool. Fleeces should be removed in a single piece with a minimum of second cuts (short pieces created by running the shears across the same area a second time). Some shearers are happy to take extra care with hobby and handspinners' flocks, but they may charge a premium price to do so. Others slow down for nothing or no one.

Shearers compute fees based on how far they must travel, the number of sheep they'll shear, and the type of wool they'll be handling. Be willing to pay extra for travel expenses or extra setup fees to obtain quality service, especially if your flock is a small one. It isn't cost effective for shearers to drive many miles to shear a flock of 50 head or fewer, so you may want to get together with other shepherds in your locale and arrange to have everyone's sheep sheared on the same day or weekend. Ask the shearer to visit each farm — don't gather sheep at a central location. Commingling sheep from several flocks contributes to the spread of disease (not to mention creating stress for animals and people).

If a sheep is tipped so its feet don't touch anything, it will remain calm for procedures like sheering or veterinary care.

The night before the shearer comes, pen your sheep in a clean, roomy area where they won't get wet. Sheep shouldn't eat for 8 to 12 hours before being shorn.

Provide a clean, level shearing area. Most shearers prefer to shear on a surface constructed of two sheets of plywood with the space between them sealed with duct tape. Have a broom handy to sweep the surface between sheep. The shearing floor should be situated in a well-lit, well-ventilated, covered area.

Arrange for additional help if you need it. You need two people at the bare minimum, one to bring sheep to the shearer and another to take each sheep away after it has been shorn. Have cool or warm drinks available for your shearer and helpers, depending on the season; wrestling sheep is thirsty work. If the shearing crew is there over lunch or supper, provide sandwiches but not a heavy meal.

As you bring pregnant, elderly, or recently ill sheep to the shearer, let him know their condition. These sheep require gentler handling. Also, a single wether in a flock of ewes could have his penis zipped off if a speedy shearer doesn't realize it's there.

SPIN YOUR OWN WOOL

You needn't buy an expensive spinning wheel to learn to spin your own wool; all you need is a drop spindle.

All wool was processed with drop spindles until the spinning wheel was invented during the Middle Ages; stone spindle whorls are routinely unearthed at Neolithic sites throughout the world. While spinning wheels made spinning faster and somewhat easier, the drop spindle never fell out of vogue. Drop-spindle spinning is still the mode of choice in many countries and with history-oriented, modern handspinners.

You can find instructions and places to buy a spindle or kit online; and online videos are especially helpful.

People have been spinning yarn with drop spindles for thousands of years.

Think like a Sheep

In the University of Tennessee paper "Applied Sheep Behavior" (see Resources and download this excellent file), world-renowned animal behaviorist Dr. Richard Kilgour is quoted describing sheep as a "defenseless, wary, tight-flocking, visual, wool-covered ruminant (cud-chewing animal) evolved from a desert or a mountain grassland habitat with low water needs and displaying a follower-type dam precocial offspring relationship with strong imitation between young and old in establishing range systems; showing seasonal breeding and a separate male sub-group structure at certain times of the year." There you have it; sheep behavior in a nutshell!

Sheep are intelligent creatures, but they very well know that they are defenseless in the face of danger, so frightened sheep don't hang around to think things over; instead, they run. This tendency has earned them an undeserved reputation for brainlessness. In fact, if you learn to think like a sheep, you'll realize there is method to their seeming madness. Here's what goes on inside sheep's heads.

Sheep Senses at a Glance

Sheep senses are similar to those of goats but with certain twists.

Vision

The average sheep has a visual field of about 270 degrees, depending on how much wool it has on its face to obstruct its vision. It has large, rectangular-shaped pupils, and its eyeballs are placed toward the sides of its head, which account for its relatively wide field of vision. Rectangular-shaped pupils provide

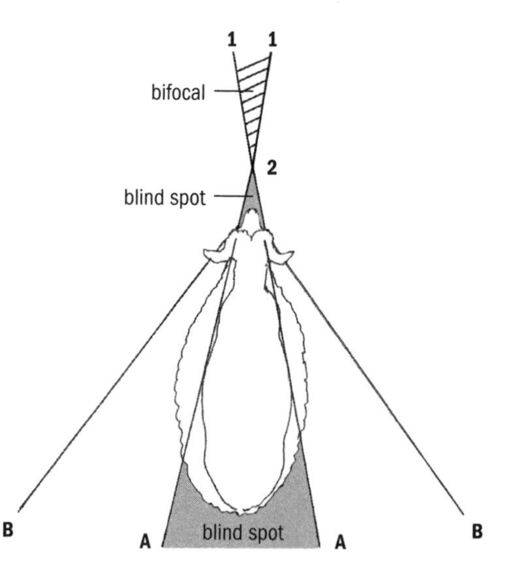

Sheep have a fairly wide field of monocular vision and a small area of bifocal vision. They also have blind spots. When sheep are sheared, they can see everything between points 1 and A on either side of their bodies. As their fleece grows out, they see less toward the back but can still see everything between points 1 and B.

SHEEPY SOCIAL ORDER

Sheep establish flock hierarchy almost entirely in the manner of goats, so turn to the chapter on goats to learn how it is done.

One major difference is that sheep don't follow a specific leader, as goats do their herd queen. When one sheep starts moving, others take note and follow, usually even in low-flocking breeds.

a wide-angle effect, giving sheep excellent peripheral vision. And by moving its head a bit, it can scan its entire surroundings. Nonetheless, sheep can't easily see objects high above their heads, and they have poor depth perception. Because of this, driven sheep avoid shadows and harsh contrasts of light and dark. They also tend to move away from darkness, toward light.

Based on research and on the number of cones and rods in sheep's eyes, scientists believe that sheep have very keen vision. They've also learned that sheep see in color, although their color acuity is less than humans'.

Smell

Sheep have a highly developed sense of smell. Ewes recognize their newborn lambs by taste and scent. Rams sniff ewes to determine which are in heat. Sheep are capable of scenting water from afar, and they sniff feed to determine if they want to eat it.

Hearing

Sheep hear well, and they are sensitive to high-pitched and sudden noises, both of which trigger a surge of stress-related hormones. They refine their hearing by moving their ears, heads, or entire body to face the thing they're focusing on.

Frantic, high-pitched *baas* are stress indicators. Medium-pitched *baas* are exchanged by adult sheep casually seeking nearby lambs, friends, or feed.

Taste

Ewes recognize themselves when licking their fluid-drenched newborn lambs, and this creates a strong maternal bond.

Sheep prefer certain feeds over others, indicating they know what they like to eat. When grazing alongside other species such as goats and cattle, each species selects different types of plants.

Touch

Because sheep are covered with wool or thick, coarse hair, they are less responsive to outward touch than are other livestock species. Sheep have very sensitive skin under their woolly covering though, so handlers should never catch or hold sheep by their wool.

Because sheep are well insulated, electric fencing is not very effective. Multiple strands of electric wire are a must, including strands placed precisely at the height of your sheep's noses.

Flocking and Driving Behavior

All sheep flock to some degree, but flocking instinct varies greatly from breed to breed. Lambs are hardwired to follow their dams and, later, other members of the flock. Sheep become agitated if separated from other sheep. Banding together in large groups protects sheep from predators that normally home in on the outliers in a flock. This flocking instinct allows shepherds to tend and move large numbers of sheep.

Fine-wool sheep such as full-size Merinos, Rambouillets, and Columbias flock closely while moving, grazing, or at rest; most other breeds flock to varying degrees when moving but spread out somewhat to graze. None of the miniature breeds are close-flocking and indeed, all except Babydoll Southdowns scatter if worked by less than topflight herding dogs. Shetlands and Soay have particularly weak flocking instincts. Herding dog trainers tend to like Miniature and Classic Cheviots because they're herdable but fast and agile, so they give good dogs a workout. It's easier to move low- and poorly flocking breeds by leading with a bucket of feed than trying to drive them.

Additional points to remember:

- If anything scary invades a sheep's personal space, or flight zone, it flees. A sheep's flight zone might be 50 feet or nothing at all, depending on the breed, gender, tameness, training, and severity of perceived threat.
- When approaching or driving sheep, don't look directly into their eyes; wolves and herding dogs do that and it instinctively makes sheep nervous.
- Calm sheep move forward; frightened sheep move backward.
- Sheep prefer not to cross water or to move through narrow openings.
- They move uphill more readily than downhill and prefer to move into the wind than against it.
- Cornered, frightened sheep will try to jump to safety, usually toward whatever is blocking their way. Be forewarned: A cornered miniature sheep can hit an adult at chest height.

- Annoyed sheep stamp their hooves, raise or nod their heads, or glare. They usually back up a few paces before charging, although ewes sometimes bash each other from a standstill. Rams or pushy ewes may rub or bump humans with their foreheads; this is a sign of early aggression and should be squelched.
- To temporarily restrain a sheep, place one hand under its jaw and raise its head while steadying its opposite flank or its hindquarters with the other hand.
- To catch a sheep, herd the flock into a corner and carefully separate the desired animal from the mob, using its flock mates' bodies to prevent it from scampering away (the smaller the pen, the easier to catch the sheep).
- Don't attempt to drive a flock faster than its usual walking speed; rushed sheep tend to scatter. Where you position your body when moving your flock or even just one sheep will either keep them moving forward or cause them to pivot back in the opposite direction. If you move too far toward the head of the flock (or the sheep), you have crossed a subtle comfort zone and they will turn backward; and if you hang back, keeping yourself toward the back third of the flock (or behind the point of its shoulder), they (it) will continue to move forward.
- Sheep remember bad experiences for up to two years. Don't lose your temper. It does not pay.

Breeding Behavior

Most wool sheep are seasonal breeders; they're most likely to breed in the fall, when days are growing shorter and temperatures are cooler.

Ewes in heat seek out and stay near the ram; some sniff, lick, or nuzzle him as well. A ram approaching a potential mate sniffs her urine, nudges the ewe, grunts, pants, and flehmens. If the ewe is in heat, she'll stand for his advances; she may also wag her tail. If disinterested, she'll simply walk away.

If he wears a marking harness, it's easy to tell if a ram is doing his job.

Mating is a very quick process. Unless you're observant you aren't likely to see it occur. For this reason, many shepherds fit their rams with marking harnesses that deposit chalk on the rumps of ewes as they're serviced.

Watch Out for Rams

The verb "to ram" is derived from the Old English word *ramm*, meaning an intact male sheep. Apart from procreating, ramming is what unaltered sheep do best.

Head butting is a national sport among rams. They make a show of backing a distance, lowering their heads, and then racing forward and smacking their opponent, poll first: *whack!*

Many rams show aggression toward humans by backing, running, then jumping and slamming to a halt just before connecting with their human target. Don't allow this! If your ram is only occasionally feisty, be ready with a bucket or a sprayer filled with water to dash or squirt in his face. Or if you're quick, grab his front legs and roll him over, then hold him down for at least five minutes.

If your guy is frequently running into humans, you need a different ram. People have been seriously injured and even killed by rams. Consider this: In 1807, President Thomas Jefferson, an avid sheep breeder, began breeding Shetlands on the South Lawn of the White House. A citizen wrote that "in passing through the President's Square I was attacked and severely wounded and bruised by your Excellency's ram — of which I lay ill for five or six weeks." Jefferson's friend Anna Maria Thornton writes in her journal that the same ram killed a young male visitor to the White House.

A great deal of butting behavior toward humans can be avoided by enforcing this rule: never fool with a ram's face or forehead. Above all, never allow anyone to play pushing games with a young ram (or buckling). This is cute when he's a lamb, but he'll grow up thinking you're an opponent, and it's not as fun when he's grown and takes you by surprise.

If cornered by an ornery ram, try to look big. Stand tall and hold your arms out to the sides. If you spy something to extend your outward reach, such as a stick, shepherd's crook, or hand tool, use it. To reward rams, scratch their

A ram shield obstructs a ram's forward vision, making it harder for him to home in on a target.

chins. This encourages the ram to raise his chin, which in turn defuses his inborn instinct to lower his head and butt.

And if he's a good ram most of the time and you want to keep him, fit your seasonally ornery ram with a ram shield during the trying period — a leather mask that allows him to see to both sides but not home in on a frontal target. Keep in mind that other sheep sometimes gang up on an individual wearing a ram shield, so remove any testy flock mates that might otherwise hurt him.

The Distaff Side of the Equation: Ewes

Ewes come in heat every 13 to 19 days throughout the fall and early winter breeding season. Mature ewes are fine with high-quality hay and a good sheep mineral until the last trimester of a typical 147-day gestation, when most fetal growth occurs. At this point they will need additional dietary consideration, such as the addition of concentrates to their diet.

If you live in a cold climate and your ewes have access to the outdoors, strongly consider shearing at least a month before lambing time. A fully fleeced ewe, being warm herself, may elect to lamb outdoors, whereas a shorn ewe realizes it's still cold outside. If ewes aren't shorn prior to lambing, they should at least be crutched. *Crutching* (an archaic spelling of crotching) is simply removing wool from around the ewe's udder, along with any wool tags a lamb might initially mistake for a teat.

Ewes display the same behaviors prior to, during, and after lambing that female goats do, so turn back to pages 309 to 312 and reread the material on kidding.

Though ewes may lamb at any time of the day or night, various studies indicate peak lambing times occur between 9 A.M. and noon, then again between 3 P.M. and 6 P.M. Ewes often lamb at dawn or dusk as well.

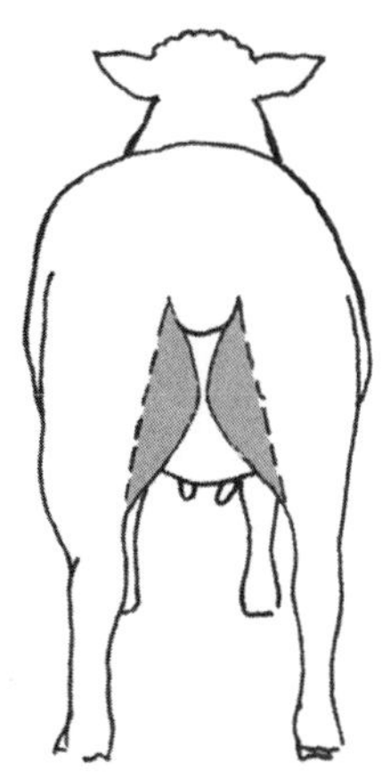

Unshorn ewes should be crutched (shaded area) prior to lambing.

In many respects, lambs behave like kids, but there are differences, too. Goats are a lying-out species; some does leave their kids and come back periodically to feed them. Kids are content to await their mother's return. Sheep are a following species; lambs follow their dams from birth.

Where kids are apt to climb to amuse themselves, lambs (who also climb but to a lesser degree) run elaborate foot races or stage "lamb stampedes." And happy lambs spronk in an all-four-feet-hitting-the-ground-at-once, Pepe le Pew gait that is sure to make the worst cynic smile.

The Breed You Need

At present, there are four established breeds of miniature sheep: Babydoll Southdowns, Classic and Miniature Cheviots, Shetlands, and Soay. All of these are historic, naturally small breeds of sheep — not scaled-down, modern-made miniatures. A number of responsible breeders are working to develop miniaturized versions of full-size breeds, such as Katadhins and Cormos, but these haven't attained breed status yet. Maybe you'd like to join them and create a miniature version of a full-size breed?

Babydoll Southdown

Registered by: Olde English Babydoll Southdown Sheep Registry (OEBSSR), North American Babydoll Southdown Sheep Association & Registry (NABSSAR)
Origin: Sussex, England
Size: OEBSSR — 24 inches (61 cm) at the shoulder, newly shorn; NABSSAR — 18–24 inches (46–61 cm) (sheep 17–18 inches [43–46 cm] and 24–26 inches [61–66 cm] are registerable but faulted)
Color: OEBSSR — Off-white, black, dilutes (paler versions of basic colors that are caused by the addition of a genetic diluting factor), spotted; NABSSAR — Off-white, black, dilutes preferred; spotted registerable but faulted (NABSSAR has specific rules regarding color, so study them before purchasing NABSSAR-registered sheep)
Ears: Moderate in length, level with the poll, neither drooping nor perpendicular to the ground
Facial profile: Straight
Fleece: According to OEBSSR, "Of fine texture, great density, and of sufficient length of staple covering the whole of the body down to the hocks and knees and right up to the cheeks, with a foretop, but not around the eyes or across the bridge of the nose."

According to NABSSAR, "Fine, tight and dense with medium to medium fine crimp that is soft and springy."
Horns: No

Babydoll Southdown

Babydoll Southdowns are arguably America's favorite hobby-farm and pet sheep. The market is strong; most breeders have waiting lists for ewe lambs. Their teddy bear faces and cute, cobby bodies are strong selling points; it's hard to resist a typical Babydoll smile.

Babydolls are today's version of the original British Southdown sheep developed by master breeder John Ellman of Glynde in Sussex, England. Southdowns are one of the oldest of the Down breeds (meat breeds that originated in the downlands of southern England). Babydoll Southdowns yield tasty, tender meat, though nowadays they're raised more for their loveable pet qualities and their wool.

John Ellman began standardizing Southdowns around 1780. He and another British breeder, Jonas Webb, exported Southdowns to the United States beginning in 1824. After World War II, consumers began demanding larger cuts of meat and these small, original Southdowns were crossed with larger New Zealand Southdowns to produce today's larger, leggier, commercial Southdown sheep.

Olde English Babydoll Southdown Registry founder Robert Mock began searching for old-type Southdowns in 1990. He found and obtained only two flocks, which he renamed Olde English Babydoll Southdowns (a.k.a. Babydolls) to differentiate them from today's commercial Southdown sheep. Additional sheep were eventually located, providing a larger gene pool.

Babydolls are easygoing, easy-keeping sheep with docile dispositions. Babydoll ewes are good mothers, and lambing problems are rare. Singles and twins are the norm, though triplets occur from time to time.

One failing: Their cute, woolly faces are prone to wool blindness, a condition in which long wool around their eyes obstructs their vision. The fix is simple: Snip it away until they can see. Also, since Babydolls have woolly faces and legs, they pick up considerably more debris than open-faced breeds that have hair instead of wool on their faces and legs. Their fleece is short-stapled, fine, and fairly greasy — some handspinners love it; others feel it's too short-stapled. If you're buying Babydolls for their fiber, order fleece and try spinning some before you buy your sheep, to be sure it's the right fleece for you.

It's worthwhile to note that these are not tiny sheep. Like Classic and Miniature Cheviots, they're low to the ground but substantially built and quite strong. They're wonderful sheep but not pocket pets and shouldn't be purchased as such.

OPEN AND SAY *BAAA*

Mature sheep have 32 teeth, including four pairs of lower incisors, but no teeth in their upper front jaw. Lambs grow eight baby incisors, which are replaced by two larger permanent teeth each year, beginning at the age of one. They start at the center and continue outward, so by simply counting teeth, it's easy to tell a sheep's age through year four. After they emerge, a sheep's teeth continue to wear and spread farther apart so that by age eight or nine, most sheep have lost or broken some of their incisors; these sheep are said to be *broken mouthed*. An elderly sheep with no incisors left in its mouth is called a *gummer*.

As with other ruminants such as goats, cattle, and deer, a hard dental pad replaces the sheep's absent upper incisors. Unless a sheep's lower teeth align properly with its dental palate, it can't efficiently grasp and rip off grass.

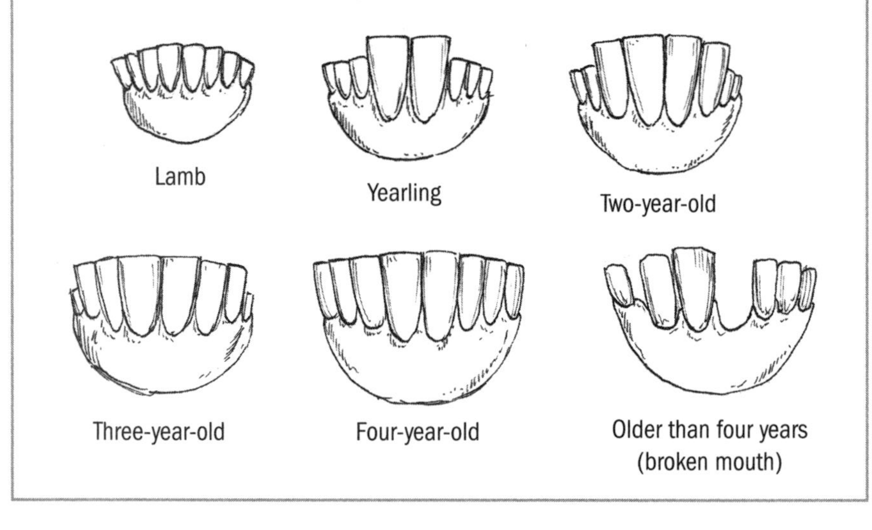

Classic Cheviot

Registered by: American Classic Cheviot Sheep Association (ACCSA)
Size: 18–24 inches (46–61 cm) at the shoulder, newly shorn. Sheep 17–18 inches (43–46 cm) or 24–26 (61–66 cm) are atypical but eligible for registration.
Color: Any
Facial profile: Arched in ewes; more strongly arched in rams

Miniature Cheviot

Miniature Cheviot

Registered by: American Miniature Cheviot Sheep Breeders Association (AMCSBA).

Size: Maximum 23 inches (58 cm) at the shoulder, newly shorn; ewes 45–85 pounds (20–39 kg), rams 55–100 pounds (25–45 kg)

Color: White, black, tan, dilute, paint; small facial markings acceptable

Facial profile: Straight or slightly arched; more strongly arched (Roman-nosed) profile acceptable in rams

Both Types of Small Cheviots

Origin: Borderlands between Scotland and Northern England

Ears: Sharp, erect, short to medium in length

Fleece: Three to seven inches (7.5–18 cm) in length, dense and even, measuring 25 to 32 microns. No wool on face or legs.

Horns: No. Males often have scur scabs but don't develop horns or true scurs.

Wee Cheviots are my breed, and I recommend them highly! They're hardy, easy-care sheep brimming with intelligence and joie de vivre. Mishandled, they live up to their reputation for flightiness, but Cheviots that are well cared for are calm and docile; many are in-your-pocket sheep.

In 1372, historical record refers to a "small, but very hardy" race of sheep grazing the bleak Cheviot Hills between Scotland and England. These sheep, ancestors of all five modern breeds of Cheviot sheep, lived on the windswept hills during summer and winter, seeking their own food to survive. Even early on, their fleece was highly prized. In 1791, Sir John Sinclair, president of the British Board of Agriculture, wrote about Cheviot wool: "The highlands of Scotland if covered with the coarse wool breeds of sheep the wool might be worth 300,000 pounds of sterling, whereas, if the same ground were covered by the Cheviot, the true mountain breed, would be worth at least 900,000 pounds sterling." The main use of Cheviot fleece was for weaving Cheviot tweed fabric, long touted as the best of Britain's tweeds.

The first Classic Cheviots came to America in the 1840s, when Thomas Laidler, a shepherd on the Cheviot Hills, sent each of his four children living in New Lisborn, New York, three Cheviot sheep. Canadians imported Cheviots shortly after that. Cheviots were a favorite breed in North America, Australia,

New Zealand, and throughout their British homeland by the beginning of the twentieth century.

Various breeds of Cheviots evolved. The first, the intrepid Border Cheviot, is the ancestor of Classic and Miniature Cheviot sheep. Other Cheviots are the North Country Cheviot, Brecknock Hill Cheviot, and Wicklow Cheviot of northern Scotland, Wales, and Ireland, respectively.

Classic and Miniature Cheviots are to Border Cheviots what Babydoll Southdowns are to old-fashioned British Southdowns — they are the original breed as it existed before it was selectively bred for longer legs and larger cuts of meat.

The original flock of Miniature Cheviots, it is said (no definitive documentation appears to exist), came to Washington from a sale barn in Canada. These sheep looked exactly the same as modern Miniature Cheviots. Breeders that founded the original registry called them American Brecknock Hill Cheviots, despite the fact that there are no records of Brecknock Hill Cheviots ever being imported to North America. There were, however, many thousands of Border Cheviots raised in British Columbia at the time. And Classic and Miniature Cheviots match historic photos and engravings of early-day British Border Cheviots.

To eliminate confusion, the breed's name was changed from American Brecknock Hill Cheviots to Miniature Cheviots when the registry was restructured in 2002. Breeders more interested in preserving historic type than breeding tiny sheep founded the Classic Cheviot registry in 2009; most Classics are registered as Miniature Cheviots, too.

Miniature and Classic sheep are longer than they are tall. They are wide, sturdy sheep with perky, upright ears; big, dark eyes; and a handsomely convex facial profile. Their faces and legs are haired instead of woolled and their soft, spongy, low-grease fleeces don't pick up a lot of debris. They are extremely easy keepers. They lamb with ease and are peerless mothers; most ewes twin, though singles and triplets occur. Many owners consider rams of this breed less aggressive than the norm (but keep in mind that no ram should ever be taken for granted).

The Miniature Cheviot registry discourages white markings and odd colors, while the Classic group encourages color diversity, including white-splashed faces and legs as well as spots in the fleece. Colored sheep in either group are often marked with irregularly shaped, white facial spots called "fairy kisses."

Ouessant (pronounced oo-shant)

Also known as: Ushant, Mouton d'Ouessant, Breton Dwarf
Registered by: Ouessant Sheep Society of Great Britain
Origin: Ouessant Island, off the coast of Brittany

Size: Rams, 19–24 inches (48–50 cm); ewes 17.5–21 inches (49–54 cm)
Color: Solid colored; most are black, but browns and whites occur
Ears: Medium size, carried slightly above horizontal
Facial profile: Straight to very slightly convex
Fleece: Double-coated; fine, soft; Ouessant undercoat averages 28 microns in diameter
Horns: Rams have big horns; ewes are polled

Ouessants are among the world's smallest sheep. A European short-tail, primitive sheep, the breed originated on the rocky, 5.8-square-mile isle of Ouessant (also called Ushant), 12 miles off the most westerly point of Brittany. As with Soay and Shetland sheep, Viking settlers may have carried the Ouessant's ancestors to their tiny island home. Two breeds were developed on the island: teensy Moribihan sheep in black, brown, and white and slightly larger black Vendeens with impressive curled horns. The breeds eventually merged to produce today's Ouessant sheep.

Ouessants are wool sheep. They are exceptionally hardy and active. Ewes have fine maternal instincts and usually produce single lambs.

Ouessant

Ouessants are an exceptionally rare breed, even in Britain and Europe. Since import regulations prohibit the importation of live Ouessants, fanciers are creating an American version by impregnating Shetland ewes with imported frozen Ouessant semen to create an F1 generation, then using imported Ouessant semen on successive generations of ewes until the breed achieves purebred status. No North American registry yet exists.

Shetland

Registered by: North American Shetland Sheepbreeders Association (NASSA)
Origin: Shetland, an archipelago off the northeastern coast of Scotland
Size: Rams, 90–125 pounds (41–57 kg); ewes, 75–100 pounds (34–45 kg)
Color: There are 11 recognized colors and 30 recognized markings, many with original Shetland dialect names
Ears: Medium size, carried slightly above horizontal
Facial profile: Prominent but not Roman

SHETLAND TAILS

The Shetland, being a member of the European short-tailed race of sheep, has a wool-covered, fluke-shaped tail, broad at the base and tapering to a flattened hairy tip. Composed of 13 vertebrae as compared with 20 or more in long-tailed sheep, it's seldom over 6 inches (15 cm) long; 4 to 5 inches 10–13 cm) is more the norm.

There is no need to dock the tails of Shetlands or any other European short-tailed breeds (including Soay).

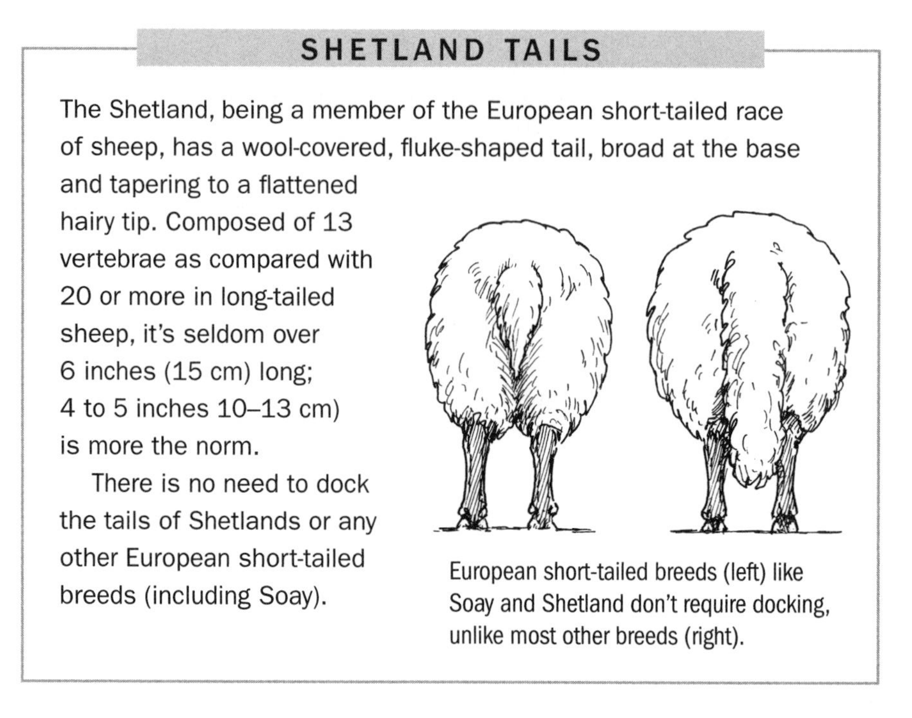

European short-tailed breeds (left) like Soay and Shetland don't require docking, unlike most other breeds (right).

Fleece: Extra fine and soft textured, longish and wavy; staple of 2–4½ inches; 20–25 microns

Horns: Rams are usually horned, ewes are not; however, there are indeed polled rams and horned ewes

Intrepid Norwegian Vikings who colonized Hjaltland (Shetland) around AD 900 brought livestock with them, including Nordic sheep that subsequently interbred with primitive sheep that had already been on the islands for thousands of years. The resulting sheep, ancestors of today's Shetlands, belong to the Northern European short-tailed group that includes, among others, Soay, Icelandics, East Friesian dairy sheep, Romanovs, and Finnsheep.

Food was scarce and in Shetland's fierce climate, smaller versions of standard livestock evolved, including scaled-down versions of ponies, cattle, pigs, poultry, and sheep.

In the early 1900s, in an effort to produce larger sheep, some shepherds on the Shetland Islands began crossbreeding their sheep with mainland sheep and selecting for white wool. So in 1923, concerned sheep raisers incorporated the Shetland Flock Book Society to monitor purebred Shetlands and make certain they didn't die out. At about the same time, hand-knit Fair Isle sweaters became the rage and cottagers on the islands began using naturally colored Shetland wool to create them, providing a ready market for purebred Shetland wool.

Shetland

Shetlands came to North America in only two importations. The first, in 1948, was of three ewes and a ram to George Flett of Fort Qu'Appelle in Saskatchewan, Canada. By carefully inbreeding their bloodlines, the Fletts maintained a purebred flock for more than 50 years.

In 1980, Colonel Dailley of the African Lion Safari in Cambridge, Ontario, Canada, imported 28 ewes and 4 rams. These sheep remained in lifelong quarantine on the Dailley farm, and no offspring could leave the farm until they were five years old. Nevertheless, nearly all North American Shetlands descend predominately, if not fully, from the Dailley importation.

Shetland sheep are hardy, thrifty, easy lambers; adaptable; and long-lived. The breed survived for centuries under difficult conditions and on a diet so poor that larger breeds would have perished. Today's Shetlands retain their ancestral hardiness and primitive survival instincts, making them easier to raise than most modern breeds.

YOU'LL ROO THE DAY

Rooing is the process of removing the fleece from a sheep by hand-plucking its wool. All Soay and a high percentage of Shetland sheep shed the current year's fleece growth in late spring or early summer. By inserting spread fingers into the fleece of a sheep that sheds and working progressively along the sheep, it's possible to remove the fleece in one piece, exactly as though it was shorn.

Sheep can remain standing while being rooed, or they can be rooed lying down with their legs tied or held together. Starting at one shoulder, the fleece is teased off until one side of the sheep is finished. The person doing the rooing goes to the other side of the sheep (if standing) or turns the sheep over (if lying down) and completes the other side.

Rooing doesn't hurt a sheep. Visit the Shetland Sheep Information Web site (see Resources) to view pictures and videos of the rooing process.

One beauty of Shetlands is that they come in an astounding array of attractive colors and patterns. Colors range from white to dark brown with nine other colors in between, most of them with traditional names such as *emsket* (dusky bluish gray), *shaela* (dark steely gray), and *moorit* (ranging from yellowish brown to dark reddish brown). Colors are further differentiated by 30 recognized patterns, such as *blettet* (having white spots on the nose and top of the head), *mirkface* (white with dark patches on the face), and *sokket* (legs a different color from the body). Another benefit is that Shetlands come in a wide array of wool types, ranging from short, crimpy fleece to longer, wavy, hairy wools with downy undercoats. There is a great diversity of color and wool types in this breed.

Soay

Registered by: Rare Breeds Survival Trust in the UK (British Soay)
Origin: Soay, an island in the St. Kilda archipelago off the west coast of Scotland
Size: About 22 inches (56 cm) tall, weighing 45 to 90 pounds (20–41 kg)
Color: Brown or blond with a white belly and rump patch (referred to as *mouflon*). A few are solid black or solid tan. Some have white patches on legs, face, and body.
Ears: To the sides, carried slightly above horizontal
Facial profile: Straight (some slightly dished, some slightly convex)
Fleece: The Soay has no wool on its face or legs. Its body fleece comprises guard hair overlaying a short, woolly undercoat. A typical Soay yields 1 or 2 pounds of fiber per year.
Horns: Soay ewes may be horned, scurred, or polled; rams are horned or scurred

Soay are a conservation breeder's dream. They're listed on the watchlist maintained by the Rare Breeds Survival Trust of Great Britain. They have an amazing history dating to ancient times, they're easy to care for, they yield marvelous wool, and everyone who tastes it raves about their delicate meat.

About 4,000 years ago, Neolithic humans who settled the St. Kilda archipelago may have put the first sheep on windswept, uninhabited Soay, essentially a granite mountain peak jutting out of the sea 41 miles off the west coast of Scotland. When Viking raiders visited in the seventh

Soay

and eighth centuries AD, they named the island So-øy, meaning "Sheep Island" probably for the tiny, feral sheep they found dwelling there.

Around 1900, a few wealthy British landowners brought small numbers of Soay to their estates and parklands in England and Scotland. Selectively bred for various characteristics, such as dark mouflon coloration and horns, these animals are sometimes referred to as Park Soay.

In 1963, Dr. Peter Jewell and his colleagues, who were studying Soay sheep on the uninhabited island of Hirta (the largest of four islands that make up the St. Kilda archipelago), brought 24 sheep back to the mainland so they could continue their studies year-round. The team selected sheep "comprised of a selection of colors, and ewes with and without horns, and even some animals with white markings, sheep that were representative of the animals as we encountered them on Hirta." These became known as Hirta Soay.

Only two groups of Soay sheep were ever exported to North America, one arriving in Canada in 1974 and another arriving there in 1990.

On December 5, 1974, Assiniboine Park Zoo in Winnipeg received four six-month-old lambs from Highland Wildlife Park in Kingussie, Scotland. Seventeen lambs were subsequently born at the zoo. These and their progeny were sold to wildlife parks and exotic breeders, and a few made their way to the United States. Breeding records were not kept, and the remaining purebred sheep were crossed with a number of different breeds. Descendants of these animals are known as American Soay.

In 1989, a Montreal-based research organization purchased six Rare Breeds Survival Trust–registered Soay sheep from a breeder in England. The company bred a closed flock of 30 purebred sheep. The records for these sheep are

SHEEP TRIVIA

- Sheep racing is a growing spectator sport in England, where sheep race around a circular course carrying floppy doll "jockeys" on their backs. Hoo Farm, a popular animal park in Shropshire, England, is famous for its sheep steeplechasing exhibitions.
- Dolly, the world's first cloned sheep, was born in 1997 at the Roslin Institute in Scotland. She gave birth to six lambs during her brief lifetime. She was euthanized when she was 6½ years old after developing a serious lung infection. She was stuffed and is now on display at the Royal Museum of Scotland.

complete, and Soay whose ancestors trace exclusively to this flock are referred to as British Soay.

Both American and British Soay are lithe, deerlike animals. They're active, sure-footed, and nimble. Ewes lamb easily; they're excellent mothers; and Soay lambs mature quickly. They're alert but gentle-natured and curious, too.

The Soay's undercoat is so cushy-soft that women on Hirta used it to knit their families' underwear. During summer, this undercoat molts, so Soay wool is gathered by *rooing* instead of shearing.

BREEDER'S STORY

Scott and Laurie Andreacci

Laurie's Lambs, Chesterfield, New Jersey

SCOTT AND LAURIE ANDREACCI raise three breeds of unusual sheep on their 10-acre farm: the Shetland, the Tunis, and the Gotland.

Laurie has been instrumental in bringing Gotland genetics to North America. Gotlands are medium-size, silvery-black sheep that were first established on the Swedish island of Gotland by Viking settlers. Like Shetlands, Gotlands are a soft-fleeced, primitive, Northern short-tailed breed. Because of strict import regulations, the import of live sheep or frozen embryos from Europe to the United States is strictly forbidden. So, an American version is being created by artificially inseminating Shetland ewes with frozen Gotland semen from Great Britain, then artificially inseminating ewes from first- and subsequent generations with Gotland semen until purebred Gotland status is achieved. Laurie is a founding member of the American Gotland Sheep Society.

When we asked Laurie about her sheep, here is what she said.

"I love the diversity of Shetland sheep. The most important thing to me is that they are soft, so I have the micron count of their fleece tested every year. If they go above a certain count, they have to go. It's sad to let some go, but when you have a goal in mind, you need to stick to the path that leads you to that goal. I can't keep them all, and my customers expect a certain level of quality when buying from me.

"My Shetlands are about the friendliest sheep on the planet because I spend so much time with them. From the moment they're born I'm

Soay are browsers; they thrive on a varied diet, making them an ideal breed for grass-fed and organic farmers or for anyone grazing marginal land. Soay meat is lean and low in cholesterol and boasts an unusually high ratio of polyunsaturated to saturated fats; its light, delectable flavor is a favorite with British chefs.

And being members of the European short-tailed race of sheep, Soay's short, skinny tails don't require docking. Their sheer-free wool and naturally short tails, combined with their hardy constitution and ease of lambing, make them easy-care sheep indeed.

scratching them, talking to them, and nuzzling them. When they grow up and I go into the field, they don't run the other way — they come running to me looking for hugs and cuddles. Friendliness is a great selling point.

"Gotland sheep also fit in very nicely with my operation. They're medium-size sheep, so they're not too much harder to handle than the Shetlands. They're friendly, docile, and inquisitive, and they have wonderful fleece. This year I'll be having 75 percent Gotland, 25 percent Shetland lambs; let's hope they're ewes!

"I do a lot of marketing of my wool. I usually sell the whole clip well before shearing, and some of the fleeces are sold to the same person for as long as I own the animal. That tells me I'm doing the right thing in making sure that softness is the first thing on my breeding agenda. I'm very proud to say I have won plenty of blue ribbons for my fleece.

"I think there is a ready market for Shetland sheep. The trend now is for smaller sheep. They're so much easier to handle, and you can put more of them on your property. With so many hobby farms and farmettes cropping up, people are looking for something easier to handle.

"There is always room for more Shetland breeders! It's like a big family. People from all over the country in all sorts of professions have one thing in common — their love for Shetland sheep. When you meet these people the enthusiasm for their sheep just spills out. The Shetland Sheep Breeders Association holds an annual gathering (the location changes every year). Anyone interested in Shetlands can come."

APPENDIX

Emergency Euthanasia

THE FIRST THING TO DO IF AN ANIMAL IS BADLY INJURED is to call your vet. The vet can assess the situation and, if necessary, humanely euthanize it using an injectable substance. Sometimes, however, no vet is available when a suffering animal needs to be put down. The best solution in this sad event is to shoot the animal — something every livestock owner should know how to do.

Make sure other animals and people are moved out of harm's way. If you don't plan to eat the animal and your vet has entrusted you with prescription drugs such as Rompun or Banamine, give it a hefty dose to sedate it (discuss this with your vet in advance, so that in an emergency you can administer the dose he recommends). If it relaxes so much that it lies down, so much the better.

Use a .22-caliber long rifle, 9 mm or .38-caliber firearm — or, if nothing else is available, a shotgun 410 gauge or larger with a rifled slug. The muzzle of the gun should be held at least 4 to 6 inches away from the skull (never directly against the skull) when fired. The use of hollow-point or soft-nose bullets will increase brain tissue destruction and reduce the chance of ricochet. When performed skillfully, euthanasia by gunshot induces immediate unconsciousness, although muscle twitching will likely continue for a little while after the animal is dead. This is a normal physiological response after death no matter what the cause.

This method should be attempted only by individuals trained in the use of firearms and who understand the potential for ricochet, not to mention the need for accurate aim. All humans and other animals should remain out of the line of fire.

Discuss with your vet ahead of time what to do with the remains; laws vary from state to state.

EQUINES AND LLAMAS **PIGS**

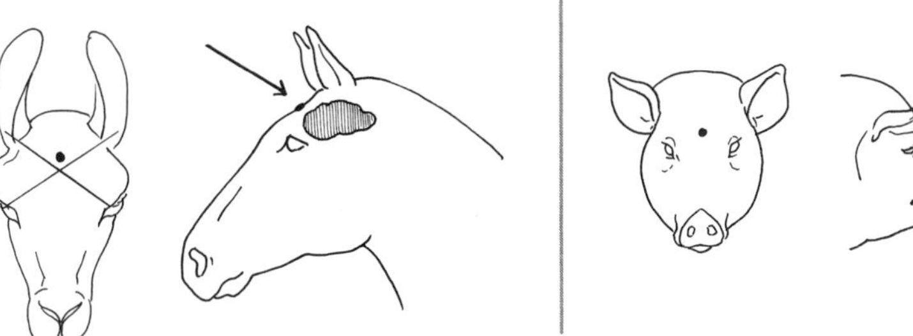

Equines, Llamas, and Pigs

The firearm should be held perpendicular to the front of the skull. Aim at a point one-half inch above the intersection of a diagonal line drawn from the base of the ear to the inside corner of the opposite eye.

Cattle

The firearm should be held perpendicular to the front of the skull. Aim at a point at the intersection of a diagonal line drawn from the base of the horn (or the lower outside edge of the poll in polled cattle) to the middle of the opposite eye.

Sheep and Goats

The firearm should be positioned behind the animal's head. (See Goats illustration at right.) Aim at the midline of the back of a horned animal's skull, just behind the bony ridge where the horns protrude, while also aiming forward toward the back of the nose.

Polled sheep (like the one at right) and polled or disbudded goats as well as kids and lambs less than four months of age may be shot from the front as indicated under Equines, Llamas, and Pigs and Cattle.

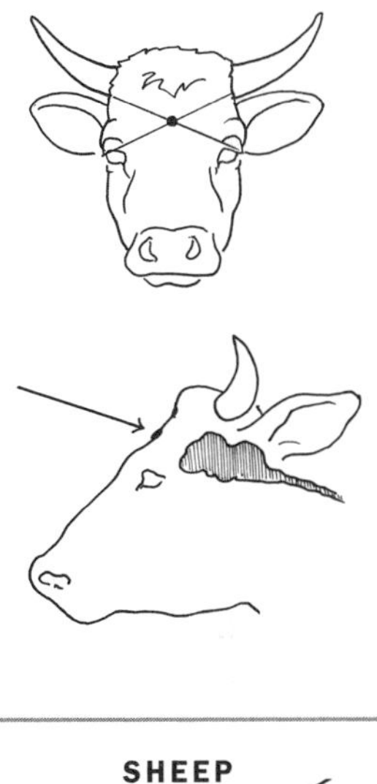

SHEEP

GOATS

Resources

We have made this section as thorough and current as possible.

Multiple Species

Organizations

Academy of Veterinary Homeopathy (AVH)
Leucadia, California
866-652-1590
www.theavh.org

American Academy of Veterinary Acupuncture (AAVA)
Glastonbury, Connecticut
860-632-9911
www.aava.org

American Association of Bovine Practitioners (AABP)
Auburn, Alabama
334-821-0442
www.aabp.org

American Association of Equine Practitioners (AAEP)
Lexington, Kentucky
859-233-0147
www.aaep.org

American Association of Small Ruminant Practitioners (AASRP)
Montgomery, Alabama
334-517-1233
www.aasrp.org

American Grassfed Association (AGA)
Denver, Colorado
877-774-7277
www.americangrassfed.org

American Holistic Veterinary Medical Association (AHVMA)
Abingdon, Maryland
410-569-0795
www.ahvma.org

The Livestock Conservancy
(formerly the American Livestock Breeds Conservancy)
Pittsboro, North Carolina
919-542-5704
www.livestockconservancy.org

American Veterinary Chiropractic Association (AVCA)
Bluejacket, Oklahoma
918-784-2231
www.animalchiropractic.org

American Veterinary Medical Association (AVMA)
Schaumburg, Illinois
800-248-2862
www.avma.org

ATTRA — National Sustainable Agriculture Information Service
Butte, Montana
800-346-9140
www.attra.org

Center for Rural Affairs
Lyons, Nebraska
402-687-2100
www.cfra.org

Equus Survival Trust
Lowgap, North Carolina
336-352-5520
www.equus-survival-trust.org

Johne's Information Center
University of Wisconsin,
School of Veterinary Medicine
Madison, Wisconsin
608-263-6920
www.johnes.org

National Institute of Food and Agriculture (NIFA)
United States Department of Agriculture
Washington, D.C.
202-720-4423
https://nifa.usda.gov
Click on Local Extension Offices to find your county extension office.

Rare Breeds Canada (RBC)
Castleton, Ontario
905-344-7768
www.rarebreedscanada.ca

Rare Breeds Conservation Society of New Zealand
Bishopdale, New Zealand
www.rarebreeds.co.nz

Rare Breeds Survival Trust (RBST)
Warwickshire, United Kingdom
+44-024-7669-6551
www.rbst.org.uk

Rare Breeds Trust of Australia
Abbotsford, Australia
+61-0408-324-346
https://rarebreedstrust.com.au

Sustainable Agriculture Research and Education (SARE)
National Institute of Food and Agriculture
Washington, D.C.
info@sare.org
www.sare.org

Information

Maryland Small Ruminant Pages
www.sheepandgoat.com
Extensive links: small farms, hay, fencing, housing, marketing, pasture, and much more

The Merck Veterinary Manual
www.merckvetmanual.com
Up to date, online, free, and searchable

Plants Poisonous to Livestock and Other Animals
Cornell University
https://poisonousplants.ansci.cornell.edu
The most comprehensive site of its kind

Raw Milk Facts
www.raw-milk-facts.com

Equipment

The BackroadHome
www.backroadhome.net
A directory of sources for manufactured buildings

Chalex Corporation
Wallowa, Oregon
www.vetslides.com
Microscopes, McMaster and nematode slides, and fecal-testing kits

Country Manufacturing
Fredericktown, Ohio
800-335-1880
www.horsestalls.com
Modular miniature stalls, barn and stable gear of all types

Hampel Corporation
Germantown, Wisconsin
800-558-8558
www.calftel.com
Calf-Tel calf hutches

Jeffers Livestock Supply
Dothan, Alabama
800-533-3377
www.jefferslivestock.com
Vaccines and equipment

Molly's Herbals
Okemos, Michigan
molly@mollysherbals.com
www.mollysherbals.com
Herbal animal care products

PolyTank & PolyDome
Litchfield, Minnesota
800-328-7659
www.polydome.com
Calf hutches

Port-A-Hut, Inc.
Storm Lake, Iowa
800-882-4884
www.port-a-hut.com
Quonset-style animal housing

Premier1 Supplies
Washington, Iowa
800-282-6631
www.premier1supplies.com
Separate sheep and goat, fencing, and clipper and shearing machine catalogs

Quality Llama Products and Alternate Livestock Supply, Inc.
Lebanon, Oregon
800-638-4689
www.llamaproducts.com
Books and equipment for llamas, miniature equines, goats

Roth Manufacturing Company, Inc.
Loyal, Wisconsin
800-472-2341
www.loyal-roth.com
Dairy, hog, poultry, small-farm, and horse equipment

Valley Vet Supply
Marysville, Kansas
800-419-9524
www.valleyvet.com
Vaccines and equipment

Woodstar Products, Inc.
Delavan, Wisconsin
800-648-3415
www.woodstarproducts.com
Modular stalls for miniatures

Periodicals

AcresUSA
800-355-5313
www.acresusa.com

Countryside
970-392-4419
www.countrysidemag.com

Hobby Farms
844-330-6373
www.hobbyfarms.com
Extensive article archives, resource lists, forum, blog

Small Farm Canada
866-260-7985
www.smallfarmcanada.ca

Books and Papers

Damerow, Gail. *Barnyard in Your Backyard.* North Adams, MA: Storey Publishing, 2002.

———. *Fences for Pasture and Garden.* North Adams, MA: Storey Publishing, 1992.

Douglass II, William Campbell. *The Milk Book: The Milk of Human Kindness Is Not Pasteurized.* South Africa: Rhino Publications S.A., 2004.

Ekarius, Carol. *Hobby Farms: Living Your Rural Dream for Pleasure and Profit.* Irvine, CA: Bowtie Press, 2005.

———. *How to Build Animal Housing: 60 Plans for Coops, Hutches, Barns, Sheds, Pens, Nest Boxes, Feeders, Stanchions, and Much More.* North Adams, MA: Storey Publishing, 2004.

———. *Small-Scale Livestock Farming: A Grass-Based Approach for Health, Sustainability, and Profit.* North Adams, MA: Storey Publishing, 1999.

———. *Storey's Illustrated Guide to Sheep, Goats, Cattle, and Pigs.* North Adams, MA: Storey Publishing, 2008.

Grandin, Temple. *Humane Livestock Handling: Understanding Livestock Behavior and Building Facilities for Healthier Animals.* North Adams, MA: Storey Publishing, 2008.

Haynes, N. Bruce. *Keeping Livestock Healthy*, 4th ed. North Adams, MA: Storey Publishing, 2001.

Macher, Ron. *Making Your Small Farm Profitable.* North Adams, MA: Storey Publishing, 1999.

Riotte, Louise. *Raising Animals by the Moon: Practical Advice on Breeding, Birthing, Weaning, and Raising Animals in Harmony with Nature.* North Adams, MA: Storey Publishing, 1999.

Robinson, Jo. *Pasture Perfect: The Far-Reaching Benefits of Choosing Meat, Eggs, and Dairy Products from Grass-Fed Animals.* Vashon, WA: Vashon Island Press, 2004.

Schmid, Ron. *The Untold Story of Milk: Green Pastures, Contented Cows and Raw Dairy Products.* Winona Lake, IN: NewTrends Publishing, 2003.

Spaulding, C. E., and Jackie Clay. *Veterinary Guide for Animal Owners.* Emmaus, PA: Rodale Press, 2001.

Speer, N. C., G. Slack, and E. Troyer. "Economic Factors Associated with Livestock Transportation," *Journal of Animal Science* 79 (E supplement) 2001, E166–E170. American Society of Animal Science.

Tresemer, David, and Peter Vido. *The Scythe Book*, 2nd ed. Chambersburg, PA: Alan C. Hood & Co., 2001.

Profiled Breeders

Please note: The following information was accurate at the time of publication (2010). When planning a trip, please confirm details by directly contacting any company or establishment you intend to visit.

Briar Patch Miniature Overo Horses
Gary and Carlene Norris
Winona, Missouri
573-325-8284
sunrae.com/briarpatchminis

Dragonflye Farms Miniature Livestock
Bev Jacobs and Bill Lanier Jr.
Goodyear, Arizona
623-451-8357
www.dragonflyefarm.com

Fancher Love Ranch
Dottie Love and Tom Sale
Ennis, Texas
www.fancherloveranch.com

The Fuzzy Farm
Richard and Gayle Dumas
Saluda, Virginia
804-815-6459
www.thefuzzyfarm.com

Laurie's Lambs
Scott and Laurie Andreacci
Chesterfield, New Jersey
609-324-0487
www.laurieslambs.com

Nissen's Lazy N Ranch
Jon, Mary, and Jay Nissen
Corwith, Iowa
515-824-3603
www.lnminidonk.com

One Goat Shy Farm
Jeanne DuBois
www.onegoatshy.com

Ridgerunners Miniature Horses
Gib and Melba Mullins
Bakersfield, Missouri
417-284-8134
sunrae.com/ridgerunnersminis

Short ASSets Ranch
Lonnie and Brenda Short
Eddy, Texas
254-859-3724
www.shortassets.com

Southern Oregon Soay Sheep Farms
Kathie Miller
Merlin, Oregon
541-955-8171
www.soayfarms.com

Stubby Acres Farm
Lauren Simermeyer, DVM
Burlington, New Jersey
609-747-1788
www.stubbyacres.com

Summer Shade Farm
Jim and Nancy Eubanks
Bethel, Ohio
513-734-4114
www.summershadefarm.net

USA KuneKunes
Jim and Lori Enright
Mira Loma, California
951-505-5230
www.usakunekunes.com

Marketing
Information

Agricultural Marketing Resource Center (AgMRC)
866-277-5567
www.agmrc.org

Instant Domain Name Search
www.instantdomainsearch.com

Internet World Stats
www.internetworldstats.com

Marketing, Business, and Risk Management
ATTRA — National Sustainable Agriculture Information Service
https://attra.ncat.org/category/topics/marketing-business-risk-management

U.S. Small Business Administration (SBA)
800-827-5722
www.sba.gov

Books

Aubrey, Sarah Beth. *Starting and Running Your Own Small Farm Business*. North Adams, MA: Storey Publishing, 2008.

Grant, Lynella, and Ted W. Parrod. *The Business Card Book: What Your Business Card Reveals about You — and How to Fix It*. Scottsdale, AZ: Off the Page Press, 1998.

Hill, Brad. *Blogging for Dummies*, 2nd ed. Hoboken, NJ: Wiley Publishing, 2006.

Popyk, Bob. *Here's My Card: How to Network Using Your Business Card to Actually Create More Business*. Los Angeles: Renaissance Books, 2000.

Winslow, Ellie. *Growing Your Rural Business from the Inside Out*. Reno: Freefall Press, 2008.

———. *Marketing Farm Products: And How to Thrive beyond the Sidewalk*. Reno: Freefall Press, 2007.

Miniature Cattle

Organizations

American Dexter Cattle Association
Watertown, Minnesota
844-588-2322
www.dextercattle.org

American Hereford Association
Kansas City, Missouri
816-842-3757
www.hereford.org

American Highland Cattle Association
Denver, Colorado
303-292-9102
www.highlandcattleusa.org

American Lowline Registry
Parker, Colorado
303-840-4343
www.usa-lowline.org

American Miniature Jersey Association & Registry
Crawford, Nebraska
308-665-1431
www.miniaturejerseyassociation.com

American Miniature Zebu Association
Ferris, Texas
972-544-3334
www.amzaonline.org

Australian Lowline Cattle Association
Agricultural Business Research Institute
Armidale, Australia
+61-02-6773-3295
www.lowlinecattleassoc.com.au

British Kerry Cattle Society
Ashbourne, United Kingdom
+44-01283-732377
www.kerrycattle.org.uk

Canadian Dexter Cattle Association
Ottawa, Ontario
613-731-7110
www.dextercattle.ca

Canadian Lowline Cattle Association
Saskatchewan, Ontario
306-397-2584
www.canadianlowline.com

Dexter Cattle Australia, Inc.
Agricultural Business Research Institute
Armidale, Australia
+61-02-6773-3471
www.dextercattle.org.au

Dexter Cattle Breeders' Society of South Africa
South African Stud Book Association
Westdene, South Africa
https://www.dextersa.co.za

Dexter Cattle Society of New Zealand
Papakura, New Zealand
+64-09-298-1789
www.dextercattle.co.nz

Florida Cracker Cattle Association
Tallahassee, Florida
850-410-0900
www.floridacrackercattle.org

International Miniature Zebu Association
Crawford, Nebraska
407-717-0084
www.imza.name

Kerry Cattle Society
Killarney, Ireland
+353-064-6631840
www.kerrycattle.ie

Miniature Hereford Breeders Association
www.mhbaonline.org

Pineywoods Cattle Registry Breeders Association
Poplarville, Mississippi
601-795-4672
www.pcrba.org

Shetland Cattle Breeders Association (SCBA)
info@shetlandcattle.org.uk
www.shetlandcattle.org.uk

Information

Weston A. Price Foundation
703-820-3333
www.westonaprice.org
Raw milk information

Books

American Kerry and Dexter Cattle Club. *American Kerry and Dexter Cattle Herd Book*, Volume I. Columbus, OH: American Kerry and Dexter Cattle Club, 1921.

Gow, R. M. American Jersey Cattle Club. *The Jersey: An Outline of Her History during Two Centuries, 1734–1935.* New York: American Jersey Cattle Club, 1936.

Grohman, Joann S. *Keeping a Family Cow*, 5th ed. Dixfield, ME: Coburn Press, 2007.

Hill, Charles L. *The Guernsey Breed*. Waterloo, IA: Fred L. Kimball, 1917.

Low, David. *On the Domesticated Animals of the British Islands.* Boston: Little, Brown and Company, 1853.

McCann, L. P. *The Battle of the Bull Runts*. Columbus, OH: L. P. McCann, 1974.

National Research Council. *Microlivestock: Little-Known Small Animals with a Promising Economic Future.* Washington, D.C.: National Academy Press, 1991.

Olson, T. A., and R. L. Willham. "Inheritance of Coat Coloration and Spotting Patterns of Cattle: A Review." Ames: Iowa State University Agriculture and Home Economics Experiment Station, 1982.

Ruechel, Julius. *Grass-Fed Cattle: How to Produce and Market Natural Beef.* North Adams, MA: Storey Publishing, 2006.

Selk, Glen. "Embryo Transfer in Cattle." Stillwater: Oklahoma Cooperative Extension Service.

Smith Thomas, Heather. *The Cattle Health Handbook*. North Adams, MA: Storey Publishing, 2009.

———. *Essential Guide to Calving: Giving Your Beef or Dairy Herd a Healthy Start*. North Adams, MA: Storey Publishing, 2008.

———. *Getting Started with Beef & Dairy Cattle.* North Adams, MA: Storey Publishing, 2005.

———. *Storey's Guide to Raising Beef Cattle: Health, Handling, Breeding.* North Adams, MA: Storey Publishing, 1998.

van Loon, Dirk. *The Family Cow.* North Adams, MA: Storey Publishing, 1976.

Miniature Donkeys and Mules

Many miniature horse resources apply to miniature donkeys and mules, too.

Organizations

American Council of Spotted Asses
Omaha, Texas
903-884-2067
www.spottedass.com

American Donkey and Mule Society
Lewisville, Texas
972-219-0781
www.lovelongears.com

The British Mule Society (BMS)
info@thebritishmulesociety.com
https://thebritishmulesociety.com

Canadian Donkey & Mule Association
www.donkeyandmule.com

Donkey All Breeds Society of Australia, Inc. (DABSA)
Kurmond, Australia
www.donkeyallbreedsaustralia.org

Donkey and Mule Society of New Zealand, Inc.
Palmerston North, New Zealand
www.donkey-mule.org.nz

Miniature Mediterranean Donkey Association (MMDA)
Stalbridge, United Kingdom
www.miniature-donkey-assoc.com

National Miniature Donkey Association
Rome, New York
www.nmdaasset.com

Information

Gotdonkeys Breeders List
www.gotdonkeys.com

LongearsMall.com
www.longearsmall.com
A free, online pedigree database

Equipment

Dinky Rugs
Harlow, United Kingdom
+44-01279-410300
www.dinkyrugs.co.uk
Finally! Turnout rugs designed to fit miniature donkeys

Packing

Northwest Pack Goats
Weippe, Idaho
888-722-5462
www.northwestpackgoats.com
Offers a high-quality wooden sawbuck pack kit for goats that is easily adapted for miniature donkeys

Periodicals

ASSET
www.nmdaasset.com/ASSET.php

The Brayer
American Donkey and Mule Society
www.lovelongears.com/brayer
Also books

Miniature Donkey Talk
719-689-2904
www.web-donkeys.com
Also tack, books, gifts

Mules and More
573-263-2669
www.mulesandmore.com
Also books and gifts

Books

Gross, Bonnie R. *Caring for Your Miniature Donkey*, 2nd ed. Westminster, MD: Miniature Donkey Talk, 1999.

Hutchins, Betsy, and Paul Hutchins; revised and edited by Leah Patton. *The Definitive Donkey: A Textbook of the Modern Ass.* Lewisville, TX: American Donkey and Mule Society, 1981. Revised 1999.

Svendsen, Dr. Elisabeth D. *The Professional Handbook of the Donkey*, 3rd ed. Wiltshire, UK: Whittet Books, 1998.

Weaver, Sue. *The Donkey Companion: Selecting, Training, Breeding, Enjoying & Caring for Donkeys*. North Adams, MA: Storey Publishing, 2008.

Miniature Goats

Organizations

American Dairy Goat Association
Spindale, North Carolina
828-286-3801
www.adga.org

American Goat Society
Pipe Creek, Texas
830-535-4247
www.americangoatsociety.com

American Nigerian Dwarf Dairy Association (ANDDA)
info@andda.org
www.andda.org

American Nigora Goat Breeders Association (ANGBA)
McMinnville, Oregon
nigoragoat_info@yahoo.com
www.nigoragoats.homestead.com

Australian Miniature Goat Association, Inc.
Muwillumbah, Australia
+61 0419 655 348
amga@australianminiaturegoat.com.au
www.australianminiaturegoat.com.au

International Goat, Sheep, Camelid Registry
Goshen, Utah
https://www.igscr-idgr.com

International Fainting Goat Association
Darlington, Pennsylvania
541-825-8580
www.faintinggoat.com

Miniature Dairy Goat Association
Woodland, Washington
www.miniaturedairygoats.net

Miniature Goat Breeders Association of Australia, Inc. (MGBA)
Boyland, Australia
https://mgba.com.au

Miniature Silky Fainting Goat Association, LLC
Lignum, Virginia
540-423-9193
https://msfgaregistry.org

Myotonic Goat Registry
Adger, Alabama
205-425-5954
www.myotonicgoatregistry.net

National Miniature Goat Association (NMGA)
Smithville, Indiana
www.nmga.net

National Pygmy Goat Association
Snohomish, Washington
425-334-6506
www.npga-pygmy.com

Nigerian Dwarf Goat Association
Wilhoit, Arizona
309-371-5896
www.ndga.org

Pygmy Goat Club (PGC)
Cornwall, United Kingdom
secretary@pygmygoatclub.org.uk
www.pygmygoatclub.org

Pygora Breeders Association
Saint Marys, Ohio
603-828-8619
https://pba-pygora.org

San Clemente Island Goat Association
The Plains, Virginia
www.scigoats.org

Information

Amber Waves Pygmy Goats
951-736-1076
https://amberwavespygmygoats.com
Great articles!

Biology of the Goat
www.goatbiology.com

Fias Co Farm
www.fiascofarm.com/goats
200+ pages of topflight information

GoatWorld
www.goatworld.com

Kinne's Minis
www.kinne.net
Scores of fantastic articles, plus extensive links

Equipment

Caprine Supply
De Soto, Kansas
800-646-7736
www.caprinesupply.com

Hamby Dairy Supply
Maysville, Missouri
816-449-1314
www.hambydairysupply.com

Hoegger Supply
Fayetteville, Georgia
800-221-4628
www.hoeggerfarmyard.com

Periodicals

Backyard Goats
https://backyardgoats.iamcountryside.com

Books

Belanger, Jerry. *Storey's Guide to Raising Dairy Goats: Breeds, Care, Dairying.* North Adams, MA: Storey Publishing, 2000.

Boldrick, Lorrie. *Pygmy Goats: Management and Veterinary Care.* Orange, CA: All Publishing Company, 1996.

Damerow, Gail. *Your Goats: A Kid's Guide to Raising and Showing*. North Adams, MA: Storey Publishing, 1993.

Leigh, Jody. *Nigerian Dwarfs: Colorful Miniature Dairy Goats.* Golden, CO: Leighstar Publications, 1993.

Sayer, Maggie. *Storey's Guide to Raising Meat Goats: Managing, Breeding, Marketing.* North Adams, MA: Storey Publishing, 2007.

Smith, Cheryl K. *Goat Health Care: The Best of Ruminations 2001–2007.* Cheshire, OR: Karmadillo Press, 2009.

Stewart, Patricia Garland. *Personal Milkers: A Primer to Nigerian Dwarf Goats*. Ashburnham, MA: Garland Stewart Publishing, 2008.

Winslow, Ellie. *Making Money with Goats*, 5th ed. Reno, NV: Freefall Press, 2007.

Miniature Horses

Organizations

American Miniature Horse Association
Alvarado, Texas
817-783-5600
www.amha.org

American Miniature Horse Registry
Morton, Illinois
309-263-4044
www.shetlandminiature.com

American Shetland Pony Club
Morton, Illinois
www.shetlandminiature.com

British Miniature Horse Society
Bridgwater, United Kingdom
+44-0-1278-685943
www.bmhs.co.uk

Caspian Horse Society (CHS)
Melton Mowbray, United Kingdom
registrar@dial.pipex.com
www.caspianhorsesociety.org.uk

Caspian Horse Society of the Americas (CHSA)
Ellsworth, Illinois
309-724-8373
www.caspian.org

Falabella Miniature Horse Association
Round Lake, Illinois
847-587-6201
www.falabellafmha.com

Independent Miniature Horse Registry, Inc. (IMHR)
Currawang, Australia
+61-02-4844-6393
www.imhr.com.au

International Miniature Horse & Pony Society Ltd. (IMHPS)
Billingshurst, United Kingdom
www.imhps.com

Miniature Horse Association of Australia, Inc.
www.mhaa.com.au

New Zealand Miniature Horse Association, Inc. (NZMHA)
Invercargill, New Zealand
+64-03-230-4666
www.nzmha.co.nz

Shetland Pony Stud-Book Society
Perth, Scotland
+44-0-1738-623471
www.shetlandponystudbooksociety.co.uk

Smallest Horse Group
www.smallesthorse.com

Information

L'il Beginnings
www.lilbeginnings.com
The best source of miniature horse information online

Little Bit's Magic Shoes
www.littlemagicshoes.com
Articles about dwarfism

The Miniature Horse
www.theminiaturehorse.com

Thumbelina: World's Smallest Horse
www.worldssmallesthorse.com

USA Miniature Horse
www.mini-horse.org

Equipment

Allen's Pony Express
Eldorado, Illinois
(618) 273-3101
www.allensponyexpress.com

Camptown Harness
www.camptownharness.com

Kingston Horse Supplies
Landsborough, Australia
www.kingstons.net.au

Mini Express
Citra, Florida
352-595-2410
www.miniexpress.com

Ozark Mountain Miniature Tack & Equine Supplies
Gassville, Arkansas
888-775-6446
www.minitack.com

Valhalla Miniature Horses & Tack
Westminster, Maryland
410-404-3365
www.valhallafarm.net

Periodicals

The Journal
American Shetland Pony Club
309-263-4044
www.shetlandminiature.com/journal.asp

Miniature Horse World
American Miniature Horse Association
817-783-5600
www.amha.org

Books

Frankeny, Rebecca L. *Miniature Horses: A Veterinary Guide for Owners & Breeders.* North Pomfret, VT: Trafalgar Square Books, 2008.

Hayes, Karen E. N. *The Complete Book of Foaling: An Illustrated Guide for the Foaling Attendant.* Indianapolis, IN: Howell Book House, 1993.

———. *Emergency! The Active Horseman's Book of Emergency Care.* Boonsboro, MD: Half Halt Press, 1995.

Naviaux, Barbara. *Miniature Horses: Their Care, Breeding, and Coat Colors.* Fort Bragg, CA: Raintree Publications, 1999.

Smith, Donna Campbell. *The Book of Miniature Horses: Buying, Breeding, Training, Showing, and Enjoying.* Guilford, CT: Lyons Press, 2007.

Miniature Llamas

Organizations

Alpaca Llama Show Association
Cypress, Texas
www.alsashow.net

American Miniature Llama Association (AMLA)
Kalispell, Montana
https://minillamalady.weebly.com

Canadian Llama & Alpaca Association
Calgary, Alberta
800-717-5262
www.claacanada.com

International Camelid Institute
Ohio State University
614-688-8160
https://icinfo.vet.ohio-state.edu

International Lama Registry
Kalispell, Montana
406-755-3438
www.lamaregistry.com

Llama Association of North America
Vacaville, California
707-447-5046
www.lanainfo.org

Pack Llama Trial Association
Grants Pass, Oregon
541-589-1406
www.packllama.org

Information

Shagbark Ridge Llamas
317-773-1201
dago@shagbarkridge.com
www.shagbarkridge.com

Southeast Llama Rescue
https://www.southeastllamarescue.org
A best-bet source of information about aberrant behavior syndrome

Periodicals

Lama Link
406-755-5473
www.lamalink.com
Back issues are archived on Lama Link's Web site as free, downloadable PDF files

Books

Birutta, Gail. *Storey's Guide to Raising Llamas: Care, Showing, Breeding, Packing, Profiting.* North Adams, MA: Storey Publishing, 1997.

McGee-Bennett, Marty. *The Camelid Companion.* Bend, OR: Raccoon Press, 2001.

Miniature Pigs

Organizations

American Guinea Hog Association, Inc. (AGHA)
Jefferson, Iowa
515-370-1021
www.guineahogs.org

American Kunekune Pig Registry
(formerly American KuneKune Breeders' Association)
Norco, California
951-505-5230
www.americankunekuneregistry.com

British KuneKune Pig Society
Saffron Walden, United Kingdom
+44-0-845-489-5863
www.britishkunekunesociety.co.uk

Miniature Pig Registry & Miniature Potbelly Pig Registry
Cleveland, Georgia
706-348-7279
www.miniaturepigregistry.com

New Zealand Kunekune Association
Fielding, New Zealand
+64-06-323-4481
www.kunekune.co.nz
Don't miss the great articles at this site!

North American Potbellied Pig Association
Gilbert, Arizona
480-899-8941
www.petpigs.com

United States Juliana Pig Registry
usjulianapigregistry@gmail.com
www.usjpr.com

Information

Pennywell Miniature Pigs
+44-01364-642023
www.pennywellfarm.co.uk

Pigs 4 Ever
352-589-1702
www.pigs4ever.com
Online info, books, and supplies

Rescue

Pig Placement Network
215-322-1539
www.pigplacementnetwork.org

Pig Breeds Available for Adoption on Petfinder
www.petfinder.com/breeds/Pig

The Pig Preserve
Jamestown, Tennessee
thepigpreserve@gmail.com
www.thepigpreserve.org
Another answer to overbreeding

PIGS, A Sanctuary
Shepherdstown, West Virginia
304-876-6766
www.pigs.org

Equipment

PA Heartland Ltd.
Pottstown, Pennsylvania
484-624-5745
www.paheartland.com
Full line of pet pig supplies

The Pig Store
Rushland, Pennsylvania
215-322-1539
http://pigstore.stores.yahoo.net

Periodicals

Lord, Linda Kay, and T. E. Wittum. "Survey of Humane Organizations and Slaughter Plants Regarding Experiences with Vietnamese Potbellied Pigs." *Journal of the American Veterinary Medical Association,* vol. 211 (1997): 562–65.

Books

Klober, Kelly. *Dirt Hog: A Hands-On Guide to Raising Pigs Outdoors . . . Naturally.* Austin, TX: 2007.

———. *Storey's Guide to Raising Pigs.* North Adams, MA: Storey Publishing, 1997.

Myers, Kathleen. *The Complete Guide for the Care and Training of Pet Potbellied Pigs,* revised edition. Ouray, CO: Ponderosa Press, 2005.

Riffle, Dale, and the staff of PIGS. *The Shelter Workers' Guide to Pigs.* Shepherdstown, WV: PIGS, Inc., 2002. Available through the PIGS, A Sanctuary Web site: *www.pigs.org*

Valentine, Priscilla. *Potbellied Pig Behavior and Training,* 2nd ed. Moscow, ID: Luminary Media Group, 2007.

Miniature Sheep

Organizations

American Classic Cheviot Sheep Association
Glenwood, Iowa
classiccheviots.com

British Coloured Sheep Breeders Association
+44-01594-529-331
www.bcsba.org.uk
Shetlands, Soays, and more. Don't miss the outstanding fleece marketing pages at this site!

Miniature Cheviot Sheep Breeders Cooperative
Silver Springs, Nevada
775-629-1211
www.minicheviot.com

North American Babydoll Southdown Sheep Association & Registry
Wamego, Kansas
502-352-7928
www.nabssar.org

North American Shetland Sheepbreeders Association
Wamego, Kansas
260-672-9623
www.shetland-sheep.org

Olde English Babydoll Southdown Sheep Registry
Graham, Washington
253-548-8815
www.oldeenglishbabydollregistry.com

Ouessant Sheep Society of Great Britain
Steyning, West Sussex, England
www.ouessantsheep.net

Soay Sheep Open Flockbook Project
Jacksonville, Oregon
541-899-1672
www.openflockbook.com

Soay Sheep Society
Macclesfield, United Kingdom
secretary.jw@soaysheep.org
www.soaysheep.org

Information

Cornell University Sheep Program
607-592-2541
www.sheep.cornell.edu

Sheep @ Perdue
Purdue University
http://ag.ansc.purdue.edu/SH

Sheep 201: A Beginner's Guide to Raising Sheep
www.sheep101.info/201

Shetland Sheep Information
www.shetlandsheepinfo.com

Periodicals

Sheep!
800-551-5691
www.sheepmagazine.com
Extensive article archives, blogs

Sheep Canada
506-328-3599
www.sheepcanada.com

Books

Dalton, Clive, and Marjorie Orr. *The Sheep Farming Guide for Small and Not-So-Small Flocks.* Christchurch, NZ: Hazard Press, 2004.

Gill, Warren. "Applied Sheep Behavior." Knoxville: Agricultural Extension Service, University of Tennessee, 2004. Available online through the University of Tennessee's Department of Animal Science: *http://animalscience.ag.utk.edu*

Parker, Ron. *The Sheep Book: A Handbook for the Modern Shepherd,* revised and updated. Athens: Swallow Press/Ohio University Press, 2001.

Simmons, Paula, and Carol Ekarius. *Storey's Guide to Raising Sheep: Breeds, Care, Facilities,* 3rd ed. North Adams, MA: Storey Publishing, 2000.

Storey Publishing, LLC. *Storey's Barn Guide to Sheep.* North Adams, MA: Storey Publishing, 2006.

Glossary

Aberrant behavior syndrome (ABS). A condition in which an adult llama (usually an uncastrated male) that was improperly imprinted on humans while it was a cria becomes dangerously aggressive toward people. Previously called berserk male syndrome (BMS).

Abomasum. The third compartment of the ruminant stomach; the compartment where digestion takes place.

Abortion. Early (often spontaneous) termination of pregnancy.

Acre. A unit of measurement: 4,840 square yards or 43,560 square feet.

Action. The manner in which an animal moves its legs.

Acute. Any process occurring over a short period of time.

Afterbirth. The placenta and fetal membranes that are expelled after giving birth.

Agouti. A color in which each hair has three or more bands of color with a definite break between each color.

Alarm call. The distinctive, whinny-like sound a llama makes when it feels it or the herd it belongs to is threatened.

ALBC. American Livestock Breeds Conservancy; a group dedicated to preserving and promoting rare and endangered breeds of American livestock and poultry.

Ammonium chloride. A mineral salt fed to male sheep and goats to inhibit the formation of bladder and kidney stones.

Amnion. One of two fluid-filled membranes enclosing an unborn fetus.

Anestrus. The period of time when a female animal is not having estrous (heat) cycles.

Anthelmintic. A substance used to control or destroy internal parasites; a dewormer.

Antibodies. Circulating protein molecules that help neutralize disease organisms.

Antitoxin. An antibody capable of neutralizing a specific disease organism.

Artificial insemination (AI). A process by which semen is deposited within a female's uterus by artificial means.

Aspirate. To pull back slightly on a syringe's plunger to draw fluid into the chamber.

Ass. Another word for donkey.

ATTRA. The National Sustainable Agriculture Information Service; managed by the National Center for Appropriate Technology (NCAT) and funded under a grant from the United States Department of Agriculture's Rural Business-Cooperative Service.

Autogenous vaccines. Vaccine made from organisms collected from a specific disease outbreak; for example, autogenous caseous lymphadenitis vaccine is manufactured using bacteria harvested from pus collected from the lanced abscess of an infected goat.

Back cross. The mating of a crossbred offspring back to one of its parental breeds.

Bag (slang). Udder

Bagging up. Enlargement of the udder prior to giving birth.

Balanced ration. Feeds having proper portions of ingredients to provide for growth, reproduction, and good health.

Banana ears. Banana-shaped llama ears that curve inward toward one another.

Banding. Castration by the process of applying a fat rubber ring to a male animal's scrotum using a tool called an elastrator.

Barren. Unable to conceive or bear young.

Barrow. A male pig castrated prior to maturity.

Bars. The gap between an equine's front and back teeth where the bit on a bridled animal rests.

Belt. A band of white around the barrel of an animal, flanked on either end by a darker color.

Billy (slang). An uncastrated male goat; the preferred term is *buck*.

Birth date. The actual date an animal was born.

Birth weight. The weight of a young animal taken within 24 hours of birth.

Bite. Occlusion; the manner in which the upper and lower teeth match up.

Bleating. Goat and sheep vocalization; in goats, also referred to as *calling*.

Blemish. A scar or deformity that diminishes an animal's beauty but doesn't affect its soundness.

Blind teat. A nonfunctional teat; it has no orifice.

Bloat. Excessive accumulation of gas in a ruminant's rumen.

Bloodlines. The ancestry of an animal.

Bloom. The healthy shine of a hair coat in good condition.

Blowing coat (shedding, molting). When a coated animal sheds its hair, usually in the spring or summer months.

Boar. An adult male pig used for breeding purposes.

Bolus. A large, oval pill; also used to describe a chunk of cud.

Booster vaccination. A second or multiple vaccinations given to increase an animal's resistance to a specific disease.

Bo-Se. An injectable prescription selenium supplement.

Bos indicus. Humped breeds of cattle descending from aurochs' ancestors domesticated in the Indus Valley of India and Pakistan.

Bos taurus. Nonhumped cattle descending from aurochs' ancestors domesticated in the Middle East, Asia, and Europe.

Bot flies. A beelike fly that lays eggs in equines' hair or in sheep and goats' nostrils.

Bot fly eggs. Minute yellow eggs deposited on the legs and chins of equines.

Bots. A type of internal parasite.

Bottom side. The dam's side of a pedigree.

Bovine. Relating to cattle.

Box stall. A roomy, four-sided stall to house livestock, particularly equines.

Bray. The loud, forceful vocalization of a donkey or mule.

Breech birth. A birth in which the rump of a baby animal is presented first, instead of its head.

Breed. Individual animals of a color, body shape, and other characteristics similar to those of their ancestors, capable of transmitting these characteristics to their own offspring.

Breeder. Generally speaking, anyone who breeds animals; more specifically, an animal's breeder is the person who owned its dam when it was foaled.

Breeding class. A livestock show class based on judging an animal's conformation and type.

Brindle. Striped.

Britchin. A contraption similar to a harness that fastens around the donkey's chest and drapes across its hindquarters.

Broken. A color term denoting white and any other color.

Broken mouth. An older animal that has lost some of its permanent incisors.

Brood mare/brood jenny. A female equine kept primarily for producing foals.

Browse. Morsels of woody plants including twigs, shoots, and leaves; also the act of consuming browse.

Buck. An uncastrated male goat.

Buck rag. A cloth rubbed on the scent glands of a buck and presented to a doe to see if she is in heat.

Buckling. An immature, uncastrated male goat.

Bull. An uncastrated male bovine.

Bunt. The act of a young animal poking its mother's udder with its head to facilitate milk letdown.

Burro. Western colloquialism referring to a donkey; donkey in Spanish is *el burro.*

Bute. Phenylbutazone, a prescription veterinary drug widely used for reducing pain.

Butt. The act of an animal bashing another animal (or a human) with its horns or forehead.

By. Short for "sired by."

CAE. Caprine arthritis encephalitis.

Calf. A baby bovine.

Calling. Vocalization.

Camelid. Members of the camel family including the old-world camelids (camels) and new-world camelids (alpacas, llamas, guanacos, vicuñas, and hybrids thereof).

Caprine. Relating to goats.

Carcass. The body of a slaughtered animal.

Castrate. Removal of a male's testes.

Catch pen. A small, well-fenced area used for catching and sometimes training livestock.

Catheter-tip syringe. A syringe with a blunt tip, used for the oral dosing of animals.

Cattle panel. A very sturdy large-gauge welded-wire fence panel; sold in various lengths and heights.

cc. Cubic centimeter; same as a milliliter (ml).

Ccara (CAR-ah). The short-woolled llama; in some places ccara refers to a working llama as opposed to a fiber llama.

CD/T. Toxoid vaccine used to protect against enterotoxemia (caused by *Clostridium perfringens* types C and D) and tetanus.

Cervix. The section of a female's uterus that protrudes into the vagina; it dilates during birth to allow the young to pass through.

Chromosome. The long DNA molecules on which genes (the basic genetic codes) are located.

Chronic. Any process occurring over a long period of time.

CL. Caseous lymphadenitis.

Classification. A system of judging within different breeds.

Claws. The two halves of a cloven hoof.

Clean legged. A sheep term denoting an animal with hair instead of wool on its legs.

Closed face. A sheep term denoting an animal with wool covering its entire head.

Coarse. Lacking refinement.

Cob. A small horse or pony with cobby conformation.

Cobby. A short, stocky body type that is close coupled and compact.

Coccidiostat. A chemical substance mixed with feed, bottle-fed milk, or drinking water to control coccidiosis.

Coggins. A blood test used to detect carriers of equine infectious anemia; also the certificate indicating that an equine has been Coggins tested.

Colostrum. The first milk a female produces after birth; high in antibodies, this milk protects newborn kids against disease; sometimes incorrectly called *colostrums.*

Colt. An uncastrated male equine under three years of age.

Come into milk. To begin lactating (producing milk).

Composite breed. A breed made up of two or more other breeds.

Concentrate. High energy, low fiber, highly digestible feed such as grain.

Condition. Amount of fat and muscle tissue on an animal's body.

Conformation. A descriptive term pertaining to the overall look of the body parts of an animal.

Congenital. A condition acquired during development in the uterus and not through heredity.

Cover. To breed (a male animal covers a female animal).

Cow. A female bovine; sometimes loosely used to refer to bovines of all ages and sexes.

Crest. The upper portion of an animal's neck, stretching from its poll to its withers.

Cria (CREE-uh). A young llama between birth and weaning age.

Crimp. Natural waviness along the length of an individual fiber or lock of fleece that allows it to stretch and then spring back into shape.

Crossbred. An animal resulting from the mating of two different breeds.

Crutch. To shear fiber from an animal's back legs and belly.

Cud. Undigested food regurgitated by a ruminant to be chewed and swallowed again.

Cull. To eliminate from a herd or breeding program; also an animal eliminated as part of the culling process.

Curaca (cur-AH-cah). A longer-woolled llama still within the ccara type.

Cush. An alternate spelling of *kush.*

Dam. The female parent.

Deccox. The brand name of Decoquinate.

Decoquinate. A coccidiostat sometimes added to feed to control coccidiosis.

Dehorning. The removal of existing horns.

Dental pad. An extension of the gums on the front part of the upper jaw of certain animals including goats, sheep, llamas, and cattle; it is a substitute for top front teeth.

Dew claws. Extra toes or vestigal hooves occurring on one or more legs.

Dewlap. A pendulous fold of loose skin that hangs below the throat.

Deworm. The use of chemicals or herbs to rid an animal of internal parasites.

Dewormer. An anthelmintic; a substance used to rid an animal of internal parasites.

Disbud. To destroy the emerging horn buds of a young animal by the application of a red-hot disbudding iron.

Dish-faced. Having a concave facial profile.

Disposition. The temperament of an animal.

Disqualification. One or more defects, deformities, or blemishes that render an animal ineligible for registration, breeding, or showing.

Dock. To shorten a lamb's tail.

Doe. A female goat.

Doeling. A young female goat.

Drench. Giving liquid medicine by mouth; also a liquid medicine given by mouth.

Dressing percentage. The percentage of a meat animal that remains as a carcass after slaughtering and eviscerating.

Dual-coat. The type of fleece produced by some primitive sheep breeds, including Shetlands, consisting of a long outer coat topping a shorter, fluffier undercoat.

Dual-purpose breeds. Animals developed for two purposes such as meat and fiber or meat and milk.

Dung piles. The areas where llamas urinate and defecate.

Dust bath. A bare, sandy, or dusty spot where animals prefer to roll (dust bathe).

DVM. Doctor of Veterinary Medicine.

Dystocia (dis-TOH-shuh). Difficulty in giving birth.

Easy keeper. An animal that easily maintains its weight.

Elastrator. A plierslike tool used to apply heavy, O-shaped rubber bands called elastrator bands to a male animal's scrotum for castration.

Emaciation. Loss of flesh resulting in extreme leanness.

Embryo. An animal in the early stages of development before birth; a fertilized egg.

Embryo transplant. Implantation of embryos into a surrogate mother.

Emu oil. A medicinal oil manufactured from emu fat, especially useful for treating cuts and abrasions.

Energy. A nutrient category of feeds usually expressed as TDN (total digestible nutrients).

Entero. A shortened, common name for enterotoxemia.

Entire. An uncastrated male animal.

Equine. All members of the family Equus, including horses, donkeys, zebras, and their hybrids.

Esophageal feeder (also called a tube feeder). A milk reservoir and flexible tube used to feed young animals that can't or won't nurse in the normal manner.

Estrogen. Female sex hormone produced by the ovaries; estrogen is the hormone responsible for the estrus portion of a female's estrous cycle.

Estrus. The period when a female animal is receptive (for example, she will mate with a male; she is "in heat") and can become pregnant.

Estrous cycle. The female reproductive cycle.

Euthanize. To humanely end an animal's life.

Ewe. A female sheep; in countries that use sheep terms to describe goats (for example, South Africa), a doe or female goat.

Ewe lamb. A female lamb.

Extra-label (also called off-label). The use of a drug for a purpose for which it isn't approved.

Eye color. The color of the iris, the circle of color that surrounds the pupil of an animal's eye.

F1. The first-generation offspring resulting from the mating of a purebred male animal to a purebred female of another breed.

Fainting goat. A common name for Myotonic goats.

Farrier. A skilled craftsperson who shoes equines and trims hooves.

Farrow. In pig terminology, to give birth.

Farrowing pen. An area set aside for farrowing, usually containing rails under which piglets can retreat to avoid being crushed by their dam.

Fatten. Feeding for increased weight gain.

Faults. Imperfections for a particular breed or variety of animal.

Favor. To limp slightly.

Fecal egg count (FEC). The number of worm eggs in a gram of feces; sometimes written as EPG (eggs per gram).

Fiber. Wool or hair.

Field shelter. A basic shelter with a roof and, usually, three sides.

Fighting teeth. Six very sharp caninelike teeth, two in the lower jaw and one in the upper jaw on both sides of a male llama's mouth.

Filly. A female equine under three years of age.

Fine fiber. Soft fiber with a low micron count.

Fineness. A measure (in microns) of the diameter of individual fibers.

Finishing. The act of feeding an animal to produce a desirable carcass for market.

Fitting. Preparing an animal for show.

Flake. One segment of a bale of hay.

Fleece weight. The weight of all usable fiber removed from a single animal.

Flehmen. Curling of the upper lip in order to increase the ability to discern scent.

Float. To file an animal's teeth to remove sharp edges.

Flock. A group of sheep.

Fly strike. A condition in which blowflies lay eggs in wounds or wet, filthy fiber; when maggots hatch out, they consume their host's flesh.

Foal. An equine less than one year of age; also the act of an equine giving birth.

Foal heat. The first estrus that occurs after foaling.

Follicle. A fluid-filled sack (on an ovary) that contains an ovum (egg).

Forage. Grass and the edible parts of browse plants that can be used to feed livestock.

Forb. A broad-leafed herbaceous plant (for example, curly dock, plantain, and dandelion).

Free choice. Available twenty-four hours a day, seven days a week; hay and mineral mixes are generally fed free choice.

Freshen. To give birth and begin lactating (producing milk).

Gait. A pattern of foot movements such as the walk, trot, canter, and gallop.

Gelding. A castrated male equine.

Genetic marker. A detectable gene or DNA fragment.

Genotype. The genetic makeup of an animal or plant.

Gestation. The length of pregnancy.

Get. The progeny of a male animal.

Gilt. A young female pig.

Grade. An unregistered animal, often of unknown breeding.

Graft. A procedure in which a newborn animal is transferred to and raised by a dam that is not its own.

Grain. Seeds of cereal crops such as oats, corn, barley, milo, and wheat.

Guard hair. Medullated (hollow) hair comprising a second, outer coat of fiber.

Gummer. An old animal that has lost most or all of its teeth.

Habitat. The place or environment where a plant or animal is normally found.

Hackles. The strip of long hair along the spine that some animals (especially goats and pigs) raise when angry or excited.

Halter. Headgear used to facilitate catching, leading, and tying.

Halter class. A class judged on soundness, style, and how well an entry physically conforms to its breed standard.

Hand. Equines are measured in hands; one hand equals 4 inches (10 cm). Equines are measured from the highest point of their withers to the ground. Fractions are shown as hands-point-inches, so that an 8.3 hand miniature mule would be 35 inches (89 cm) tall.

Handspinner. A person who spins fiber by hand using a spinning wheel or drop spindle; a fiber hobbyist.

Hard keeper. An animal that requires more than the usual amount of feed to maintain weight.

Haunches. Hindquarters.

Hay. Grass mowed and cured for use as off-season forage.

Heart girth. Circumference of the chest immediately behind the front legs.

Heat. See Estrus.

Heifer. A female cow less than three years of age.

Herbivore. A plant-eating animal.

Heritability. The degree to which a trait is inherited.

Heterosis. The increased performance of hybrids over purebreds; hybrid vigor.

Hinny. The sterile hybrid offspring of a stallion and a jenny.

Hog. A pig.

Horns. Solid, bony cores covered by a sheath of hard, fibrous material. They form part of an animal's skull, grow throughout its lifetime, and are never shed.

Huarizo (h'whar-EE-soh). An alpaca-llama hybrid; some say only a male alpaca bred to a female llama produces a huarizo and that the mating of a male llama to a female alpaca produces a misti.

Humming. The droning sound llamas make under a variety of circumstances.

Hybrid vigor. See Heterosis.

Hypothermia. A condition characterized by low body temperature.

IgG (Immunoglobulin G). Antibodies in the colostrum of near-term female animals and those that have just given birth.

IM (Intramuscular). The injecting of a solution, usually a vaccine or drug, into muscle mass.

Immunity. A natural or acquired resistance to a specific disease.

IN (Intranasal). The spraying of a solution, usually a vaccine or drug, into the nostrils.

In foal (kid, lamb, calf, pig). Pregnant.

In milk. Lactating.

In season. In heat; see Estrus.

Inbreeding. Mating closely related individuals such as father and daughter, mother and son, and full or half siblings.

Induced ovulator. A female animal that ovulates after, instead of before, being bred.

IU (International Unit). A unit of measurement used in labeling vitamins and drugs.

IV (Intravenous). The injection of a solution into a vein.

Jack. An uncastrated male donkey.

Jackass. An uncastrated male donkey.

Jennet (JEN-et). The correct term for a female donkey.

Jenny. A colloquial word for jennet (and the one generally used in this book).

John. A gelded male mule or hinny.

Jug. An approximately 4 × 5-foot-pen (1.2 × 1.5 m) where a doe or ewe and her newborns are kept for the first 24 to 72 hours after giving birth.

Kemp. Coarse, medullated hair fibers scattered throughout a fleece.

Ketones. Substances found in the blood of late-term pregnant females suffering from pregnancy toxemia.

Killed vaccine. Being or containing a virus that has been inactivated (as by chemicals) so that it is no longer infectious.

Kush (also spelled cush). The act of a camelid lying down sternally with its legs tucked under it. It is also the name of the position as well as the command given to an animal to signal it to *kush*.

Lactation. The period when a female is giving milk.

Lamb. A baby sheep. In countries that use sheep terms to describe goats, a kid or baby goat. The meat of young sheep. The act of a female sheep giving birth.

Lamb fleece. Fleece obtained from a lamb's first shearing, usually the softest and finest it will ever produce.

Lame. A condition in which an animal does not carry weight equally on all four legs, due to disease or injury.

Lanuda (lah-NOO-dah). A long-woolled llama with fringes on its ears and abundant wool on its legs.

Larvae. Immature stages of adult parasite; the term applies to insects, ticks, and worms.

Lead. A rope or strap used for leading livestock.

Lead rope. A sturdy 7 to 10-foot (2–3 m) rope with a snap on one end, used for leading livestock.

Legume. Plants such as alfalfa, clover, and lespedeza.

Libido. Sex drive; the desire to copulate.

Line. A group of related individuals.

Linebreeding. The mating of individuals sharing a common ancestor.

Litter. Two or more young born from the same mating, to the same dam.

Live vaccine. One in which live virus is weakened in order to produce an immune response without causing the severe effects of the disease.

Longear. Colloquialism for a donkey or mule.

Luster (spelled lustre in Britain). The natural sheen of certain fibers.

Maiden. An animal that has never given birth.

Malocclusion. An inherited defect whereby the upper and lower jaws do not allow the upper teeth or dental pad to meet correctly with the lower teeth.

Marbling. Fat distributed within muscle mass.

Mare. A female horse three years of age or older; in Britain, female donkeys three years of age or older are also called mares, rather than jennies or jennets.

Mare mule. A female mule.

Marking harness. A simple harness incorporating a colored crayon or colored chalk, which is sometimes worn by rams and goat bucks during breeding season.

Mastitis. Inflammation of the udder.

Meconium. The sticky, usually blackish fecal matter that a baby animal passes within a few hours after birth.

Mediterranean donkey. A miniature donkey.

Medullated. Hollow.

Medullation. The degree to which a fleece contains medullated fiber.

Micron. A measurement of fiber diameter, equal to 1/25,000 of an inch or 1/1,000 of a millimeter. Used to refer to the fineness of a fiber: a smaller micron equates with finer fiber.

Milk letdown. Release of milk by the mammary glands.

Miniature donkey. The American Donkey and Mule Society/Miniature Donkey Registry records donkeys standing up to 36 inches (91 cm) tall, measured at the withers; the International Miniature Donkey Registry registers miniature donkeys up to 38 inches (97 cm) tall.

Miniature donkey (oversized or Class B). The American Donkey and Mule Society/ Miniature Donkey Registry records donkeys 36.01 to 38 inches (91.5–97 cm) tall in their Class B studbook.

Miniature horse. In American Miniature Horse Association terminology, a horse standing 34 inches (86 cm) or less, measured at the last hairs of the mane; in American Miniature Registry terminology, a horse standing less than 38 inches (96.5 cm), measured at the last hairs of the mane.

Miniature mule. Miniature mules are the offspring of a miniature donkey jack bred to a miniature horse mare. The American Miniature Mule Society registers them in two sizes: Class A — Under 38 inches (97 cm), and Class B — 38 inches to 48 inches (97–122 cm).

Mitochondrial DNA. Genetic material inherited from one's mother, contained within the mitochondria of each cell.

ml (milliliter). A unit of liquid measure; the same as a cubic centimeter (cc).

Molly. A female mule or hinny.

Molly mule. A female mule.

Molt (moult). To shed hair, wool, or fur.

Monensin. A coccidiostat sometimes added to feed to control coccidiosis; marketed under the brand name Rumensin; monensin is highly toxic to equines.

Monkey mouth. See Underbite.

Monogastric. A digestive system containing a simple stomach system (found in humans, equines, pigs).

Mothering pen. See Jug.

Motility. The ability of sperm to move themselves.

Mule. The sterile hybrid offspring of a male donkey (jack) and a female horse (mare).

Multispecies grazing (mixed grazing). Grazing two or more species of animals on the same unit of land.

Mutation. A spontaneous change in characteristics that is inherited by an individual's progeny.

Myotonia congenita. The inherited neuromuscular condition that causes Myotonic goats' major muscles to temporarily seize up.

Myotonic goat. The preferred name for a "fainting goat"; a muscular goat carrying the gene for myotonia congenita.

Nanny (slang). A female goat; the preferred term is doe.

Needle teeth. Eight tiny teeth present at a piglets' birth that some people clip to prevent injury to the dam's teats and littermates' ears.

Nematode. A type of internal parasite; a worm.

Nick. A fortuitous mating in which offspring are superior to their parents.

Off feed. Not eating as much as usual.

Omasum. The third part of the ruminant stomach; it's sandwiched between the reticulum and the abomasum.

Omnivorous. An animal that eats both flesh and plant food.

Oocyst. A minute pouch or saclike structure containing the fertilized cell of a parasite

Open. Not pregnant.

Open show. Shows that are open to exhibitors of all ages; shows that are open to all breeds.

Open-faced. A sheep term denoting having little or no wool on the face.

Orifice. The opening in the end of a functional teat.

Ovary. One of a pair of egg- and hormone-producing glands in a female animal.

Over the counter (OTC). Nonprescription drugs.

Overconditioned. Fat.

Overshot or parrot mouth. When the lower jaw is shorter than the upper jaw and the teeth hit in back of the dental pad.

Ovine. Referring to sheep.

Ovulation. When the follicle ruptures and the ovum (egg) is released from the ovary.

Ovum. An egg; also called an ova or oocyte.

Ox (plural: oxen). Castrated adult male cattle used for riding and driving or for draft purposes.

Oxytocin. A naturally occurring hormone that plays a role in milk letdown and muscle contraction during the birthing process.

Paddock. A small, enclosed area used for grazing.

Palatable. Agreeable in flavor.

Palpation. Examining something with one's hands.

Papered. Registered.

Papers. A registration certificate.

Parturition. The act of giving birth.

Passive transfer of immunity. Acquiring protection against infectious disease from another animal; this occurs when a newborn consumes antibody-rich colostrum from its dam.

Pasture breeding. When a male animal runs loose with a group of females and breeding occurs without human intervention.

Pathogen. An agent that causes disease, especially a living microorganism such as a bacterium or virus.

Pecking order. The social hierarchy within a group of animals.

Pedigree. A certificate documenting an animal's line of descent.

Pen breeding. The breeding system by which one male and one female animal are released in a small enclosure for mating purposes.

Percentage. A crossbred animal that is at least 50 percent of a specific breed.

Perennial. A plant that doesn't die at the end of its first growing season but returns and regrows from year to year.

pH. A measure of the activity of hydrogen ions in a solution and therefore its acidity or alkalinity.

Pharmaceutical. A substance used in the treatment of disease: a drug, medication, or medicine.

Phenotype. An individual's observable physical characteristics.

Piglet. A baby pig.

Pizzle. The urethral process, a stringy-looking structure at the end of some male animals' penises.

Placenta. See Afterbirth.

Pneumonia. Infection in the lungs.

Polled. Naturally hornless.

Pony. Members of the *Equus equus* (horse) subspecies standing, depending on breed and/or local custom, less than 14 hands or sometimes 14.2 hands tall.

Porcine. Relating to pigs.

Postpartum. After giving birth.

Predator. An animal that lives by killing and eating other animals.

Prepartum. Before giving birth.

Prepotency. The ability of an individual to sire or produce uniform offspring.

Prey animal. An animal belonging to a species preyed upon by predators.

Probiotic. Living organisms used to influence rumen health by assisting in the fermentation process.

Processing. Slaughtering an animal and preparing its meat for home use or market.

Produce. A female animal's offspring.

Progeny. Offspring.

Progesterone. A hormone secreted by the ovaries and produced by the placenta during pregnancy.

Proliferate. To vastly multiply in numbers, usually over a short span of time.

Prolific. Producing more than the usual number of offspring.

Protein. A nutrient category of feed used for growth, milk, and repair of body tissue.

Proven. An animal that has successfully sired or produced live offspring.

Puberty. When an animal becomes sexually mature.

Pulpy kidney. Another name for enterotoxemia.

Pureblood. See Purebred.

Purebred. An individual whose ancestors are of the same breed for a predetermined number of generations.

Put down. Euphuism for the euthanization or humane destruction of an animal.

Quarantine. To isolate or separate an individual from others of its kind.

Ram. An uncastrated male sheep; in countries that use sheep terms to describe goats, an uncastrated male goat.

Ration. Total feed given an animal during a 24-hour period.

Raw fiber. Unwashed fiber.

Recipient (doe, ewe, cow, mare, and so on), also called a *recip*. A female into which one or two flushed embryos, depending on species, are inserted and which she carries to term.

Registered animal. An animal that has a registration certificate and number issued by a breed association.

Rehydrate. To replace body fluids lost through dehydration.

Replacement. An animal retained for future breeding purposes.

Reticulum. The second chamber of a ruminant's stomach.

Ring. A metal ring sometimes placed in the edge of a pig's nose to keep it from rooting; also the act of inserting the ring. A ring placed in the septum of a bovine's nose to aid in controlling it or to provide attachment for bridle reins.

Roman-nosed. A convex facial profile.

Rooting. When pigs use their snouts to unearth food.

Rotational grazing (or browsing). Moving grazing or browsing animals from one paddock to another before plant growth in the first paddock is fully depleted; allows pasture regrowth.

Roughage. Plant fiber.

Roundworm. A parasitic worm with an elongated round body.

Rumen. The first compartment of the stomach of a ruminant, in which microbes break down the cellulose in plants.

Rumensin. The brand name for monensin.

Ruminant. An animal with a multicompartmented stomach and that chews cud.

Rumination. The process whereby a cud or bolus of rumen contents is regurgitated, rechewed, and reswallowed; "chewing the cud."

Runt. The smallest of the young in a litter; sometimes used to describe any unusually small or underdeveloped animal.

Rut. The period during which a male sheep, or goat is interested in breeding females.

Sanctioned. Events held under the direction of a registry, organization, or breeder's group.

Sardinian donkey. A miniature donkey.

Scours. Diarrhea.

Scrapie. The goat and sheep version of "mad cow disease"; sometimes incorrectly spelled scrapies.

Scrotum. The external pouch in which a male animal's testicles are suspended.

Scur. A small, misshaped horn or horn button.

Seasonal breeders. Female animals that only come in heat during part of the year.

Selection. Choosing superior animals as parents for future generations.

Self-colored. The same color over the entire body; solid colored.

Septicemia. An infection of the bloodstream that affects the entire body.

Settle. Get pregnant.

Sexing. Determining the sex of an animal.

Sheath. The outer skin covering protecting a male animal's penis.

Shoat. A young pig.

Show. A female animal is "showing" when she indicates she's receptive to being bred.

Silent heat. In heat (estrus) but showing no outward signs.

Sire. The male parent.

Sound. Having no defects that affect serviceability.

Sow. An adult female pig; usually one that has given birth to one or more litters.

Sow-mouth. See Underbite.

Spin. To twist fiber into yarn; this can be done using commercial machinery, a spinning wheel, or a drop spindle.

Spit test. Exposing a previously bred female llama to a male llama to see if she will "spit him off," indicating that she has conceived.

Spooky. Easily startled.

Square. Having a boxy appearance with "a leg in each corner"; a very desirable trait in most miniature species.

Stallion. An uncastrated male horse three years of age or older. In Britain and a few other parts of the world, uncastrated male donkeys three years of age or older are also called stallions (instead of jacks).

Standard (also Standard of Perfection). The desirable characteristics of a breed of animal as approved and written down by its registration organization.

Standing heat. The period during estrus (heat) when a female animal allows a male to breed her.

Staple. A group or lock of individual fibers.

Staple length. The length of a group or lock of individual fibers.

Steer. A castrated male bovine.

Stocking rate. The number of animals that can be pastured on one acre or the number of acres required to pasture one animal.

Straw. The stems of plants (oats, wheat, barley) that are cut and baled to be used for animal bedding.

Stud fee. The charge for breeding to a male animal.

Studbook. A compilation of information about individual breeding animals, maintained by a registry.

Subcutaneous (SQ). Under the skin.

Substance. A strong, stockily built animal with good bone has substance.

Suri (SIR-ee). A type of llama characterized by individual locks of fleece hanging in ringlets.

Sustainable agriculture. An approach to producing profitable farm products while enhancing natural productivity and minimizing adverse effects to the environment.

Sweet feed. A commercial mixture of grains with molasses added.

Swine. A term used to describe all pigs.

Switch. The long-haired lower section of a donkey's tail (also called a *swish*).

Symmetry. The relationship of all body parts when viewed as a whole on a given animal.

Systemic. Affecting the entire body.

Tack. Equipment used for riding, driving, and caring for animals.

Tags. Unusable bits of felted or dirty fleece removed from the lower legs.

Tail wrap. Material (nowadays usually 4-inch wide, self-stick, disposable bandage) used to wrap around and protect an animal's tail during trailering, breeding, and giving birth.

Tampada (tam-PAH-dah). A long-wool llama, but not as woolly as a lanuda.

Tampuli (tam-POO-lee). A catch term for woolly llamas (tampadas and lanudas).

Tapeworm. A segmented, ribbonlike, intestinal parasite.

TDN. Total Digestible Nutrients; energy value of feeds can be assayed in a laboratory.

Teaser. A male animal used to verify the readiness of a female for breeding purposes, often one that has had its spermatic cords cut or tied (has had a vasectomy).

Testosterone. A hormone that promotes the development and maintenance of male sexual characteristics.

Three-in-one package. A pregnant female animal sold with her unweaned baby at her side.

Top side. The sire's side of a pedigree.

Topline. The area between an animal's withers and the beginning of its tail.

Total Digestible Nutrients (TDN). A standard system for expressing the energy value of feed.

Toxin. Any poisonous substance of biological origin.

Trace minerals. Minerals needed in only minute amounts.

Trachea. Windpipe leading from the throat to lungs.

Trimester. One-third of a pregnancy.

Tusks. Long upper and lower canine teeth.

Type. The body conformation of an animal or the shape of a particular body part, as in "head type."

Uc. Urinary calculi; mineral salt crystals ("stones") that form in the urinary tract and sometimes block the urethras of male animals.

Udder. The female mammary system.

Ultrasound. A procedure in which sound waves are bounced off tissues and organs; widely used to confirm pregnancy in females.

Underbite. When an animal's lower jaw is longer than its upper jaw and its lower teeth extend forward past its upper teeth or the dental pad on upper jaw; also known as monkey mouth or sow mouth.

Underconditioned. Thin.

Urethral process. The pizzle; a stringy-looking structure at the end of a male goat's penis.

Urinary calculi (UC). Stones formed in the urinary tract.

USDA. United States Department of Agriculture.

Uterus. The female organ in which fetuses develop; the womb.

Vagina. The passageway from the female uterus to the outside of her body; the birth canal.

Vascular. Pertaining to or provided with vessels; usually refers to veins and arteries.

Vegetable matter (VM). Sticks, burrs, hay chaff, and so forth in a raw, uncleaned fleece.

Vulva. The external female genital organ.

Wallow. The act of lying in mud or water; a muddy spot made for wallowing.

Weanling. A young animal less than one year of age that has been weaned from its dam.

Wether. A castrated male goat or sheep; also the act of castrating male goats or sheep.

Withdrawal period. After administering drugs, the amount of time during which an animal must not be sent to market to ensure no drug residues remain in its meat.

Wool. Sheep fiber.

Wool blindness. A condition caused by excessive wool growth around sheep's eyes.

Yearling. An animal of either sex that is one to two years of age.

Index

Page references in *italics* indicate photos or illustrations; page references in **bold** indicate charts.

B

E

F

M

S

Photography Credits

Page 151: © Lynn Stone, top left; © Dusty Perin/www.DustyPerin.com, top right and bottom
Page 152: © Jason Houston, top left; © Bob Langrish, top right; © Kit Houghton Photography, middle left; © Design Pics/SuperStock, bottom
Page 153: © Kit Houghton Photography, top; © Robert Dowling, middle; © Jean Paul Ferrero/ardea.com, bottom left; Courtesy of American Miniature Horse Association, www.amha.org, bottom right
Page 154: Courtesy of American Miniature Horse Association, www.amha.org, top; © Bob Langrish, middle; © Stanislas Merlin/Jupiterimages, bottom left; © Sue Weaver, bottom right
Page 155: © Dusty Perin/www.DustyPerin.com, all
Page 156: © Robert Dowling, top and bottom right; © Pauline S. Mills/iStockphoto, bottom left
Page 157: © Jason Houston, top; © Robert Dowling, bottom
Page 158: © Robert Dowling, top; © Elise Toups, bottom
Page 159: © Judy McPhail/iStockphoto, top left; © Jason Houston, top right and middle; © Robert Dowling, bottom left; © Bart van den Dikkenberg/iStockphoto, bottom right
Page 160: © Terry Swartz/iStockphoto, top; © Jared Waddams, bottom
Page 161: © Jared Waddams, top; © Nancy Nehring/iStockphoto, bottom left; © Linda Davis, Davis Farmland, bottom right
Page 162: © Robert Dowling, top and middle right; © AP/Wide World Photos, middle left; © www.ToplineLowlines.com, bottom left; © Patti Adams, Wakarusa Ridge Ranch, bottom right
Page 163: © Jeff Greenberg/age footstock, top; © Patti Adams, Wakarusa Ridge Ranch, middle; © Lynn Stone, bottom
Page 164: © Sue Weaver, top left and right; © Linda Davis, Davis Farmland, middle; © Mark Boulton/ardea.com, bottom
Page 165: © Robert Dowling, top; © Dusty Perin/www.DustyPerin.com, bottom left; © Sue Weaver, middle right; © Mark Boulton/ardea.com, bottom right
Page 166: © Gayle Dumas, all

Build up Your Barnyard Knowledge with More Books from Storey

Butchering Poultry, Rabbit, Lamb, Goat, and Pork by Adam Danforth
Learn how to slaughter and butcher small livestock humanely with the help of hundreds of detailed step-by-step photos. This award-winning guide includes in-depth coverage of food safety, freezing and packaging, tools and equipment, butchering methods, and preslaughter conditions.

The Donkey Companion by Sue Weaver
From foaling to first aid and from grooming to professional showing, this guide offers you everything you need to know about selecting, breeding, training, enjoying, and caring for these friendly, dependable animals.

The Encyclopedia of Animal Predators by Janet Vorwald Dohner
Learn to identify threatening species through tracks, scat, and the type of damage they leave behind. In-depth profiles of more than 30 animal predators will help you prevent your livestock, poultry, and pets from becoming prey.

Farm Dogs by Janet Vorwald Dohner
Anyone interested in dogs, as well as farmers and ranchers looking for the right four-legged partner, will fall in love with this beautiful and accessible introduction to 93 highly intelligent, independent, and energetic working breeds.

How to Build Animal Housing by Carol Ekarius
This all-inclusive guide contains illustrated diagrams and in-depth explanations about building shelters that meet animals' individual needs: barns, windbreaks, and shade structures, plus watering systems, feeders, chutes, stanchions, and more.

Temple Grandin's Guide to Working with Farm Animals
Keep your animals calm and safe with Dr. Grandin's groundbreaking methods, tailored for small farms. This in-depth manual describes the behavior, fears, and instincts of herd animals and teaches you how to set up the most humane and productive facilities on your farm.